T0339866

Electrochemical Water Treatment Methods

Methods

Fundamentals, Methods and Full Scale Applications

Electrochemical Water Treatment Methods

Fundamentals, Methods and Full Scale
Applications

Mika Sillanpää and Marina Shestakova

Butterworth-Heinemann
An imprint of Elsevier

Butterworth-Heinemann is an imprint of Elsevier
The Boulevard, Langford Lane, Kidlington, Oxford OX5 1GB, United Kingdom
50 Hampshire Street, 5th Floor, Cambridge, MA 02139, United States

Notices
Knowledge and best practice in this field are constantly changing. As new research and experience broaden our understanding, changes in research methods, professional practices, or medical treatment may become necessary.

Practitioners and researchers must always rely on their own experience and knowledge in evaluating and using any information, methods, compounds, or experiments described herein. In using such information or methods they should be mindful of their own safety and the safety of others, including parties for whom they have a professional responsibility.

To the fullest extent of the law, neither the Publisher nor the authors, contributors, or editors, assume any liability for any injury and/or damage to persons or property as a matter of products liability, negligence or otherwise, or from any use or operation of any methods, products, instructions, or ideas contained in the material herein.

Library of Congress Cataloging-in-Publication Data
A catalog record for this book is available from the Library of Congress

British Library Cataloguing-in-Publication Data
A catalogue record for this book is available from the British Library

ISBN: 978-0-12-811462-9

For information on all Butterworth-Heinemann publications visit our website at https://www.elsevier.com/books-and-journals

Working together
to grow libraries in
developing countries

www.elsevier.com • www.bookaid.org

Publisher: Matthew Deans
Acquisition Editor: Ken McCombs
Editorial Project Manager: Jennifer Pierce
Production Project Manager: Mohanapriyan Rajendran
Designer: Greg Harris

Typeset by TNQ Books and Journals

Contents

3.   Emerging and Combined Electrochemical Methods
Mika Sillanpää, Marina Shestakova

Chapter 1

Introduction

Mika Sillanpää, Marina Shestakova
Lappeenranta University of Technology, Lappeenranta, Finland

NOMENCLATURE

Latin alphabet

Q	Electric charge	C
I	Current	A
t	Time	s
m	Mass of the substance liberated or deposited at an electrode	
M	Molar mass of a substance	g/mol
F	Faraday constant	96485.33289(59) C/mol
N_A	Avogadro's number	$6.022141 \cdot 10^{23}$ mol^{-1}
e	Elementary charge of an electron	$1.6021766 \cdot 10^{-19}$ C
z	Number of electrons participating in reaction	
E	Cell potential at a given temperature	V
E^0	Standard cell potential	V
R	Universal gas constant	8.314472 (15) J/K mol
T	Temperature	K
Q_r	Reaction quotient	
K	Electrochemical equivalent of a substance	
Vs	Solution volume	dm^3
ΔCOD	COD decay	g/dm
$E^0_{Red, cathode}$	Standard reduction potential for the reduction half reaction occurring at the cathode	V
$E^0_{Red, anode}$	Standard reduction potential for the oxidation half reaction occurring at the anode	V
ΔG	Change in the Gibbs energy	J/mol
W	Work by the galvanic cell for the chemical transformation of 1 mol of the reactant	J/mol
ΔH	Change in process enthalpy	J/mol

Electrochemical Water Treatment Methods. http://dx.doi.org/10.1016/B978-0-12-811462-9.00001-3

ΔS	Change in process entropy	J/K mol
γ	Activity coefficient	
a	Activity	mol/cm^3
a_e/a_s	Ion activities in the near-electrode layer/the bulk solution, respectively	mol/cm^3
A	Electrode area	cm^2
j	Current density	A/cm^2
j_0	Exchange current density at redox reaction equilibrium	A/cm^2
j_L	Limiting current density	A/cm^2
k^0	Heterogeneous electron transfer rate constant	cm/s
k_m	Mass transfer coefficient	cm/s
J	Diffusional flux	mol/s cm^2
α_a/α_c	Anodic/cathodic charge transfer coefficients, respectively	
ΔE_C	Overpotential due to concentration polarization	V
ΔE	Overpotential	V
D	Diffusion coefficient	cm^2/s
δ	Thickness of the diffusion layer	cm
l	Interelectrode distance	cm
ρ	Solution-specific conductance	S/cm
V_R	Reactor volume	L
P	Power consumed during the electrolysis process	kW
s_n	Normalized space velocity	
$\Delta G^{\ddagger}$	Gibbs free energy required for the reaction activation	J/mol
υ	Rate of the first-order chemical reaction under constant pressure (concentration)	cm/s
R_p	Charge transfer resistance	
u_i	Mobility of species	cm^2 mol/J s
z_i	Charge number of species	eq/mol
ϕ	Electrostatic potential	V
υ_s	Electrolyte solution velocity	cm/s

Abbreviations

An$^-$	Anion
COD	Chemical oxygen demand
BOD	Biological chemical demand
DOM	Dissolved organic matter
CE	Current efficiency
EO	Electrochemical oxidation
US	Ultrasound/ultrasonication
SHE	Standard hydrogen electrode
EMF	Electromotive force
ICCP	Impressed current cathodic protection
HOMO	Highest occupied molecular orbital
LUMO	Lowest unoccupied molecular orbital
E_F	Fermi level

Water is unique indispensable component of all living things playing the role of solvent, metabolite, temperature buffer, and living environment. It is a well-known fact that human body consists of 50–75 wt.% water, where the upper value of 75% is the water content in newborn babies and the lower value of 50% is in elderly people. People can survive without food for about 45–60 days when drinking freshwater, but the absence of water intake (or loss of 20% moisture) leads to invertible death within a week.

Water is the most abundant compound on the Earth. About 71% of the Earth's surface is covered by water accumulated, for example, in oceans, seas, lakes, rivers, and ice. The total volume of water resources is 1386 million m^3 where 97.5% is salt water and 2.5% is freshwater [1]. Most of the freshwater is contained in glaciers and ice caps (about 68.7%) and only around 30.7% of fresh ground, river, and lake water are available for human use, from which around 70% is used by agriculture, 23% is consumed by industry and only 7% is left for domestic use [2]. In sum, from the total volume of water accumulated on the planet, only about 0.8% of freshwater is available for human use. A graphical visualization of the distribution of water on Earth is shown in Fig. 1.1.

The world population is growing rapidly and the need for fresh drinking water only increases. Earth's population is equal approximately to 7.4 billion people, and the annual growth of the world population is around 83 million people. It should be noted that freshwater use has increased 17-fold for the period when the world population has tripled. Moreover, according to some forecasts, freshwater consumption is expected to triple in 20 years. One out of

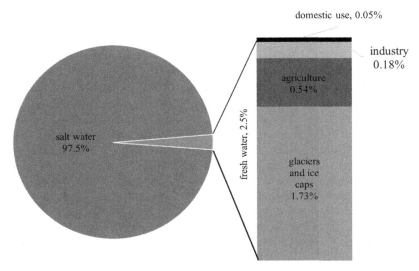

FIGURE 1.1   Visualization of the distribution of water on Earth.

every six people on the planet does not have access to clean drinking water. The growth of cities, the rapid development of industry, intensification of agriculture, a significant expansion of irrigated land, improving the level of life, climate change, and other factors increase problems of water supply. Along with freshwater scarcity the main problem significantly reducing existing freshwater resources is anthropogenic contamination. This is facilitated by industrial emissions and discharges, runoffs of fertilizers from the fields, as well as saline water intrusion into freshwater aquifers near to coastal areas due to overpumping of groundwater.

New directions of rational use of natural water resources such as better exploitation and widening of freshwater resources as well as development of new technological solutions preventing pollution of water bodies and minimizing the water consumption are under continuous development at the present time.

1.1 CLASSIFICATION OF POLLUTANTS AND WATER TREATMENT METHODS

Sources of FreshWater Supply

Nearly 20% of freshwater resources are *groundwater*. Usually groundwater is less polluted than surface water and sometimes can be used for water supply without purification and disinfection especially in the case of deep water basins. This makes the groundwater sources more preferable compared to surface water. However, due to confined groundwater volumes, their instability, dependence on atmospheric sediments, deep occurrence of clean water and contamination of the first layer from the Earth's surface groundwater in the areas of high-populated cities, the use of groundwater for public water supply is often possible only for rural areas.

Groundwater accumulates in aquifers above impermeable-confining beds. Aquifers are composed of water-bearing sedimentary rocks that form the top layer of the Earth's crust and contain free water. This water is able to pass through the rocks under capillary and gravity forces. The most common water-bearing rocks are sands, sandstones, limestones, pebbles, etc. Impermeable-confining rocks either hold water or slightly let it pass through their thickness. Clay, heavy loam, granite, rock salt, etc. can be considered as impermeable-confining rocks. While passing through the soil and deeply lying rocks water becomes filtered and clarified from turbidity, contaminants, bacteria, smell, color, etc. The deeper the groundwater level, the cleaner the water.

Aquifers can be divided into two categories (Fig. 1.2), which are confined or unconfined. Unconfined aquifer has a free water surface (water table) and is located on the first solid impermeable layer from the Earth surface. The pressure at the water table of unconfined aquifer is equal to the atmospheric pressure. Soil below the water table is completely saturated with water, which

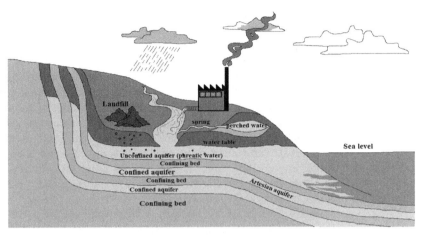

FIGURE 1.2 Types of underground aquifers.

is called unconfined or phreatic water. Phreatic waters are fed by infiltration of atmospheric precipitations and surface water, condensation of water vapors, soil water evaporation, and inflows from deeper aquifers. The level of phreatic water table is not constant and mainly depends on the amount of precipitations. The depth of the phreatic water level, water temperature, salinity, and water flow are subjected to systematic fluctuations occurring as a rule in a range of a day to a year or few years. Soil above the phreatic water table is unsaturated and called aeration zone. When it has inclusions of low impermeable rocks, perched water bodies are formed above these inclusions. Perched water formation is similar to phreatic waters and significantly influenced by the terrain. When comparing with phreatic water, perched water bodies have a limited distribution area and no hydraulic connection with the surface water. They are usually not suitable for continuous water supply and easily polluted. The quality of phreatic water may also vary. In general, due to the absence of an impermeable layer above the unconfined aquifer, phreatic water is vulnerable to pollution infiltrating with rain and meltwater containing, for example, landfill and agricultural field leachates or highways water runoffs. Bacterial composition in unconfined aquifer depends on soil contamination of aeration zone.

Damaged sewer pipes and drainage fields, located near the water wells, can be a source of contamination of aquifers by pathogens. The higher the filtration capacity of the soil, the slower the spread of biological contamination of groundwater. Confined aquifer is located between two impermeable layers (confining beds) and therefore is well protected against contamination. Confined aquifers can recharge in places of resurgence often rather far away from the water accumulation places. So even if the water reserves are located under a built-up area, they are replenished with clean water at a great distance from that area. Confined water may come to the surface in the form of water springs. Usually water-bearing rocks between confining bed are completely saturated with water, which is confined under the pressure. If confining beds have inclined position, then water inside the confined aquifer has high hydrostatic pressure. Such deep pressurized groundwater is called artesian. As a rule, artesian water has high organoleptic properties such as transparency, absence of odor, high-quality taste, and usually it does not contain bacteria. The constancy of the artesian water quality in an artesian well is determined by the proximity of recharge zone (the further it is, the higher and more constant the water quality). The natural water cycle for confined water accumulated at the second impermeable layer is usually in a range of centuries, while confined water formation located at the third layer of impermeable rocks from the Earth's surface can last for millennia.

Surface water sources, which serves the most of freshwater supply needs, are slightly above 1.2% from the total amount of the Earth's freshwater. All surface water sources can be divided into two categories, which are flow water sources such as rivers and artificial canals and standing or slow-moving water bodies such as lakes, ponds, or reservoirs. Surface water sources are the least reliable in respect to sanitary water quality sources; however, they are the only possible sources for large urban areas.

In contrast to the groundwater, surface water always has to be treated and disinfected. Surface water has poor organoleptic properties and contains large amounts of microbes and pollutants. Water from surface sources contains much less mineral salts in comparison with groundwater. Contamination of surface waters is particularly intense during floods, when dirt, bacteria, and organic matter are washed away from the soil surface. Pollution of surface water bodies are also often caused by industrial wastewater spill and discharges.

Flow water sources are more suitable for water supply than standing water, as they have significant water capacity, can undergo a self-cleaning, and usually do not have eutrophication problems, which is the case with standing water. Rivers are the most common source of centralized water supply followed by large lakes and reservoirs.

Classification of Water Pollutants and Water Treatment Methods

Freshwater is consumed by many sectors of people life, such as processing, transporting, and extracting of energy, for irrigation of crops, for livestock, domestic use, health, ecosystems, mining, and production of goods and services [3]. Spent industrial and domestic water cannot be directly released back to the environment or returned to a technological cycle because it is contaminated with mechanical impurities and organic and inorganic pollutants. The quality of natural waters is closely linked to the state of the soil, climate, and human activities. Water pollution can be determined by any changes to the physical, chemical, and biological properties of water in the water bodies caused by inflow of liquid, solid, and gaseous compounds that endanger aquatic biota, health, and public safety and lead to the detriment of the national economy. There are the following ways changing the natural water quality:

- direct discharge of industrial and municipal wastewaters to water bodies and groundwater sources (including mine activity, spills from tankers and oil platforms, leakage of sewer lines, timber rafting, and discharges from water and rail transport);
- surface runoff caused by precipitations (including runoffs from landfills and agricultural fields);
- with contaminated precipitations;
- infiltration of pollutants through contaminated soils.

All water pollutants can be divided into few categories depending of their origin, nature, size, toxicity, etc. Classification by source divides pollutants to anthropogenic or natural ones. Anthropogenic pollutants are caused by industrial, agricultural, mining, domestic, and traffic activities. Natural pollutants formation is caused by geological processes such as volcanoes activity, epithermal deposits, leaching of pollutants from rocks and soils, and decomposition products of biological decay of dead plant and animal materials. [4]. Depending on the nature, pollutants can be classified into organic and inorganic compounds, pathogens, and thermal and radioactive pollution.

Industrial factories such as sulfate, phosphate, and nitrogen fertilizer factories and lead, zinc, and nickel ore enrichment factories discharge mainly wastewaters containing inorganic pollutants, including acids, alkalis, metal ions, and others. Refineries, petrochemical, organic synthesis, and coke plants discharge effluents containing various petroleum products, ammonia, aldehydes, resin, phenols, and other harmful substances. Wastewater of pulp and paper industry adversely affects the flora and fauna of water bodies. Oxidation of the pulp is accompanied by consumption of significant quantities of oxygen, which leads to the death of eggs, fry, and adult fish. The fibers, leached tannins, and other water insoluble substances contaminate water and affect its physicochemical properties. Timber rafting contaminates the bottom of water bodies that in turn deprives fish spawning and feeding areas. Tannery, food industry, and agriculture are among the sources of organic pollution of water bodies with plant fibers, animal and vegetable fats, manure contamination, etc.

Anthropogenic activity causes the following changes of natural water quality:

- Increase of organic matter content due to discharge of wastewaters containing surfactants, pesticides, herbicides, polychlorinated biphenyls (PCBs), polycyclic aromatic hydrocarbons (PAHs), dyes, etc., which in turn increases biological oxygen demand (BOD) and chemical oxygen demand (COD) in water bodies. Pesticides are also accumulated in plankton, benthos, and fish and can fall into the human body through the food chain. Moreover, many organic pollutants are toxic for aquatic habitants and harmful to humans.

- Decrease of dissolved oxygen content due to consumption of oxygen by biogenic elements (phytoplankton, algae, and plants) and for self-cleaning processes of organic compounds mineralization, thermal pollution, and pollution with hydrophobic compounds preventing oxygen diffusion from the atmosphere.
- Contamination with nutrients such as phosphates, nitrates, nitrites, and ammonium, which causes the water bodies eutrophication and increases of toxic metal ions content such as As, Cd, Pb, and Zn.
- Decrease of pH of freshwater bodies due to their contamination with acid precipitations caused by emissions of sulfur dioxide, nitric oxide, and nitrogen oxide into the atmosphere. Formation of sulfuric and nitric acids in the atmosphere is described by the following reactions:

$$SO_2 + 2HO^{\bullet} \rightarrow H_2SO_4 \tag{1.1}$$

$$2SO_2 + O_2 \rightarrow 2SO_3 \tag{1.2}$$

$$SO_3 + H_2O \rightarrow H_2SO_4 \tag{1.3}$$

$$2NO + O_2 \rightarrow 2NO_2 \tag{1.4}$$

$$NO + HO^{\bullet} \rightarrow HNO_2 \tag{1.5}$$

$$NO_2 + HO^{\bullet} \rightarrow HNO_3 \tag{1.6}$$

$$2NO_2 + H_2O \rightarrow HNO_3 + HNO_2 \tag{1.7}$$

$$N_2O_5 + H_2O \rightarrow 2HNO_3 \tag{1.8}$$

- Increase of calcium, magnesium, silica, aluminum, and other ions in surface and groundwater as a result of their leaching and dissolution from rocks that are in contact with acidified precipitations.

$$CaCO_3 \text{ (limestone)} + 2H^+ \rightarrow Ca^{2+} + H_2O + CO_2 \tag{1.9}$$

$$4KAlSi_3O_8 \text{ (feldspar)} + 4H^+ + 2H_2O \rightarrow 4K^+ + Al_4Si_4O_{10}(OH)_8 \\ + 8SiO_2 \tag{1.10}$$

$$Mg_2SiO_4 \text{ (forsterite)} + 4H^+ \rightarrow 2Mg^{2+} + 2H_2O + SiO_2 \tag{1.11}$$

- Increase of total dissolved solids content as a result of surface runoff from the landfills and direct discharge of wastewaters.

- Contamination of natural waters with radioactive isotopes of chemical elements (examples of Fukushima and Chernobyl nuclear power plants). Radioactive substances are accumulated in plankton and fish and then transmitted to other animals through the food chain. Moreover, the radioactivity of planktonic inhabitants is thousands of times greater than the water they live in.
- Changes of organoleptic properties of water such as decrease of water clarity, increase of turbidity, color, odor, and taste contamination.
- Bacterial and biological contamination of water by a variety of pathogenic microorganisms, fungi, and small algae.

Rivers and other water bodies are characterized by a natural process of water self-purification. However, the process is slow and water bodies cannot cope with extensive pollution. Therefore, there is a need for wastewater neutralization, purification, and recycling.

Wastewater treatment methods can be divided into mechanical, chemical, physical—chemical, biological, and combined methods in a case of joint use of few methods. The use of a particular method in a particular case depends on the pollutants' nature and their extent of hazard. Mechanical treatment is used usually as a pretreatment step for pollutants' removal [5]. The essence of mechanical water treatment methods is removal of mechanical impurities by gravity separation methods and filtration. Suspended solids, depending on the size and density, are trapped by grids, sieves, grit chambers, septic tanks, grease traps, clarifiers, filters, etc. Mechanical treatment allows to remove about 60%—75% of insoluble impurities from the municipal wastewater and up to 95% from industrial wastewaters [6]. The chemical treatment is characterized by addition of chemical reagents, which react with contaminants and in the majority methods precipitate in the form of insoluble precipitates. It can be used as a pretreatment, primary treatment, or secondary treatment. The chemical methods of wastewater treatment include neutralization, oxidation, and reduction. Physical—chemical methods of water treatment of wastewater remove fine-dispersed and dissolved inorganic impurities, decompose organic matter and recalcitrant compounds, recover metals, etc. Coagulation, flotation, crystallization, electrochemical treatment, sorption, extraction, ultrasound, and ion exchange are the most widespread physical—chemical methods. Biological treatment is the primary treatment method and removes dissolved nutrients such as nitrogen and phosphorus and organic pollutants by using microorganisms (bacteria and protozoa) or earthworms, which are called activated sludge or biofilm. As a result of the treatment the COD and BOD content of water decreases. Both aerobic and anaerobic microorganisms can be used in the treatment. From a technical point of view, there are several methods for biological treatment. At the moment the main ones are activated sludge (aeration tanks), biofilters, digestion tanks, membrane bioreactors, and constructed wetlands.

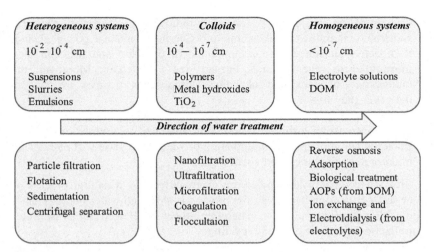

Heterogeneous systems	Colloids	Homogeneous systems
$10^{-2} - 10^{-4}$ cm	$10^{-4} - 10^{-7}$ cm	$< 10^{-7}$ cm
Suspensions Slurries Emulsions	Polymers Metal hydroxides TiO_2	Electrolyte solutions DOM

Direction of water treatment

Particle filtration Flotation Sedimentation Centrifugal separation	Nanofiltration Ultrafiltration Microfiltration Coagulation Floccultaion	Reverse osmosis Adsorption Biological treatment AOPs (from DOM) Ion exchange and Electroldialysis (from electrolytes)

FIGURE 1.3 Classification of water pollutants by particle size and the main water treatment methods depending on the size of the pollutants. *AOPs*, advanced oxidation processes; *DOM*, dissolved organic matter.

Classification of pollutants by size the mostly defines the treatment methods to be chosen for particles removal. Depending on the size pollutants can form heterogeneous ($10^{-2}-10^{-4}$ cm), colloidal ($10^{-4}-10^{-7}$ cm), or homogeneous solutions ($<10^{-7}$ cm) with water (Fig. 1.3). Heterogeneous systems include suspensions, emulsions, and slurries. Polymers and metal hydroxides are colloidal particles. Electrolytes and dissolved organic matter are homogeneous pollutants. Water treatment always begins with a larger particle size removal, which is achieved in the mechanical phase separation processes such as flotation, sedimentation, particle filtration, and centrifugal separation. Coagulation, flocculation, and micro- and ultrafiltration, are usually used in the removal of colloidal pollutants from water. Dissolved organic matter (DOM) and inorganic ions with the size below 1 nm form homogeneous or true solutions with water. The most common water treatment methods allowing removal of small organic molecules are adsorption, biological treatment, reverse osmosis, advanced oxidation processes (AOPs), and rarely ion exchange methods. From the first glance, conventional methods such as biological treatment, membrane technologies, or adsorption seem to be the most reasonable for homogeneous pollutants removal because they are efficient, low cost, and widespread. However, all of them have environmental issues with the secondary waste production such as spent adsorbent, retentate in adsorption and membrane treatment, and waste-activated sludge in biological treatment. Due to the tightening of environmental legislation, there is a need for new waste-free technologies.

AOPs are modern methods using reactive radicals (mostly hydroxyl radicals) for the mineralization of organic pollutant into simple inorganic compound such as carbon dioxide and water. AOPs are a wide group of treatment methods including electrochemical oxidation (EO), ultrasonication (US), Fenton process, photocatalysis and photodegradation, ozonation, persulfate oxidation, and radiolysis. Advantages and disadvantages of AOPs are described in Table 1.1 [7–9]. The most commonly used methods for removal

TABLE 1.1 Advantages and Disadvantages of Different Methods Used for Removal of Small Organic Molecules [7]

Treatment Methods		Advantages	Disadvantages
Adsorption		Complete removal of pollutants is achievable Ease of implementation Widespread systems for different flow rates and concentrations Low operation costs	Regeneration of saturated and utilization of spent adsorbents Efficiency dependence on temperatures and pH Utilization of retentate is often required
Nanofiltration		Simultaneous water softening Systems are available for different flow rates Low investment costs	Membrane fouling Concentration polarization Short service life of membranes Utilization of retentate and spent membranes is required
Biological treatment		Possibility to recycle anaerobic sludge High efficiencies of aerobic treatment Widespread method Low-cost method	Large working areas Waste-activated sludge of aerobic treatment should be disposed Efficiency depends on temperature Postdisinfection is often required Toxic compounds are dangerous for microorganisms Addition of nutrients is required
AOPs	Electrocatalysis	Complete decomposition and high mineralization is achievable Ease of implementation and automation No chemicals required in industrial wastewater applications	Energy-intensive process Electrodes polarization and contamination Electrodes corrosion High cost of electrodes
	US	Fast decomposition reactions	Low mineralization and degradation rates

Continued

TABLE 1.1 Advantages and Disadvantages of Different Methods Used for Removal of Small Organic Molecules [7]—cont'd

Treatment Methods	Advantages	Disadvantages
	No chemicals required Ease of implementation and automation Multicomponent degradation mechanism by ˙OH radicals, pyrolytic reactions, physical effect of bubble collapse	Energy-intensive process Strong reproducibility dependence on reactor types Cooling is required Formation of toxic by-products is possible
Fenton	Complete mineralization is achievable Fast decomposition reactions No energy input required	Addition of chemicals is required pH-dependent degradation Ferric hydroxide sludge should be utilized
Photodegradation and photocatalysis	Complete mineralization is achievable No chemicals required	Nonapplicable for turbid and colored wastewaters Separation of photocatalyst is required pH-dependent degradation
Ozonation	High degradation efficiencies are achievable Fast decomposition reactions No chemicals required No pH changes during the treatment Simultaneous disinfection effect	Energy-intensive process Ozone toxicity and fire hazards issues Formation of toxic by-products
Radiolysis	Fast decomposition reactions Nonselective process	High investment cost of electron accelerator Hazards of radioactive personnel exposure High-resistant materials are required

AOPs, advanced oxidation processes; *US*, ultrasonication.

of ions from water are ion exchange and electrodialysis methods. The latter will be discussed in more detail in Chapter 2. As it is seen from Table 1.1 environmental issues of secondary water pollution due to the sludge formation can be arisen while considering water treatment by Fenton reaction. Nevertheless, for the possible complete mineralization of compounds, this method requires addition of chemicals, which would require additional working areas for chemical storage and equipment for their dosage. Radiolysis treatment requires high investment costs for electron accelerator. Moreover, in respect of possible radioactive exposure of personnel and contamination of the environment in the case of equipment breakage this method is unlikely to be accepted by the society, especially after a series of accidents at nuclear facilities worldwide. Ozonation and ultrasonic treatment are cost-effective methods, but the formation of toxic by-products is possible [10,11]. Moreover, ultrasonic treatment is not effective in mineralization of organic pollutants [12,13]. Photodegradation and electrocatalysis are AOPs, which can provide high mineralization of refractory organic pollutants. Both methods are easily automated, implemented in compact reactors, do not require addition of chemicals, and no secondary waste produced during the treatment. Nevertheless, both methods are energy intensive. The methods are suitable for both centralized and decentralized water treatment applications [14]. In contrast to photodegradation, electrocatalysis is suitable for the treatment of turbid and colored wastewaters. Moreover, there is no need to separate catalysts, which is a necessary step in photocatalysis.

Classification of Electrochemical Water Treatment Methods

Electrochemical water treatment is related to the physical—chemical water treatment methods. Electrochemical treatment is characterized by multistage and relative complexity of physical and chemical phenomena occurring in electrochemical reactors (electrolyzers). The mechanism and rate of occurrence of the individual reaction steps are dependent on many factors, which have to be identified to determine the optimal reactor design and conditions for its operation. Based on the physical—chemical properties, electrochemical methods for wastewater treatment applications can be divided into three main categories. They are conversion methods, separation methods, and combined methods (Fig. 1.4).

Conversion method provides a change of physical—chemical and phase characteristics of dispersed pollutants in regard of their neutralization, conversion, and removal from the wastewater. The transformation of impurities can undergo a series of successive stages, starting with the electronic interaction of soluble compounds and ending with the change of electrosurface and volume characteristics of suspended substances contained in the wastewater. Electrooxidation, electroreduction, and electrocoagulation are common conversion methods. Electrooxidation and electroreduction are used for water treatment from dissolved impurities such as cyanides, thiocyanates, amines, alcohols, aldehydes, nitro compounds, azo dyes, sulfides, and thiols. Electrochemical oxidation processes lead either to complete decomposition of compounds present in the wastewater to CO_2, NH_3, water or formation of simple nontoxic substances, which can be removed by other

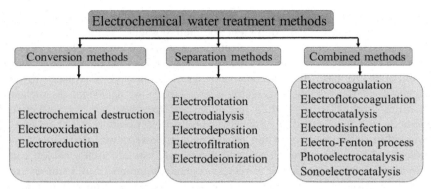

FIGURE 1.4 Classification of electrochemical methods by the mechanism of treatment.

methods. Different insoluble conductive materials such as graphite or mixed metal oxide electrodes (lead dioxide, manganese, ruthenium, iridium, etc., applied to a titanium substrate) are used as anodes. Cathodes are usually made of molybdenum, tungsten alloy with iron or nickel, graphite, stainless steel, and other metals and metal alloys.

Separation methods are used for pollutants concentration in the local volume of the solution without significant changes of their phase or physical—chemical properties. Separation of impurities from water is conducted mainly by electrogenerated gas bubbles in electroflotation or by the power of the electric field, which provides transport of charged particles in the water during electrodialysis.

Combined methods of electrochemical wastewater treatment incorporate one or more conversion and separation methods in one reactor. Electrocoagulation method is based on electrolysis process using steel or aluminum anodes subjected to electrolytic dissolution. Dissolved iron or aluminum cations pass to the electrolyte solution where they react with pollutants forming flakes and causing flake precipitation. In general, during electrocoagulation pollutants lose their aggregative stability and as a result undergo sedimentation, i.e., phase separation. However, in addition to phase separation, coagulant can cause pollutant transformation. For example, dissolved Fe(II) ions reduce Cr(VI) ions to Cr(III) ions during the coagulation process.

All electrochemical processes take place at the electrodes while passing direct electric current through the electrolyte solution. Electrochemical methods can be used for drinking water and wastewater treatment. In the case of wastewater treatment there is a possibility to extract valuable products from water by a relatively simple technological scheme without the use of chemicals. The main disadvantage of these methods is the high energy consumption. Depending on the required effect of water treatment suitable electrochemical method and reactor can be chosen.

Electrolyzers for the electrochemical water treatment can be classified by the following features:

- flow kinetics (continuous or batch);
- hydrodynamic operation (pressurized or nonpressurized);
- reactor type (open, close, diaphragm, or membrane cell);
- movement of water in the interelectrode distance (horizontal, angled, vertical with the ascending and descending water flow);
- type of the impact on pollutants (electric field, electrode process, electric discharge or complex effects).

The effectiveness of electrochemical methods is evaluated by a number of parameters such as current density, overpotential, current efficiency, and energy consumption, which will be discussed in this chapter.

1.2 FUNDAMENTALS OF ELECTROCHEMICAL PROCESSES IN WATER TREATMENT

Electrochemical Phenomena

Because chemical reactions are usually associated with the rearrangement of charged particles such as electrons and atomic nuclei, chemical and electrical phenomena are closely linked. Chemical transformation occurring owing to the external applied electrical current or leading to generation of electrical current is studied in electrochemistry. Electrochemistry studies the mutual conversion of chemical and electrical forms of energy. Electrochemical reactions are of great practical importance. Chemical current sources such as batteries are used in everyday life. Electrolysis is used in metallurgy of nonferrous metals, chemical industry, water treatment, etc. Electroplating of silver, gold, chromium, brass, bronze, and other metals and alloys are widely used for the protection of steel products from corrosion, for decorative purposes, and for the manufacture of electrical connectors and printed circuit boards in the electronics industry. Electrochemical processes are the basis of many modern methods of analysis. The developing branch of electrochemistry is chemotronics, developing electrochemical and optical data converters. Electrochemical methods are used for high-precision dimensional processing of workpieces made of metals and alloys, particularly those that cannot be processed by conventional mechanical methods as well as for formation of parts of complex profile. Aluminum and titanium protective oxide films are formed on metal surface by anodization. Such films are made on the surface of workpieces of aluminum, tantalum, and niobium in the manufacture of electrolytic capacitors. Large-scale production of many substances is based on the electrochemical synthesis. While conducting the electrolysis of brine in chloralkali process, chlorine and hydroxides are formed, which are later used for the production of organic compounds and polymers, as well as in the pulp and paper industry. Sodium chlorate, persulfate, and sodium permanganate are products of electrolysis. Important in industry metals such as aluminum, magnesium, lithium, sodium, and titanium are obtained using electroextraction.

So what is the difference between chemical and electrochemical reactions? Let us consider the following chemical reaction:

$$Zn + 2Fe^{3+} \rightarrow Zn^{2+} + 2Fe^{2+} \tag{1.12}$$

If this reaction is a chemical process, it will be characterized by some peculiarities. The chemical reaction is only possible in a case of collision of the reactants with each other. Hence, the need for contact between reacting species is the first feature of the chemical transformation. The transfer of electrons from one particle to the other or from the reducing agent (Zn) to the oxidant (Fe^{2+}) is possible only at the time of the collision. The electron path will be very short and this is the second feature of the chemical process. Collisions can occur at any point of the reaction volume and in all relative positions of reacting species, so the electronic transitions can be performed in any direction in space. Randomness of particle collisions and electron transfer are the third feature of the chemical reaction. As a result of these peculiarities the energy effect of the chemical reaction is expressed in the form of release or absorption of heat. It is necessary to set certain conditions in the system to convert energy from chemical reactions to electrical energy, i.e., to create an electrochemical process.

In electrochemical processes, the transfer of electrons from one reactant to another is performed over a significantly long path. It is explained by the fact that production and consumption of electric energy is always associated with the passage of electric current, which is a stream of electrons traveling along the same path. Therefore, spatial separation of the reactants (reducing and oxidizing agents) is required for electrochemical processes to keep electrons flow from reducing to oxidizing agents. In this regard, direct contact between reactants should be replaced with the two metal plates connected to each other by a metallic conductor. To ensure the continuous passage of electric current through the reactionary space, charge carriers having a high ionic conductivity should be present or added in the reactionary solution. Thus, a system called an electrochemical cell is required to conduct electrochemical reactions.

Wastewater treatment by electrochemical methods is based on conducting the electrolysis process. To conduct the electrolysis an external source of electrical energy is required to generate and maintain a proper potential and as a result electrochemical reactions at anode and cathode, which are placed into electrochemical cell (for example, into industrial electrolyzer). Michael Faraday was the first scientist who investigated the relationship between amount of electric charge Q (current I multiplied by time t) passed through the electrode/electrolyte solution interface and chemical reactions caused by this charge. In 1832 Faraday reported that the amount of electricity required to produce a given quantity of substances does not depend on the electrode size, number of working electrodes, and the distance between electrodes. It was stated that the mass m of the substance liberated at an electrode is directly proportional to the electric charge Q, passed through the electrolyte and directly proportional to the equivalent weight (M/z) of the element for a given amount of electricity (1.13):

$$m = k \cdot Q = k \cdot I \cdot t = \left(\frac{I \cdot t}{F}\right) \cdot \left(\frac{M}{z}\right) \tag{1.13}$$

where k is electrochemical equivalent of a substance, $k = \frac{M}{F \cdot z}$, M is molar mass of a substance, $1\,F = 1\,\text{mol} \cdot e^- = 6.02 \times 10^{23}\ e^- = e \cdot N_A = 26.8\,\text{A h/mol} = 96{,}485.33289(59)$ C/mol is Faraday constant, z is the number of electrons participating in the reaction (valency of ion of the substance).

Michael Faraday together with his friend William Whewell developed a new terminology in electrochemistry. He called conductors immersed in the solution such as the electrodes (earlier they were called poles), introduced the concept of electrolysis (chemical changes associated with the current passage), electrolyte (conductive liquid in electrochemical cells), anode (electrode with oxidation reaction on it), and cathode (an electrode with the reduction reaction on it). The charge carriers in liquids were called ions (from the Greek wanderer); the ions moving to the anode (positive electrode) were called anions and to the cathode (negative electrode) cations. Colloids and suspended solids also can participate in the transfer of electric current; however, due to low mobility, they can carry only insignificant part of electric current.

Electrochemical equivalent can be used to calculate the amount of reactive substance in the anodic and cathodic processes, such as anodic dissolution of metal, gas evolution at the cathode, and products of EO. Electrochemical equivalent value for the same substance may differ depending on the electrochemical process, in which the substance participates. Let us consider three different reactions of chlorine, hypocholite, and chlorate electrolytic evolution and anode's half-reactions (1.14, 1.16, 1.17).

$$\text{NaCl} + H_2O \rightarrow \text{NaOH} + \tfrac{1}{2}\ H_2 + \tfrac{1}{2}\ Cl_2 \tag{1.14}$$

Anode reaction:

$$Cl^- \rightarrow \tfrac{1}{2}\ Cl_2 + e^- \tag{1.15}$$

$$\text{NaCl} + H_2O \rightarrow \text{NaClO} + H_2 \tag{1.16}$$

Anode reaction:

$$Cl^- \rightarrow Cl^+ + 2e^-$$

$$\text{NaCl} + 3H_2O \rightarrow \text{NaClO}_3 + 3H_2 \tag{1.17}$$

Anode reaction:

$$Cl^- \rightarrow Cl^{5+} + 6e^-$$

As it can be seen from anode's reaction the number of electrons participating in electrolytic formation of chlorine, hypochlorite, and chlorate are equal to 1, 2, and 6 electrons, which means that $z = 1$, 2, or 6, respectively. Thus, electrochemical equivalents of NaCl for chlorine, hypochlorite, and chlorate formation are equal to $z = 58.44/(1 \cdot 96{,}485) = 6.1 \cdot 10^{-4}$ g/C $= 0.61$ mg/C; 0.3 mg/C; and 0.1 mg/C, respectively.

Faraday's laws are strictly observed. Observed deviations from Faraday's laws often associated with the presence of unaccounted parallel electrochemical reactions, such as oxygen evolution reactions, hydrogen peroxide formation, or products recombination reactions. Deviations from Faraday's law in industrial systems are associated with Faradic current losses appearing as heat or unwanted by-products, loss of material by spraying the solution, etc. The ratio of the actual amount of product (charge/electrons) obtained/spent in electrolysis to the theoretical amount of product (charge/electrons) calculated based on Faraday's law is typically below one in the technological processes. This relation is called current efficiency (CE; faradaic or coulombic efficiency; 1.18).

$$CE = \frac{m_{\text{actual}}}{m_{\text{theoretical}}} = \frac{Q_{\text{actual}}}{Q_{\text{theoretical}}} \tag{1.18}$$

Current efficiency in EO processes of organic pollutants can be monitored through the COD decay values at a constant current using Eq. (1.19) [15].

$$CE = \frac{\Delta(\text{COD})FV_s}{8It} \tag{1.19}$$

where ΔCOD is the COD decrease during degradation of pollutants at time t and 8 is the oxygen equivalent mass (q equiv^{-1}).

J. Gibbs and W. Nernst contributed to the development of electrochemical thermodynamics and particularly to the determination of the nature of electrical potential (voltage) in the electrochemical cell and the balance between electrical, chemical, and thermal energies. The electrochemical potential is determined by the energy of the chemical processes occurring in electrochemical cell and also depends on their kinetics.

Electrode Potentials

By the end of the XIX century it was clear that the main reason of the generation of electric current by the electrochemical cell is the emergence of the electrode potential, i.e., the difference in electrostatic potential between the electrode and a solution surrounding the electrode. The theory of the electrode/solution interface structure (electrical double layer theory) and the mechanism of the electrode potential were developed in the works of H. Helmholtz (1879), L. G. Gouy (1910), and O. Stern (1924).

The general concept of the generation of electrode potential is explained by the fact of spontaneous particle exchange between the metal (electrode) surface and surrounding solution. If an active metal such as Zn, Fe, or Ca is immersed in an aqueous solution of its salt, the negative poles of H_2O molecule affect the positive ions of the crystal lattice "withdrawing" them into electrolyte solution, which is becoming charged positively. When metal cations pass into solution, free electrons are left at the electrode surface. This leads to the formation of electrode's negative charge and formation of electrostatic potential. The more cations leave the electrode, the more free electrons accumulate in the electrode and the greater the electrostatic potential of the electrode.

While entering the solution, metal cations tend to be stabilized and undergo solvation, i.e., surrounding by solvent molecules. When the solvent is water, electronegative oxygen atom is attracted electrostatically to the positive metal cation. As a result, the cation becomes surrounded with water molecules. Solvation process takes place spontaneously with the release of energy. Simultaneously, the formation of the negative charge on the electrode prevents the further exit of cations in the solution, as well as keeps passed into solution ions in the near-electrode region. At the same time, a part of the cations from the solution interacts with the electrons and returns to the metal lattice nodes. When dynamic equilibrium is established (1.20), the transition of cations in the solution stops.

$$M + nH_2O \leftrightarrow \underset{\text{in solution}}{M^{m+}(H_2O)_n} + \underset{\text{on metal}}{me^-} \tag{1.20}$$

where M is the active metal.

When immersing less active metal such as Cu, Ag, or Pt in the electrolyte solution, reverse process occurs. The ions pass from solution in the crystal lattice, charging metal positively, and electrolyte solution is charged negatively because of excess anions.

The major part of the released cations into solutions (Fig. 1.5A; the case of active metals) or left in the solution anions (Fig. 1.5B; the case of less active metals) accumulates near the surface of the electrode. As the distance from the surface of the electrode increases, the concentration of cations/anions decreases and becomes equal to the concentration in solution. As a result the electrical double layer (EDL) is formed.

While EDL is formed, the potential difference between the electrode surface (ϕ_0) and solution at distance x from the electrode surface (ϕ_x) is called electrode potential (E, $E = \phi_0 - \phi_x$). Potential behavior in the EDL allows to separate a dense (I) and diffuse (II) parts in it. The dense part of the EDL (Helmholtz/Stern layer) is formed by ions located at a distance of ions radius (about 1 Å) to the surface of the electrode. The diffuse part of the ELD (the Gouy–Chapman layer) contains ions distributed deep into the solution due to their thermal motion. The thickness of the diffusion layer is about 1–10,000 nm. Potential in the dense layer changes linearly. Electrode potential describes the equilibrium between the unpolarized (nonoperated) electrode and electrolyte solution system. The value of electrode potential depends on

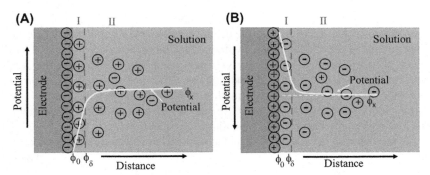

FIGURE 1.5 Schematic representation of the electrode potential development in the case of active (A) and less active (B) metal electrodes. Electrical double layer. ϕ_0 is the surface potential, ϕ_δ is the Stern potential, and ϕ_x is the potential at distance x from the surface, I is the dense part of the EDL (Helmholtz/Stern layer), and II is the diffuse part of the ELD (the Gouy-Chapman layer).

the material of electrode, the nature of the solvent, temperature, and concentration of the electrode exchange ions. Therefore, the electrode potential is measured by comparing with the reference electrode potential. Typically, this is the standard hydrogen electrode (SHE), whose potential is stable and is assumed to be zero $\left(E^0{}_{2H^+/H_2} = 0\ \mathrm{V}\right)$.

The theoretical value of electrode potential can be calculated using the Nernst equation.

$$E = E^0 - \frac{RT}{zF}\ln\frac{[a_{\mathrm{ox}}]}{[a_{\mathrm{red}}]} \tag{1.21}$$

where E is the electrode potential at a given temperature, E^0 is the standard electrode potential, R is the universal gas constant, T is the temperature, z is the number of electrons participating in the process, a_{ox} and a_{red} are activities of oxidized and reduced forms of the substance, respectively. Activities are often approximated by concentrations for dilute electrolyte solutions. In the case of the concentrated solutions activities have to be used since the calculation errors can be significant.

$a = \gamma \cdot C$, where γ is the activity coefficient, C is the concentration of the solution. Activity coefficients can be calculated theoretically from thermodynamic models, determined experimentally, or found from the literature.

Substituting appropriate values of constants in equation and converting natural logarithm in more frequently used decimal logarithm the following expression for the Nernst equation can be obtained.

$$E = E^0 - \frac{0.0002T}{z}\lg\frac{[a_{\mathrm{ox}}]}{[a_{\mathrm{red}}]} \tag{1.22}$$

The Nernst equation can be further simplified for a half-cell operating at room conditions (25°C).

$$E = E^0 - \frac{0.05916}{z} \lg \frac{[a_{ox}]}{[a_{red}]} \tag{1.23}$$

The Nernst equation for a full cell can be expressed as follows:

$$E_{cell} = E^0_{cell} - \frac{RT}{zF} \ln Q_r \tag{1.24}$$

where $Q_r = \frac{[a_{products}]}{[a_{reactants}]}$ is the reaction quotient; $E^0_{cell} = E^0_{red, cathode} - E^0_{red, anode}$ is the standard cell potential. The electrode's potential difference E_{cell} in galvanic cell is called the cell potential of the electromotive force (EMF). $E^0_{Red, cathode}/E^0_{Red, anode}$ is the standard reduction potential for the half-reaction occurring at the cathode/anode, respectively. $E^0_{Red} = E^0_{Ox}$.

A number of standard electrode potentials can be obtained by measuring the relative value of the metal potential toward hydrogen electrode under standard conditions (25°C, 1 M effective concentration of all reactive specious, 1 atm pressure for all gaseous specious). Standard electrode (reduction) potential can be found from the literature. Important standard electrode (reduction) potentials are listed in Table 1.2.

TABLE 1.2 Standard Electrodes Potentials and Reduction Half-Reactions [16]

Electrode	Electrode Reaction	E^0, V
Li^+/Li	$Li^+ + e^- = Li$	−3.04
K^+/K	$K^+ + e^- = K$	−2.936
Ba^{2+}/Ba	$Ba^{2+} + 2e^- = Ba$	−2.906
Ca^{2+}/Ca	$Ca^{2+} + 2e^- = Ca$	−2.868
Na^+/Na	$Na^+ + e^- = Na$	−2.714
La^{3+}/La	$La^{3+} + 3e^- = La$	−2.522
Mg^{2+}/Mg	$Mg^{2+} + 2e^- = Mg$	−2.36
Be^{2+}/Be	$Be^{2+} + 2e^- = Be$	−1.968
Al^{3+}/Al	$Al^{3+} + 3e^- = Al$	−1.677
Ti^{2+}/Ti	$Ti^{2+} + 2e^- = Ti$	−1.628
Zr^{4+}/Zr	$Zr^{4+} + 4e^- = Zr$	−1.529
V^{2+}/V	$V^{2+} + 2e^- = V$	−1.186

Continued

TABLE 1.2 Standard Electrodes Potentials and Reduction Half-Reactions [16]—cont'd

Electrode	Electrode Reaction	E^0, V
Mn^{2+}/Mn	$Mn^{2+} + 2e^- = Mn$	−1.180
WO_4^{2-}/W	$WO_4^{2-} + 4H_2O + 6e^- = W + 8OH^-$	−1.05
Se^{2-}/Se	$Se + 2e^- = Se^{2-}$	−0.77
Zn^{2+}/Zn	$Zn^{2+} + 2e^- = Zn$	−0.762
Cr^{3+}/Cr	$Cr^{3+} + 3e^- = Cr$	−0.744
Ga^{3+}/Ga	$Ga^{3+} + 3e^- = Ga$	−0.529
S^{2-}/S	$S + 2e^- = S^{2-}$	−0.51
Fe^{2+}/Fe	$Fe^{2+} + 2e^- = Fe$	−0.44
$Cr^{3+}, Cr^{2+}/Pt$	$Cr^{3+} + e^- = Cr^{2+}$	−0.408
Cd^{2+}/Cd	$Cd^{2+} + 2e^- = Cd$	−0.402
$Ti^{3+}, Ti^{2+}/Pt$	$Ti^{3+} + e^- = Ti^{2+}$	−0.369
Pb^{2+}/Pb	$PbSO_4 + 2e^- = Pb(s) + SO_4^{2-}$	−0.3588
Tl^+/Tl	$Tl^+ + e^- = Tl$	−0.3363
Co^{2+}/Co	$Co^{2+} + 2e^- = Co$	−0.277
Ni^{2+}/Ni	$Ni^{2+} + 2e^- = Ni$	−0.236
Mo^{3+}/Mo	$Mo^{3+} + 3e^- = Mo$	−0.20
Sn^{2+}/Sn	$Sn^{2+} + 2e^- = Sn$	−0.141
Pb^{2+}/Pb	$Pb^{2+} + 2e^- = Pb$	−0.126
$Ti^{4+}, Ti^{3+}/Pt$	$Ti^{4+} + e^- = Ti^{3+}$	−0.04
$D^+/D_2, Pt$	$D^+ + e^- = 1/2\ D_2$	−0.0034
$H^+/H_2, Pt$	$H^+ + e^- = 1/2\ H_2$	
Ge^{2+}/Ge	$Ge^{2+} + 2e^- = Ge$	+0.01
$Br^-/AgBr/Ag$	$AgBr + e^- = Ag + Br^-$	+0.0732
$Sn^{4+}/Sn^{2+}, Pt$	$Sn^{4+} + 2e^- = Sn^{2+}$	+0.142
$Cu^{2+}, Cu^+/Pt$	$Cu^{2+} + e^- = Cu^+$	+0.161
Cu^{2+}/Cu	$Cu^{2+} + 2e^- = Cu$	+0.339
$Fe(CN)_6^{4-}, Fe(CN)_6^{3-}/Pt$	$Fe^-(CN)_6^{3-} + e^- = Fe^-(CN)_6^{4-}$	+0.36

TABLE 1.2 Standard Electrodes Potentials and Reduction Half-Reactions [16]—cont'd

Electrode	Electrode Reaction	E^0, V
OH^-/O_2, Pt	$\frac{1}{2} O_2 + H_2O + 2e^- = 2OH^-$	+0.401
Cu^+/Cu	$Cu^+ + e^- = Cu$	+0.521
I^-/I_2, Pt	$I_2 + 2e^- = 2I^-$	+0.535
Te^{4+}/Te	$Te^{4+} + 4e^- = Te$	+0.56
MnO_4^-, MnO_4^{2-}/Pt	$MnO_4^- + e^- = MnO_4^{2-}$	+0.564
Fe^{3+}, Fe^{2+}/Pt	$Fe^{3+} + e^- = Fe^{2+}$	+0.771
Hg_2^{2+}/Hg	$Hg_2^{2+} + 2e^- = 2Hg$	+0.788
Ag^+/Ag	$Ag^+ + e^- = Ag$	+0.799
Hg^{2+}/Hg	$Hg^{2+} + 2e^- = Hg$	+0.854
Hg^{2+}, Hg^+/Pt	$Hg^{2+} + e^- = Hg^+$	+0.915
Pd^{2+}/Pd	$Pd^{2+} + 2e^- = Pd$	+0.987
Br^-/Br_2, Pt	$Br_2 + 2e^- = 2Br^-$	+1.078
Pt^{2+}/Pt	$Pt^{2+} + 2e^- = Pt$	+1.18
Mn^{2+}, H^+/MnO_2, Pt	$MnO_2 + 4H^+ + 2e^- = Mn^{2+} + 2H_2O$	+1.23
Cr^{3+}, $Cr_2O_7^{2-}$, H^+/Pt	$Cr_2O_7^{2-} + 14H^+ + 6e^- = 2Cr^{3+} + 7H_2O$	+1.3
Tl^{3+}, Tl^+/Pt	$Tl^{3+} + 2e^- = Tl^+$	+1.25
Cl^-/Cl_2, Pt	$Cl_2 + 2e^- = 2Cl^-$	+1.36
Pb^{2+}, H^+/PbO_2, Pt	$PbO_2 + 4H^+ + 2e^- = Pb^{2+} + 2H_2O$	+1.458
Au^{3+}/Au	$Au^{3+} + 3e^- = Au$	+1.498
MnO_4^-, H^+/MnO_2, Pt	$MnO_4^- + 4H^+ + 3e^- = MnO_2 + 2H_2O$	+1.692
Ce^{4+}, Ce^{3+}/Pt	$Ce^{4+} + e^- = Ce^{3+}$	+1.61
SO_4^{2-}, $H^+/PbSO_4$, PbO_2, Pb	$PbO_2 + SO_4^{2-} + 4H^+ + 2e^- = PbSO_4 + 2H_2O$	+1.682
Au^+/Au	$Au^+ + e^- = Au$	+1.69
H^-/H_2, Pt	$H_2 + 2e^- = 2H^-$	+2.2
F^-/F_2, Pt	$F_2 + 2e^- = 2F^-$	+2.89

Electrochemical Cells

Basic electrical parameters of electrochemical cells are current (it is measured in amperes, A) and potential (measured in volts, V). The current is determined by the rate of electrode reactions and the potential is determined by the energy of chemical processes occurring in the cell. The potential is equal to the energy (measured in joules, J) and referred to the amount of electric charge (measured in coulombs, C), i.e., 1 V = 1 J/C. Consequently, the potential of the cell (EMF) is a measure of the energy generated during the reactions occurring therein.

The electrochemical cell typically consists of two electrodes immersed in an electrolyte. The electrodes are made of conductive material (metal or carbon) and more rarely of a semiconductor. The charge carriers in the electrodes are electrons (an exceptional type of conductivity is the hole conductivity of semiconductors) and in the electrolyte are ions. For example, the electrolyte solution of common salt sodium chloride (NaCl) contains charged particles of sodium cations (Na^+) and chlorine anions (Cl^-). If this solution is placed in an electric field, the Na^+ ions would move to the negative cathode and Cl^- ions would move to the positive anode. Molten salt such as NaCl and some solids such as β-alumina (sodium polyaluminate) containing mobile ions of sodium or ion-exchange resins can also be electrolytes [17]. The most common charge carriers in natural wasters are cations such as K^+, Ca^{2+}, Na^+, Mg^{2+}, and H^+ and anions such as HCO_3^-, SO_4^{2-}, Cl^-, and OH^-. In the case of multiple electrolytes, electrodes can be divided by a cell separator, which prevents electrolytes mixing but does not interfere with the movement of ions. Salt bridges, ion exchange membranes, and porous glass plates could serve the role of such partitions.

There are two types of electrochemical cells. They are galvanic (voltaic) cells and electrolytic cells (electrolyzers). Spontaneous chemical reactions at the electrode/electrolyte interface take place in galvanic cells. Several galvanic cells connected in series form a battery or, in other words, the chemical source of energy. In the electrolytic cell reaction at the electrode/electrolyte interface occur due to external electrical energy source; the latter is converted into chemical energy of reactions products occurring at the electrodes. Schematic drawings of electrochemical and galvanic cells are shown on Fig. 1.6 and 1.7.

It is worth to notice that the same cell can operate either as a galvanic or electrolytic cell depending on the operating mode. As an example, lead-acid batteries in cars can act as a galvanic element when used for engine start (in this case it is discharged) and as an electrolytic cell when charging from a car generator or a battery charger. According to the IUPAC rules the schematic representation of electrochemical cells (irrespective of whether they are galvanic or electrolytic) shows that left-hand half-cell must be one with an electrochemical oxidation process and a right-hand one is with a process of electroreduction. In the galvanic cell, electrode in the left-hand half-cell is

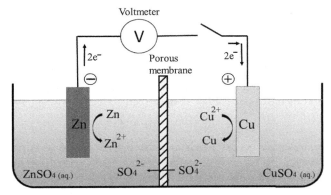

FIGURE 1.6 Galvanic cell. Every two electrons moving through an external circuit cause an oxidation of one zinc atom and deposition of one copper atom.

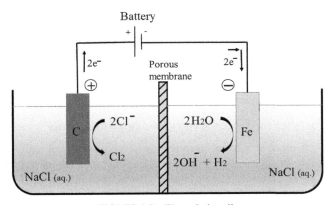

FIGURE 1.7 Electrolytic cell.

signed with the minus ($-$) sign and in the right-hand half-cell is with the plus ($+$) sign.

Simple galvanic cell, which is similar to that shown in Fig. 1.6, was invented in 1836 by J. F. Daniell. The only difference between the cell in Fig. 1.6 and the Daniell cell is the partition. In the case of Daniell cell, salt bridge played the role of such partition. Simple galvanic cell in Fig. 1.6 consists of zinc and copper electrodes. Zinc electrode (anode) is immersed in an aqueous solution of zinc sulfate and copper electrode is in an aqueous solution of copper sulfate. When external circuit is closed, zinc atoms are oxidized to ions on the surface of zinc electrode releasing electrons.

$$Zn \rightarrow Zn^{2+} + 2e^- \qquad (1.25)$$

These electrons move through an external circuit on the copper electrode where copper ions are reduced to atoms

$$Cu^{2+} + 2e^- \rightarrow Cu \qquad (1.26)$$

The flow of electrons into an external circuit is the current produced by the cell. The overall balanced reaction leading to the chemical transformation and the generation of electrical energy is described by the following equation:

$$Zn + Cu^{2+} \rightarrow Cu + Zn^{2+} \qquad (1.27)$$

The same reaction occurs when metallic zinc is added to copper sulfate solution; however, in that case chemical energy is converted into heat. When the external circuit is open, electrode reactions do not occur.

To describe any galvanic cell diagram in words, specific rules for the line notation have been adapted. In the line notation anode is written to the far left and cathode is written to the far right. The phase boundaries in a half-cell are shown by a vertical line ($|$ or $/$), and the boundary between two half-cells connected with a salt bridge or porous membrane is shown by a double vertical line ($\|$ or $//$). Electroactive species of a single half-cell participating in oxidation or reduction reaction are separated by comma. The concentrations or pressures of all reactive species are given within brackets right after the species notation if their values are different from the standard state values (1 atm. for gases or 1 M for solutions). Thus, the line notation for the cell shown in Fig. 1.6 is written as follows:

$$Zn_{(s)} \big| Zn^{2+}{}_{(aq)} \big\| Cu^{2+}{}_{(aq)} \big| Cu_{(s)} \qquad (1.28)$$

Passage of electric current through the electrochemical cell is always accompanied by the occurrence of electrochemical processes at the electrodes. Fig. 1.7 shows an electrolytic cell, which is similar to industrial electrolyzers for the production of chlorine gas and sodium hydroxide from a brine (a saturated aqueous solution of sodium chloride). Each of the two electrons moving in external circuit leads to the oxidation of two chloride ions to one molecule of chlorine gas at carbon electrode and decomposition of two water molecule to one hydrogen atom and two hydroxide ions on iron electrode. Electric potential different from the equilibrium potential is established during the electrolysis. This potential forces electrochemical reactions to occur. Because potential is a measure of energy, the mechanism of the initiation of electrochemical reactions can be considered on the example of the Fermi level (E_F) of an electrode metal (M) and molecular orbitals of reactant in electrolyte solution near to the electrode surface (Fig. 1.8). The Fermi level is the highest energy level occupied by electrons in atom. If the Fermi level is lower than the lowest unoccupied molecular orbital (LUMO) of a molecule, then according to thermodynamics, electrons cannot jump from the electrode to the molecule. When applying external electrical energy to the electrode, the Fermi level can be raised as it is not fixed. If the magnitude of applied potential difference is

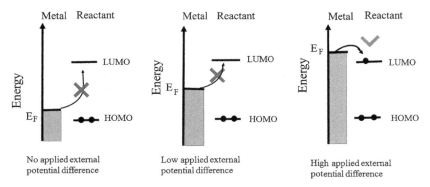

FIGURE 1.8 Schematic representation of electron transfer from an electrode to a molecule of electrolyte solution. *LUMO*, lowest unoccupied molecular orbital; *HOMO*, highest occupied molecular orbital.

low, it cannot raise the Fermi level of the metal sufficiently and electron transfer still remains thermodynamically unfavorable. When the magnitude of the applied external potential difference is sufficient, the Fermi level can be lifted above the LUMO level of a molecule and electron transfer from the electrode to the molecule becomes thermodynamically favorable. The probability of further electrochemical reaction will depend on the rate of electron transfer and will be discussed in more detail under polarization part. In electrolytic cell negative ions undergo oxidation (giving off electrons to the circuit) at anode. In the example of Fig. 1.7 chloride ions are oxidized to chlorine gas at a graphite electrode, and water is reduced to hydrogen and hydroxide ions at an iron electrode. Electrolytes remain electrically neutral due to the movement of sodium ions through the ion exchange membrane.

$$2Cl^- + 2H_2O \rightarrow Cl_2 + H_2 + 2OH^- \tag{1.29}$$

Compounds, participating in the reactions caused by electrolysis, can be divided into three groups. They are compounds of electrolyte solution, material of electrode, and extraneous substances.

Compounds of Electrolyte Solution

While carrying out the electrolysis, reactions, which cannot occur spontaneously according to the laws of thermodynamics, can be conducted. Chemical transformations can involve both compounds of electrolyte and solvent. In the case of aqueous solutions, water plays the role of the solvent and water molecules dissociate with formation of H^+ and OH^- ions. Thus, the electrolyte cations and hydrogen ions will be reduced at the cathode and electrolyte anions and hydroxide ions will be oxidized at the anode. The overall reaction of water decomposition is described as follows [18]:

$$2H_2O \rightarrow 2H_2 + O_2 \tag{1.30}$$

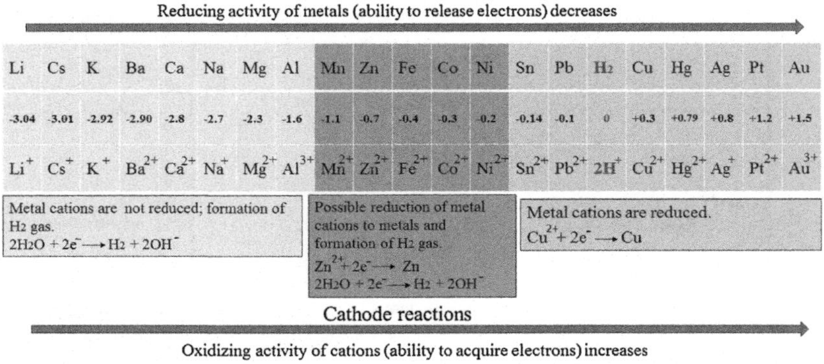

Reducing activity of metals (ability to release electrons) decreases

Li	Cs	K	Ba	Ca	Na	Mg	Al	Mn	Zn	Fe	Co	Ni	Sn	Pb	H_2	Cu	Hg	Ag	Pt	Au
-3.04	-3.01	-2.92	-2.90	-2.8	-2.7	-2.3	-1.6	-1.1	-0.7	-0.4	-9.3	-0.2	-0.14	-0.1	0	+0.3	+0.79	+0.8	+1.2	+1.5
Li^+	Cs^+	K^+	Ba^{2+}	Ca^{2+}	Na^+	Mg^{2+}	Al^{3+}	Mn^{2+}	Zn^{2+}	Fe^{2+}	Co^{2+}	Ni^{2+}	Sn^{2+}	Pb^{2+}	$2H^+$	Cu^{2+}	Hg^{2+}	Ag^+	Pt^{2+}	Au^{3+}

Metal cations are not reduced; formation of H_2 gas. $$2H_2O + 2e^- \longrightarrow H_2 + 2OH^-$$	Possible reduction of metal cations to metals and formation of H_2 gas. $$Zn^{2+} + 2e^- \longrightarrow Zn$$ $$2H_2O + 2e^- \longrightarrow H_2 + 2OH^-$$	Metal cations are reduced. $$Cu^{2+} + 2e^- \longrightarrow Cu$$

Cathode reactions

Oxidizing activity of cations (ability to acquire electrons) increases

FIGURE 1.9 Reactivity series of some metals.

The implementation of each specific reaction is associated to a minimum energy consumption. Therefore, at first the oxidized form of the compound with a higher value of the electrode potential is reduced at the cathode and the reduced form of the compound with a lower electrode potential is oxidized at the anode.

The data on standard electrode potentials allow constructing the reactivity series (Fig. 1.9), which give the information on the metal reactivity with aqueous solutions of salts and acids as well as on the cathodic and anodic processes at electrolysis can be compiled. For example, if an anode is made of metals such as Cu, Ag, Ni, Fe, or Zn, it will be oxidized and dissolved into solution in the form of cations. The cations will gradually move to the cathode where they are reduced and deposited.

$$M^0 - ze^- \rightarrow M^{z+} \quad \text{if } E^0_M < 1.23 \text{ V}.$$

As it is seen in Fig. 1.9 reducing activity of metals (ability to donate electrons) decreases and oxidizing activity (ability to gain electrons) of metal cations increases in the reactivity series from left to right. In other words, the smaller the value of the metal electrode potential, the more prone the metal atoms to oxidation, that is more active the metal. More active metal (which is located more left in the series) displaces a less active from a salt solution. For example, interaction of metallic Zn^0 and Cu(II) ions can flow only in the forward direction (see R. 1.27). Zinc displaces copper from an aqueous solution of its salt. Thus zinc metal is dissolved and copper metal is liberated out of solution. Another example of displacement of less active metal from the solution of its salt is the displacement of mercury by copper (1.31).

$$Cu + Hg(NO_3)_2 \rightarrow Hg + Cu(NO_3)_2 \quad (1.31)$$

Metals, standing to the right of hydrogen, do not interact with the solutions of nonoxidizing acids (i.e., acids that cannot act as oxidizing agent; for example, HCl, HBr, HF, HI, and H_3PO_4) under normal conditions. Metals located up to hydrogen in the reactivity series displace hydrogen gas from dilute acids (see 1.32).

$$Zn + 2HCl \rightarrow H_2 + ZnCl_2 \qquad (1.32)$$

Cathodic processes in the electrolysis of aqueous solutions of electrolytes are subjected to certain rules of discharge. In the presence of metal cations in the solution, standing at the beginning of reactive series (approximately up to Al), only hydrogen is released at the cathode due to the water (neutral or alkaline solutions) or H^+ ions reduction (acid medium).

At cathode

$$2H^+ + 2e^- \rightarrow H_2, \qquad E^0 = 0.00 \text{ V} \qquad pH < 7$$
$$2H_2O + 2e^- \rightarrow H_2 + 2OH^- \quad E^0 = -0.83 \text{ V} \quad pH \geq 7$$

If the cation of the electrolyte has a higher electrode potential (approximately from Sn to Au), then only metal cations are discharged at the cathode. If the electrolyte solution contains the metal cations of average potential in the reactive series (from Al up to $\sim$ Ni), water molecules and metal cations can simultaneously be reduced at the cathode. In the presence of multiple species in the electrolyte solution, cations with a higher potential value are reduced first.

Metal cations in the solution of simple salts are discharged at the cathode with a small overpotential, i.e., at potentials close to the equilibrium values of $E_{Me^+/Me}$. Cathodic overpotential is close to zero for ions of metals such as Zn, Cd, Cu, Ag, Au, and Hg.

Electrode Material Itself

Anodic processes in the electrolysis of aqueous solutions of electrolytes are determined by the material of the anode and the anion types. All anodes can be divided into two groups. They are inert and active. Inert or insoluble anodes (for example, graphite, Au, Pt, Ir, or Ta) are not oxidized during electrolysis. Such anodes can promote chemical reactions without participating in them only through the supply or withdraw of electrons from an electrolyte. So they are used for the electrolysis of the metal salts, which are located from the beginning of reactive series up to Al inclusive. Active or soluble anodes are oxidized during electrolysis.

Depending on the nature of the anion, different processes may occur on inert anodes during electrolysis. Anions (A^-) of hydrohalic acids (Cl^-, Br^-, I^-, except F^- and including S^{2-}), which are present in the solution are oxidized at the anode (see Fig. 1.10A). Oxidation of oxygen-containing anions can take place at very high potentials.

$$An^{z-} - ze^- \rightarrow An^0$$

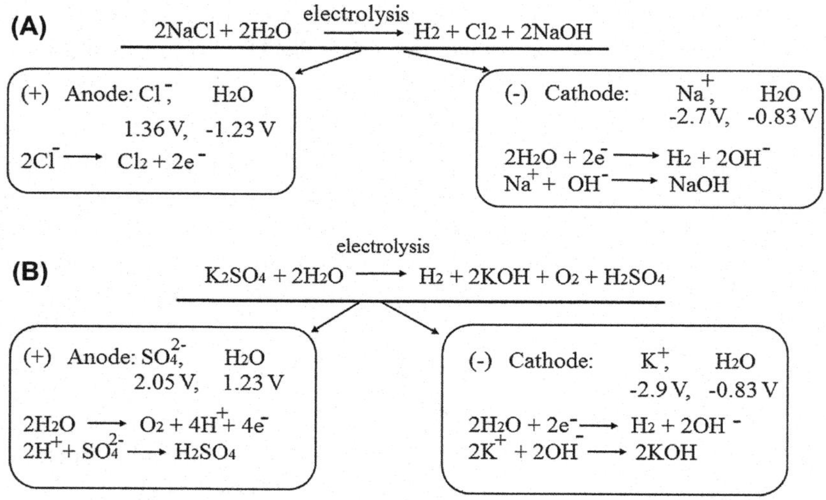

(A)

$$2NaCl + 2H_2O \xrightarrow{electrolysis} H_2 + Cl_2 + 2NaOH$$

> (+) Anode: Cl^-, H_2O
> 1.36 V, -1.23 V
> $2Cl^- \longrightarrow Cl_2 + 2e^-$

> (-) Cathode: Na^+, H_2O
> -2.7 V, -0.83 V
> $2H_2O + 2e^- \longrightarrow H_2 + 2OH^-$
> $Na^+ + OH^- \longrightarrow NaOH$

(B)

$$K_2SO_4 + 2H_2O \xrightarrow{electrolysis} H_2 + 2KOH + O_2 + H_2SO_4$$

> (+) Anode: SO_4^{2-}, H_2O
> 2.05 V, 1.23 V
> $2H_2O \longrightarrow O_2 + 4H^+ + 4e^-$
> $2H^+ + SO_4^{2-} \longrightarrow H_2SO_4$

> (-) Cathode: K^+, H_2O
> -2.9 V, -0.83 V
> $2H_2O + 2e^- \longrightarrow H_2 + 2OH^-$
> $2K^+ + 2OH^- \longrightarrow 2KOH$

FIGURE 1.10 An example of electrolysis of aqueous solutions of (A) NaCl and (B) K_2SO_4 in acidic media at inert anodes.

Water molecules or hydroxide ions are discharged at the anode in acidic and neutral or alkaline media, respectively, in the presence of oxygen containing anions $(SO_4{}^{2-}, PO_4{}^{3-}, NO_3{}^-, OH^-,$ etc.$)$ and the fluoride ion F^- (see Fig. 1.10B).

At anode

$$2H_2O - 4e^- \rightarrow O_2 + 4H^+ \quad E^0 = 1.23 \text{ V} \quad pH \leq 7$$
$$4OH^- - 4e^- \rightarrow O_2 + 2H_2O \quad E^0 = -0.41 \text{ V} \quad pH > 7$$

If the electrolyte solution contains several kinds of anions, the anions with a lower electrode potential are oxidized primarily at the inert anode. In the case of active (soluble) anode, the anode metal undergoes oxidation (dissolution) if its standard electrode potential is lower than a standard electrode potential of the oxygen electrode (+1,229 V).

Let us now compare the electrolysis of $NiSO_4$ aqueous solution in neutral media on an inert anode (Fig. 1.11A) such as carbon anode and compare it to the electrolysis on an active Ni anode (Fig. 1.11B).

Ni metal is located in the middle of the reactive series. This means that there are two simultaneous processes on the cathode. They are reduction of metal cation and water molecules.

At cathode

$$Ni^{2+} + 2e^- \rightarrow Ni \quad E^0 = -0.25 \text{ V}$$
$$2H_2O + 2e^- \rightarrow H_2 + 2OH^- \quad E^0 = -0.83 \text{ V}$$

(A)

Inert anode

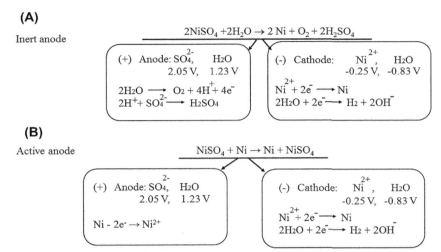

(B)

Active anode

FIGURE 1.11 Electrolysis of aqueous solutions of $NiSO_4$ in neutral media at (A) inert and (B) active anodes.

Two above mentioned reactions cannot be summarized in the overall reaction and have no relationship between the amount of nickel and hydrogen produced. The amount of hydrogen depends on the concentration, temperature of the cathode material, and other factors. Sulfate anion is not oxidized at anode, but water molecule is oxidized.

At anode

$$2H_2O - 4e^- \rightarrow O_2 + 4H^+ \quad E^0 = 1.23 \text{ V}$$

The overall reaction of the direct electrolysis taking into account the balance of electronic factor is as follows:

$$Ni^{2+} + 2e^- \rightarrow Ni \qquad |2$$
$$2H_2O - 4e^- \rightarrow O_2 + 4H^+ \qquad |1$$
$$2Ni^{2+} + 2H_2O \rightarrow 2Ni + O_2 + 4H^+$$
$$2NiSO_4 + 2H_2O \rightarrow 2Ni + O_2 + 2H_2SO_4$$

The overall side reaction is as follows:

$$2H_2O + 2e^- \rightarrow H_2 + 2OH^- \qquad |2$$
$$2H_2O - 4e^- \rightarrow O_2 + 4H^+ \qquad |1$$
$$6H_2O \rightarrow 2H_2 + 4OH^- + O_2 + 4H^+$$
$$2H_2O \rightarrow 2H_2 + O_2$$

Let us now consider the electrolysis of $NiSO_4$ using active Ni anode. Cathode reactions will be similar to those considered in the electrolysis with inert anode. Anode material (Ni) will undergo the oxidation with the formation of nickel cations, which further move to the cathode and undergo reduction.

There are two reactions at cathode, which are the main reaction of Ni deposition and the side one of water molecules reduction; therefore, there are two overall reactions. The main electrolysis reaction, the process of which is used for electrochemical Ni purification, is as follows:

$$Ni^{2+} + 2e^- \rightarrow Ni$$
$$\underline{Ni - 2e^- \rightarrow Ni^{2+}}$$
$$NiSO_4 + Ni \rightarrow Ni + NiSO_4$$

The overall side reaction is as follows:

$$2H_2O + 2e^- \rightarrow H_2 + 2OH^-$$
$$\underline{Ni - 2e^- \rightarrow Ni^{2+}}$$
$$Ni + 2H_2O \rightarrow H_2 + Ni(OH)_2$$

Extraneous Substances

Extraneous substances are not included in the composition of electrodes or electrolytes (gases, solids), but able to get to the surface of electrode and leave it.

The power (P) consumed during the electrolysis process directly influences the cost of the treatment and can be found by Eq. (1.33) [19]

$$P = \frac{k_m \cdot A \cdot E_{cell} \cdot Q}{2.303 \cdot V_R} \tag{1.33}$$

where k_m is the mass transfer coefficient, V_R is the reaction volume, A is the electrode area, E_{cell} is the cell potential, and Q is the electric charge.

Important parameter of electrochemical reactors, which should be considered at the planning stage is the normalized space velocity (s_n). Normalized space velocity shows the volume of effluent, which can be treated in the reactor of volume V_R per unit time (1.34).

$$s_n = \frac{k_m \cdot A}{2.3 \cdot V_R} \tag{1.34}$$

Briefly the difference between galvanic (electrochemical) and electrolytic cells (electrolyzer) are summarized in Table 1.3.

Thermodynamics

Thermodynamic phenomena play an important role in electrochemical processes. The work (W) done by the galvanic cell and spent for the chemical transformation of 1 mol of a reactant on the electrodes is determined as follows:

$$W = E_{cell} \cdot Q = z \cdot F \cdot E_{cell}, \tag{1.35}$$

where Q is electric charge spent for the transformation of 1 mol of reactant.

TABLE 1.3 Difference Between Galvanic and Electrolytic Cells

Galvanic Cell	Electrolytic Cell
Galvanic cell converts chemical energy into electrical one.	Electrolytic cell converts electrical energy to chemical one.
Redox reactions at the electrode surface are spontaneous.	Redox reactions are not spontaneous and initiated by applied external electrical energy.
Electrons are generated by oxidized compounds at anode and then transformed to the cathode through the external circuit.	Electrons are applied to the cathode from external battery.
Anode is a negatively charged electrode, and cathode is a positively charged electrode.	Anode is a positively charged electrode, and cathode is a negatively charged electrode.

To estimate the spontaneity of chemical reactions on electrodes the cell potential (E_{cell}) and the change in the Gibbs energy (ΔG; $W = -\Delta G$) can be used. E_{cell} is positive and ΔG is negative in galvanic cells with spontaneous reactions and under constant temperature and pressure. The dependence of E_{cell} on ΔG is expressed by Eq. (1.36):

$$\Delta G = -z \cdot F \cdot E_{cell} \quad \text{or} \quad \Delta G^0 = -z \cdot F \cdot E^0_{cell} \quad \text{under standard conditions}$$

(1.36)

Analogous to ΔG, E_{cell} is used to determine the direction of chemical reactions in electrochemical cells. The following can be concluded from Eq. (1.36):

- If $E_{cell} > 0$ ($\Delta G < 0$), then spontaneous reaction proceeds in the forward direction (from left to right ion the balanced chemical reaction) and system is analogous to a galvanic cell.
- If $E_{cell} < 0$ ($\Delta G > 0$), then spontaneous reaction proceeds in the reverse direction (from right to left in the balanced chemical reaction) and to conduct the reaction in the forward direction an external source of electric energy (EMF) is required. In that case the system will be similar to an electrolytic cell (electrolyzer).
- If $E_{cell} = 0$ ($\Delta G = 0$), then the redox system is in equilibrium state.

The change of the Gibbs energy in an electrochemical process is the sum of the Gibbs energies of reaction products with deduction of the sum of the Gibbs energies for the reactants of spontaneous reaction. Let us consider the following reaction:

$$aA + bB \rightarrow cC + dD \pm ze^-$$

(1.37)

where a, b, c, d are stoichiometric coefficients.

The change in the Gibbs energy for R. (1.37) can be calculated as follows:

$$\Delta G = (c \cdot \Delta G_C + d \cdot \Delta G_D) - (a \cdot \Delta G_A + b \cdot \Delta G_B) \tag{1.38}$$

Using the similar approach and knowing the values of enthalpy for the products and reactants, the thermal effect of electrochemical reaction can be calculated:

$$\Delta H = \sum \Delta H_{\text{products}} - \sum \Delta H_{\text{reactants}} \tag{1.39}$$

Entropy allows to determine the direction of the thermodynamic processes in insulated systems. According to the Gibbs–Helmholtz equation:

$$\Delta G = \Delta H + \frac{T\Delta G}{\Delta T} = \Delta H - T\Delta S \tag{1.40}$$

$$z \cdot F \cdot E_{\text{cell}} = -\Delta H + z \cdot F \cdot T \frac{dE_{\text{cell}}}{dT} \tag{1.41}$$

$$E_{\text{cell}} = -\frac{\Delta H}{zF} + T \frac{dE_{\text{cell}}}{dT} \tag{1.42}$$

$$\Delta S = z \cdot F \cdot \frac{dE_{\text{cell}}}{dT} \tag{1.43}$$

The temperature coefficient $\frac{dE_{\text{cell}}}{dT}$ can be calculated from the approximate Eq. (1.44) using the measured values $E_{\text{cell}1}$ and $E_{\text{cell}2}$ at the corresponding temperatures $T_{\text{cell}1}$ and $T_{\text{cell}2}$.

$$\frac{dE_{\text{cell}}}{dT} \approx \frac{\Delta E_{\text{cell}}}{\Delta T} = \frac{E_{\text{cell 2}} - E_{\text{cell 1}}}{T_{\text{cell 2}} - T_{\text{cell 1}}} \tag{1.44}$$

As it is seen from Eq. (1.41) that if $\frac{dE_{\text{cell}}}{dT} < 0$, then electrical work is smaller than thermal effect of the reaction. It means that under isothermal condition, galvanic cell gives its heat to the environment or is heated in thermal insulation. Pb, $PbCl_2$ | Cl^- | AgCl, Ag can be mentioned as an example of such galvanic cell.

If $\frac{dE_{\text{cell}}}{dT} > 0$, then electrical work is higher than thermal effect of the reaction. It means that such system compensates the lack of energy from the environment and cools down in thermal insulation. Pb | $Pb(CH_3COO)_2$ || $Cu(CH_3COO)_2$ | Cu can serve as an example of such system.

The equilibrium constant (K) of a chemical reaction can be found using the following equation:

$$\ln K = -\frac{\Delta G^0}{R \cdot T} = \frac{z \cdot F \cdot E^0_{\text{cell}}}{R \cdot T} \tag{1.45}$$

Polarization

Electrode potential of operating cell is always different than the calculated equilibrium value corresponding to the reversible electrochemical reaction. The reason for this is the polarization. Polarization is a change of potential from its equilibrium value under the flow of electrical current. The difference between equilibrium potential (E) and applied potential (E_j) required to initiate the initiate the electrode reactions is called overpotential (ΔE).

$$\Delta E = E_j - E \tag{1.46}$$

Polarization curves, showing the dependence of electrode potential on the flowing through the electrode electrical current, are used to determine polarization. From the Faraday's law it is known that current is proportional to the amount of substance reacted at the electrode per unit time, i.e., the electrochemical reaction rate. Therefore, the current value may be used to quantify the rate of the electrochemical reaction. Because the electrodes can be different in size, the same potential can generate different currents. Therefore, the reaction rate is usually referred to a unit of electrode surface area. Current (I) to electrode area (A) relation is called the current density (j).

$$j = I/A \tag{1.47}$$

Fig. 1.12 shows an example of polarization curve for a specific electrode–electrolyte combination.

The magnitude of electrode polarization can be determined by the difference between the potential at the passage of current density E_j and the equilibrium potential E. It can be seen that higher current densities generate the greater deviation of potential from the equilibrium potential. Thus, the higher the polarization of the electrode, the greater the current density; in other words, the electrochemical reaction rate can be increased by increasing the polarization.

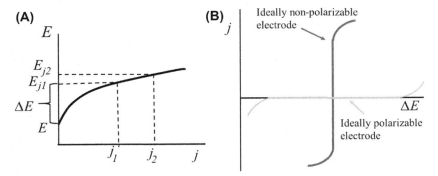

FIGURE 1.12   Experimental (A) and ideal (B) polarization curves.

The polarization is caused by the inhibition of the individual stages of the electrochemical process such as the rate of electrode reactions, supply and removal of reactants, and reaction products to and from the surface of electrode, respectively. The rate of an electrode reaction depends on the displacement value of electrode potential from its equilibrium potential. The rate of electrochemical reactions is determined by the rate of charge (electron) transfer between the electrode and electrolyte species, which becomes critical in the near-electrode layer (few Å from the surface of the electrode). The rate of electrochemical reactions controlled by the movement of reactants and reaction products in diffuse layer near the electrode is referred to the mass transport mechanism. Diffuse layer is a layer near to the electrode surface where there are changes in solution concentration; however, electrical neutrality is preserved. This layer must be distinguished from the diffuse layer of EDL (Fig. 1.5), which is located closer to the electrode and where the total charge of cations and anions differ in sign and absolute value. Therefore, polarization is mainly divided into activation polarization, dealing with the inhibition of electron transfer in the near-electrode layer, and concentration polarization, considering the inhibition of reactants mass transport in the diffuse layer. Sometimes the ohmic polarization caused by the ohmic resistance of electrolyte solution is also distinguished.

Ohmic Polarization

Ohmic polarization or ohmic drop or *IR* drop (ΔE_{ohm}) is referred to the resistance of the media during the flow of electrical current through the cell. Ohmic polarization is related to the rate of ion flow between electrodes and described by the magnitude of current, electrolyte conductivity, and the distance between the two electrodes. According to the Ohm's law, ohmic potential can be represented as follows:

$$\Delta E_{ohm} = -I \cdot R_{cell} = -j \cdot R_{cell}, A \tag{1.48}$$

where $R_{cell} = \frac{l}{\rho \cdot A}$ is the ohmic resistance of a cell [20]; l is the interelectrode distance, ρ is solution specific conductance, and A is the working electrode area.

The minus sign is attributed to the fact that electrical current flows toward more negative potentials in the external circuit. To initiate the electrochemical reactions in a cell, potential value, which includes the ohmic drop ($-IR$) due to the cell media resistance, should be applied to the cell. In other words, while applying potential to a cell equal to the equilibrium potential, no chemical reactions and as a result no current will flow in the cell. Therefore, the value of applied cell potential, which considers the ohmic drops and loss of potential due to activation and concentration overpotentials, can be expressed as follows [19]:

$$E_{cell} = E^0{}_{cell} - \sum |\Delta E| - \sum IR \tag{1.49}$$

Activation Polarization

Activation polarization is referred to the kinetics of all electrochemical reactions at electrodes. Reaction rates depend on many factors such as temperature, the rate of electron transfer from metal electrodes to ions carriers, and the rate of electron transfer within metals, for example, evolution of hydrogen gas $2H^+ + 2e^- \rightarrow H_2$ and its accumulation at the electrode/electrolyte interface, which prevents the passage of charge carriers to the surface of the electrode. Electron transfer occurs at a finite rate in any electrochemical reaction. Potential shift during the passage of electric current caused by the slowness of the charge transfer stage is called activation polarization.

The relation between activation polarization and current density (see Fig. 1.13) at first was obtained empirically and expressed by the Tafel equation [21] (Eq. 1.50).

$$\Delta E = \pm(a + b \ lg j) \tag{1.50}$$

The minus sign in the Tafel equation refers to the anodic overpotential and the plus sign refers to the cathodic overpotential. Afterward, the Tafel equation was confirmed theoretically by the Butler–Volmer equation.

The general electrochemical reaction at an electrode can be written as follows:

$$Ox + ze^- \underset{K_a}{\overset{K_c}{\rightleftarrows}} Red$$

According to the Arrhenius equation the rate (υ) of the first-order chemical reaction under constant pressure (concentration, C) is described as follows:

$$\upsilon = k_x \cdot C \cdot \exp\left(-\frac{\Delta G^{\ddagger}}{R \cdot T}\right) \tag{1.51}$$

where $\Delta G^{\ddagger}$ is the Gibbs free energy required for the reaction activation and k is the rate constant.

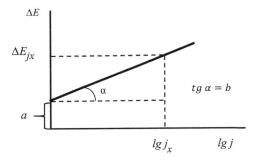

FIGURE 1.13 Tafel plot.

Because electrode potential influences the rate of electrochemical reaction, it should be taken into account in the evaluation of electrochemical reaction rates. As mentioned earlier the current density can be considered as a measure of electrochemical reaction rate. Based on these assumptions, current densities for anodic and cathodic reactions can be found using the following equations:

$$j_c = z \cdot F \cdot k_c \cdot C_{Ox} \cdot \exp\left(-\frac{\alpha_c \cdot z \cdot F \cdot \Delta E}{R \cdot T}\right) \qquad (1.52)$$

$$j_a = z \cdot F \cdot k_a \cdot C_{Red} \cdot \exp\left(\frac{\alpha_a \cdot z \cdot F \cdot \Delta E}{R \cdot T}\right) \qquad (1.53)$$

Where k_c and k_a are the rate constants of cathodic and anodic reactions, respectively; α_a and α_c are anodic and cathodic charge transfer coefficients, respectively ($\alpha_c + \alpha_a = 1$; therefore, it is possible to assume that $\alpha_c = \alpha$ and $\alpha_a = 1 - \alpha$).

At the equilibrium $j_c = j_a = j_0$, where $j_0 = z \cdot F \cdot k^0 \cdot \left[C_{ox}^{\alpha}\right] \cdot \left[C_{red}^{(1-\alpha)}\right]$ is the exchange current density (current density of the redox reaction at equilibrium when cathodic current of reduction reaction is equal in absolute value to the anodic current of oxidation reaction); k^0 is heterogeneous electron transfer rate constant.

The final equation for the calculation of cell current density can be written as follows.

$$j = j_0 \cdot \left\{\exp\left[-\frac{\alpha \cdot z \cdot F \cdot \Delta E}{R \cdot T}\right] - \exp\left[\frac{(1-\alpha) \cdot z \cdot F \cdot \Delta E}{R \cdot T}\right]\right\} \qquad (1.54)$$

Let us consider the limiting conditions. At low overpotential of $|\Delta E| \ll \frac{RT}{zF}$, Eq. (1.54) can be simplified to a linear current-potential dependence (1.55):

$$j = -\frac{j_0 \cdot z \cdot F \cdot \Delta E}{R \cdot T} \qquad (1.55)$$

Thus $\Delta E = -\frac{R \cdot T \cdot j}{z \cdot F \cdot j_0} = R_p \cdot j$, i.e., a linear dependence between ΔE and j, is observed, which is similar to the Ohm's law. $R_p = \frac{R \cdot T}{j_0 \cdot z \cdot F}$ is called the charge transfer resistance. It is seen that the higher the j_0 value, the smaller the potential deviation from its equilibrium value at a given current density and reaction is more reversible and vice versa. In this regard definitions for ideally polarizable and ideally nonpolarizable electrodes can be introduced (Fig. 1.12B). Ideally polarizable electrode is an electrode characterized by exchange current density equal to zero ($j_0 = 0$), i.e., no faradic current at the electrode/electrolyte interface. Ideally nonpolarizable electrode is an electrode characterized by infinitely large exchange current: $j_0 = \infty$. Real current exchange always has some finite value. But if j_0 is much greater than the current j flowing through the electrode, the change of the electrode potential can be neglected. In other words, electrodes potential is not dependent on the passing current. Therefore, electrodes with sufficiently large exchange currents are usually selected as reference electrodes.

At high overpotentials when $|\Delta E| \gg \frac{RT}{zF}$

$$j = -j_0 \cdot \exp\left[\frac{(1-\alpha)\cdot z \cdot F \cdot \Delta E}{R \cdot T}\right] \tag{1.56}$$

Overpotential can be reduced using catalytic electrodes. For example, platinum and palladium can facilitate the hydrogen evolution due to the decreased overpotential. Electrochemical polarization decreases with increasing temperature and reactant concentration and is independent of the stirring solution. Overpotential can be reduced by increasing the area of the electrodes.

Concentration Polarization

Change of the electrode potential due to changes of reactants concentration in the near-electrode layer under electrical current flow is called the concentration polarization. In turn, the change in concentration of the reactants in the near electrode layer slows down the supply of reactants to the electrode or the removal of reaction products from the electrode. The transfer of the reactants in the electrolyte solution may be carried out by three mechanisms. They are diffusion, migration, and convection. Molecular diffusion is characterized by the movement of electrochemical species under the influence of a concentration gradient and it is the main mechanism of mass transfer. Diffusion is observed in all electrode processes, whereas other mechanisms can be overlapped with the diffusion process or even be absent. Migration is the movement of charged particles in an electric field, which occurs in the diffusion layer because of the differences in potential between electrode and electrolyte solution. Convection can be caused for example by removal of gaseous products from the solution, stirring of the electrolyte solution, or changes in solution densities, which lead to the concentration gradient changes.

The processes of reactants diffusion are described by Fick diffusion equations. In a simple case, diffusional flux ($J_D = -D_i \cdot \frac{\partial C_i}{\partial x}$, flux along the coordinate x, i.e., normal to the electrode surface) determines the rate of material movement and is characterized by the linear change of compounds concentration (C) while passing through a unit of planar surface area (A) per unit time (t).

$$\frac{\partial C_i}{\partial t} = D_i \cdot \frac{\partial^2 C_i}{\partial x^2} \tag{1.57}$$

where D is the diffusion coefficient [22].

If diffusional flux should be evaluated in 3D environment, i.e., Cartesian directions (x, y, z), previous equation can be generalized to the following equation:

$$\frac{\partial C_i}{\partial t} = D_i \cdot \frac{\partial^2 C_i}{\partial x^2} + D_i \cdot \frac{\partial^2 C_i}{\partial y^2} + D_i \cdot \frac{\partial^2 C_i}{\partial z^2} \tag{1.58}$$

If electrochemical reactions occur at stationary conditions, i.e., $\frac{\partial C_i}{\partial t} = 0$, then $\frac{\partial C_i}{\partial x} = \frac{C_{ib}-C_{is}}{\delta_i}$, where C_{ib} and C_{is} are concentrations of electrochemical species in the bulk solution and on the surface of the electrode, respectively; δ is the thickness of the diffusion layer. Therefore, the equation for diffusion flux can be written as follows:

$$J_D = -D_i \cdot \frac{C_{ib}-C_{is}}{\delta} \tag{1.59}$$

To simplify calculations the value of mass transfer coefficient $k_m = \frac{D}{\delta}$ is introduced. At first approximation, current density can be used to describe the rate of electrochemical reactions; then for the steady-state of electrode reaction the following can be concluded:

$$j_i = \pm z \cdot F \cdot k_m \cdot (C_{ib} - C_{is}) \tag{1.60}$$

where minus sign is attributed to the cathodic reaction and plus sign is to anodic reactions [23].

When concentration of electroactive species near to the electrode surface is equal to zero due to their complete reduction or oxidation, the maximum current density called the limiting current density (j_L) is produced [24].

$$j_L = \frac{z \cdot F \cdot C_i \cdot D}{\delta} = z \cdot F \cdot C_i \cdot k_m \tag{1.61}$$

Overpotential for anodic concentration polarization (E_{Ca}) can be calculated as follows:

$$\Delta E_{Ca} = \frac{R \cdot T}{z \cdot F} \cdot \ln\left(\frac{j_L}{j_L - j_i}\right) \tag{1.62}$$

Consequently, for the cathodic concentration polarization (E_{Cc}), overpotential can be found from the following equation.

$$\Delta E_{Cc} = \frac{R \cdot T}{z \cdot F} \cdot \ln\left(\frac{j_L - j_i}{j_L}\right) \tag{1.63}$$

The greater the current density, the greater the difference between concentrations of ions in the near-electrode layer and the greater the concentration polarization.

In the case if concentration changes along the normal to the electrode surface, the migratory flux (J_M) caused by the potential (ϕ) difference between solution and electrode in electric field is proportional to the species mobility (u) and described as follows:

$$J_m = -z_i \cdot u_i \cdot F \cdot C_i \cdot \frac{\partial \phi}{\partial x} \tag{1.64}$$

where z_i is the charge of i species.

The change of species concentration during convection mass transport depends on the electrolyte solution velocity (v_s). In this regard, convective flux can be calculated as follows.

$$J_C = -v_s \cdot \frac{\partial C_i}{\partial x} \tag{165}$$

The total flux (J) of a reactant is composed of diffusion, convection, and migration fluxes as follows:

$$J = J_D + J_M + J_C \tag{1.66}$$

In the case of concentrated solutions and absence of stirring, migration and convection fluxes can be neglected. Concentration polarization decreases with an increase in the diffusion coefficient and reactant concentration and reduction in thickness of the diffusion layer. Diffusion layer is a thin layer near the surface of the electrode, wherein the fluid mixing is absent, i.e., there is no convection and thus molecules are transferred only by diffusion. While stirring the solution, the thickness of the diffusion layer and, as a consequence, the concentration polarization decrease. Stirring also influences the limiting current density.

Corrosion

Corrosion is a spontaneous destruction of metallic materials under the influence of the chemical environment. Corrosion is a redox process, accompanied by the transition metal in the ionic state.

Chemical corrosion occurs when the metal is in contact with dry gases at elevated temperatures or nonelectrolytes (liquids, nonconductive electrical current such as raw oil). Chemical corrosion occurs without the occurrence of an electric current.

Electrochemical corrosion of metals is more common. It includes all cases of corrosion in aqueous solutions or in a humid atmosphere. Mechanism of electrochemical corrosion is similar to the galvanic cell operation. Anodic oxidation of the metal and the cathodic reduction of oxidant (medium) occur during electrochemical corrosion. Formation of electric current during the contact of metal (or alloy) with an electrolyte solution activates the metal disruption. Seawater, freshwater, water films on the metal surface in soils, and atmosphere containing conductive impurities serve as an electrolyte during metal corrosion. Both metal and impurities can be an electrode. Impurities with a higher electrode potential play the role of cathode and metal itself is anode. During anodic oxidation, metal ions leave the metal lattice and pass into solution and excess electrons are left in the lattice. Then these electrons are moved to the cathode and cause the chemical reactions on its surface. All aqueous media contains hydrogen ions, which are transformed to hydrogen gas.

$$2H^+ + 2e^- \rightarrow H_2 \; E^0 = 0 \text{ V} \tag{1.67}$$

$$CO_2 + H_2O \rightarrow H_2CO_3$$

If the aqueous media contains the oxygen, the following transformation can occur at the cathode.

$$O_2 + 2H_2O + 2e^- \rightarrow 4OH^- \; E^0 = 0.4 \text{ V} \tag{1.68}$$

Formed hydroxyl ions can further react with the metal anions discharged at the anode forming metal hydroxides.

$$M^{z+} + zOH^- \rightarrow Me(OH)_z$$

As it is seen from the R. (1.67, 1.68), the increase of H^+ ions and oxygen molecules leads to the increased corrosion. However, reduction of oxygen is more preferable than reduction of hydrogen ions because of higher electrode potential value. On contacting two metals, the corrosion process proceeds in the direction of the dissolution of metal with a lower electrode potential. Corrosion of alloy containing two or more metals is more complicated because the surface of the metal alloy has different kinds of galvanic pairs. Composition and metal structure, temperature, pH, fluid flow, deformations, and stress of the surface and surface treatment influence the corrosion. Higher temperatures usually enhance corrosion.

Methods of Metals Protection From Corrosion

The most common way to protect metals from corrosion is to coat the surface with insulating coatings. There are nonmetallic and metallic protective coatings. Nonmetallic coatings (enamels, plastics, ceramics, etc.) are effective as long as coating layer is leakproof. Metal coatings are applied either thermally (for example, zinc and tin coatings) or by electroplating (for example, nickel, silver, gold, or chrome plating). Another way to create a protective layer on the metal surface is to subject metal products to chemical treatment to obtain a surface layer of a chemical compound resistant to corrosion such as metal oxides and phosphates as well as nitrogen compounds.

Metallic coating can be divided into cathodic and anodic coating by the type of protective effect. Covering metal of anodic coating has a more negative electrode potential than covered metal. As an example, let us consider the galvanized iron (protective layer of zinc on iron). Because zinc is more active (has lower electrode potential) than iron, as long as there is zinc on the surface of iron, zinc undergoes the dissolution on the open/leaky parts of coatings.

In the case of cathodic coatings (for example, Ni-plated Fe), coating tightness is a very important characteristic. While the protecting metal layer is leakproof, the coated metal (Fe) is well preserved, but metal outcrop, at least in one point, facilitates the vigorous corrosion. Another type of cathodic corrosion protection is impressed current cathodic protection (ICCP). The essence of the ICCP is that protected product is connected to the negative terminal of the external DC power supply, i.e., it becomes the cathode. The anode is an auxiliary electrode, usually steel electrode. The auxiliary electrode (anode) is dissolved, and hydrogen is released onto the protected metal product. ICCP is used for corrosion protection of underground pipelines, offshore pipelines, hulls of submarines, etc.

1.3 SUMMARY

The amount of freshwater available among the total volume of water resources in use is equal to about 2.5%. Because of the intensive development of industry and agriculture, the quality of freshwater decreases and it becomes unsuitable for crop irrigation, good production, and drinking purposes without deep pretreatment. The load on water resources also increases due to population growth and global warming. In addition to the surface water, the quality of ground water is also deteriorating, especially in the upper unconfined layer of the aquifer. Contamination of water occurs through the direct discharge of wastewater into water bodies, accidental spills, infiltration from the contaminated soil, and with atmospheric precipitates.

There are different classifications of pollutants depending on their origin, size, toxicity, etc. Human activity causes significant changes in the environment that can threaten the survival of world population in a long run. The most significant adverse effects of anthropogenic activity on natural water sources are increase of toxic organic and inorganic compounds, nutrients, COD and BOD content, radioactive pollution of natural waters causing inhibition of aquatic organisms, depletion of oxygen, drastic changes in pH, increase of turbidity, and biological contamination of water. Because the load on water bodies exceeds their self-purification capacity, it is necessary to maintain the practice of wastewater and drinking water treatment. Depending on the nature and size of pollutants, different water treatment techniques can be used for pollutant remediation. In general, all water treatment methods can be divided into mechanical, biological, chemical, and physical—chemical methods. Mechanical treatment allows removing insoluble coarse particle from the water. Biological water treatment is used for the degradation of nonrefractory organic compounds and some inorganic salts. Chemical water treatment can be successfully used for the removal of both toxic organic and inorganic compounds, and physical—chemical methods are intended to remove a wide range on contaminants from water starting from organic and inorganic compounds and ending with biological pollution.

To a large extent, the choice of water treatment method depends on the size of the contaminants. Heterogeneous pollutants with the particle size of $10^{-2}-10^{-4}$ cm such as suspensions, emulsions, and slurries are removed using, for example, flotation, centrifugation, or particle filtration. Colloidal particles, such as polymers or metal hydroxides, with the size varying between 10^{-4} and 10^{-7} cm can be removed using, for example, ultra- and nanofiltration, coagulation, and flotation techniques. Homogeneous systems containing pollutants in the form of electrolyte solution and dissolved organic matter are treated up using reverse osmosis, biological and advance oxidation methods, etc. If solution contains particles of different size category, then water treatment is conducted into few steps. At first, more coarse particles are removed, then medium-sized particles, and finally fine particles.

Among the various water treatment methods, electrochemical treatment starts to gain more attention. They allow adjusting physical and chemical properties of the treated water, concentrating and recovering valuable chemical products and metals from water, providing deep mineralization of organic contaminants and disinfection effect, and greatly simplifying technological treatment schemes. In many cases, electrochemical methods are environmentally friendly, excluding secondary water pollution by anionic and cationic residues, which are a side effect of chemical methods.

The main law describing electrochemical transformations is Faraday's law, which is strictly observed. All electrochemical processes are based on redox reaction occurring at the electrode surface because of established potential difference between these electrodes. To initiate the desired electrochemical reactions of degradation or separation of contaminants from water, it is necessary to conduct the electrolysis process. That means it is necessary to pass the electric current through a conductive solution containing these dissolved compounds. When passing the current through solution, potential difference discrepant in the standard electrode potential is established and forces electrochemical reaction to occur. The theoretical value of established electrode potential can be determined using the Nernst equation. The real value of electrolysis cell potential is calculated taking into account potential drops due to polarization in the system. The solvent, dissolved in electrolyte compounds, the material of electrodes, and other substances present in the electrolyte solution can undergo chemical transformations, and the sequence of these transformations depends on the magnitude of the potential and is explained by the minimum energy consumption and described by the laws of thermodynamics. Therefore, these parameters should be taken into account when planning electrolysis.

SELF-CONTROL QUESTIONS

1. What are the major sources of freshwater used for public supply?
2. What is the difference between galvanic and electrolytic cell? What are the differences and similarities between the anode and cathode in galvanic cell and electrolyzer?
3. What is the anode? What reactions do occur at the anode?

4. What is the cathode? What kind of reactions do occur at the cathode?
5. What is the difference between chemical and electrochemical reactions?
6. What types of underground aquifers do you know? How do you characterize unconfined and confined aquifers? What is the pollution level of different aquifers?
7. What kind of classification of pollutants do you know? How are pollutants classified by size? What are the main water treatment methods used for different size categories of pollutants?
8. Which values are used to determine the direction of chemical reactions in electrochemical cells?
9. How is the dependence between entropy, enthalpy, and cell potential expressed?
10. What reactions do occur at electrolysis at inert and active anodes?
11. What is the dependence of cations reduction reactions in electrolytic cells on the standard potential of electrodes? What does the reactive species show?
12. What is corrosion and its mechanism?
13. What metals can be used for the anodic protection of Al, Fe, Ni, and Cu products from corrosion?
14. What kinds of corrosion protection of metals do you know? Which one is the most effective and why?
15. What does polarization of electrodes mean? What is the mechanism of polarization? What types of polarization do you know?
16. What are the main mechanisms of reactants mass transfer?
17. What are advantages and disadvantages of electrocatalytic water treatment comparing to other AOPs?
18. How can electrochemical water treatment methods be characterized?

REFERENCES

[1] U.S. Department of the Interior, The USGS water science school. The World's water, U.S. Geolo. Surv. (2 May 2016) [Online]. Available: http://water.usgs.gov/edu/earthwherewater. html.

[2] Managing our future water needs for agriculture, industry, human health and the environment. Discussion document for the world economic forum meeting 2008, World Econ. Forum (1 January 2008) [Online]. Available: http://www3.weforum.org/docs/WEF_ ManagingFutureWater%20Needs_DiscussionDocument_2008.pdf.

[3] M. Beck, R. Vallarroel Walker, On water security, sustainability, and the water-food-energy-climate nexus, Front. Environ. Sci. Eng. 7 (2013) 626–639.

[4] M. Tarzia, B. De Vivo, R. Somma, R. Ayuso, R.A.R. McGill, R. Parrish, Anthropogenic vs. natural pollution: an environmental study of an industrial site under remediation (Naples, Italy), Geochem. Explor. Environ. Anal. 2 (1) (2002) 45–56.

[5] K. Kolmetz, A. Sidney Dunn, A.M. Som, C. Phaik Sim and Z. Mustaffa, Benchmarking Waste Water Treatment Systems. [Online]. Available from: http://www.klmtechgroup.com/ PDF/Articles/articles/WWT-Paper-Expanded.pdf.

[6] R. Arntzen, Gravity Separator Revamping, Norwegian University of Science and Technology, Trondheim, 2001.

[7] M. Shestakova, Ultrasound-assisted Electrochemical Treatment of Wastewaters Containing Organic Pollutants by Using Novel Ti/Ta2O5-sno2 Electrodes, Lappeenrannan teknillinen yliopisto-Digipaino 2016, Lappeenranta, 2016.

[8] M. Shestakova, M. Sillanpää, Removal of dichloromethane from ground and wastewater: a review, Chemosphere 93 (2013) 1258–1267.

[9] I. Levchuk, A. Bhatnagar, M. Sillanpää, Overview of technologies for removal of methyl tert-butyl ether (MTBE) from water, Sci. Total Environ. 476–477 (2014) 415–433.

[10] S. Luster-Teasley, J. Yao, H. Herner, J. Trosko, S. Masten, Ozonation of chrysene: evaluation of byproduct mixtures and identification of toxic constituent, Environ. Sci. Technol. 36 (2002) 869–876.

[11] B. Upham, J. Yao, J. Trosko, S. Masten, Determination of the efficacy of ozone treatment systems using a gap junction intercellular communication bioassay, Environ. Sci. Technol. 29 (1995) 2923–2928.

[12] W. Chen, Y. Su, Removal of dinitrotoluenes in wastewater by sonoactivated persulfate, Ultrason. Sonochem. 19 (4) (2012) 921–927.

[13] S. Na, C. Jinhua, M. Cui, J. Khim, Sonophotolytic diethyl phthalate (DEP) degradation with UVC or VUV irradiation, Ultrason. Sonochem. 19 (5) (2012) 1094–1098.

[14] J. Radjenovic, D. Sedlak, Challenges and opportunities for electrochemical processes as next-generation technologies for the treatment of contaminated water, Environ. Sci. Technol. 49 (2015) 11292–11302.

[15] C. Martínez-Huitle, E. Brillas, Decontamination of wastewaters containing synthetic organic dyes by electrochemical methods: a general review, Appl. Catal. B Environ. 87 (3–4) (2009) 105–145.

[16] S. Bratsch, Standard electrode potential and temperature coefficients in water at 298.15 K, J. Phys. Chem. Ref. Data 18 (1989) 1–21.

[17] L. Niedrach, Sensor with ion exchange resin electrolyte. USA Patent US3714015 A, 30 January 1973.

[18] P. Atkins, J. de Paula, Physical Chemistry, 8th ed., Oxford University Press, 2006, p. 947.

[19] F. Walsh, Electrochemical technology for environmental treatment and clean energy conversion, Pure Appl. Chem. 73 (12) (2001) 1819–1837.

[20] M. A. B.V, Ohmic Drop. Part 2-Measurements, 1 July 2011 [Online]. Available: http://www.ecochemie.nl/download/Applicationnotes/Autolab_Application_Note_EC04.pdf.

[21] J. Tafel, Über die polarisation bei kathodischer wasserstoffetwicklung, Z. Phys. Chem. Bd. 50 (1905) 641–712.

[22] L. Austin, Polarization at diffusion electrodes, in: Symposium on Fuel Cells, 3–8 September 1961.

[23] F.M. Donahue, Fundamentals of electrochemical engineering, in: Engineering Summer Conferences, pp. VIII-1-VIII-15, 12–16 August 1985.

[24] J.-Y. Chen, C.-L. Hsieh, N.-Y. Hsu, Y.-S. Chou, Y.-S. Chen, Determining the limiting current density of vanadium redox flow batteries, Energies 7 (2014) 5863–5873.

Chapter 2

Electrochemical Water Treatment Methods

Mika Sillanpää, Marina Shestakova
Lappeenranta University of Technology, Lappeenranta, Finland

NOMENCLATURE

Latin alphabet

α	Efficiency of the treatment process	%
A_H	Hamaker constant	J
A	Surface area	m^2
b	Air solubility in water at atmospheric pressure and a given temperature	cm^3/L
β	Degree of saturation	
C_0	Initial concentration of the dispersed phase in water	mol/m^3
C_t	Final concentration of the dispersed phase in water after time t of the treatment	mol/m^3
C_i	Concentration of i-species	mol/m^3
C_{TDS}	Salt content	g/L
C	Constant	
c	Distance between particle centers	m
c_d	Drag coefficient	
g	Gravitational acceleration	m/s^2
d_{gs}	Equivalent diameter of gas bubble—solid particle complex	
D_i	Diffusion coefficient	m^2/s
E, E_{cell}	Cell potential at a given temperature	V
E_a, E_c	Anodic and cathodic potential, respectively	V
ε_r	Relative dielectric permittivity of the medium containing particles	
ε_0	Electrical permittivity of vacuum	$8.854 \cdot 10^{-12}$ F/m

Electrochemical Water Treatment Methods. http://dx.doi.org/10.1016/B978-0-12-811462-9.00002-5

F	Faraday constant	96485.33289(59) C/mol
F_A	Archimedes buoyancy	J
F_d	Drag force	J
G_{el}	Electrostatic repulsion forces	J
G_A	Van der Waals attraction forces	J
h	Distance between particle surfaces	m
I	Current	A
j	Applied current density	A/m^2
j_{lim}	Limiting current density	A/m^2
J_i	Total flux of ions	mol/m$^2 \cdot$s
k	Inverse of the thickness of diffuse electrical double layer	m^{-1}
m	Mass of i-species	kg
M	Molar mass	g/mol
μ	Ionic strength of solution	mol/m^3
N	Number of cell pairs	
η	Dynamic viscosity of the liquid medium	kg/m s
p	Gas storage pressure	Pa
P	Absolute pressure at which water is saturated with the air	Pa
Q_d	Flow rate of dilute	m^3/s
Q_l	Amount of water-saturated air	m^3/h
Q	Water flow	m^3/h
R_1, R_2, R_{12}	Particle radii	m
R	Gas constant	8.314 J/K mol
ρ	Specific conductance	1/Ω cm
ρ_l	Liquid density	g/L
ρ_x	Electron density of particles	
T	Temperature	K
t	Time	s
t_i	Transport number of i-species	
$\overline{t^{CEM}}, \overline{t^{AEM}}$	Transport number of cation- and anion-exchange membranes	
U_{cell}	Cell voltage	V
u_s, u_g	Particles and bubbles velocities, respectively	m/s
V	Volume of electrolyte	m^3
V_{mH_2}	Molar volume of 1 mol hydrogen	m^3
V_g, V_s	Volumes of gas bubble and particle, respectively	m^3
ω	Probability of formation of the "bubble-particle" complex	%
z_i	Charge number of species	
$\gamma_{sg}, \gamma_{lg}, \gamma_{sl}$	Surface tension at the interface of solid–gas, liquid–gas, and solid–liquid phases, respectively	J/m^2
ϕ_δ	Stern potential	V
θ_c	Contact angle	degrees
δ_i	Stoichiometric coefficient of i gas	
ϑ	Velocity of ion in solution due to convection	m/s

Abbreviations

BDD	Boron-doped diamond electrode
BOD	Biological oxygen demand
CE	Current efficiency
COD	chemical oxygen demand
DLVO	Derjaguin, Landau, Verwey, and Overbeek theory
DSA	Dimensionally stable anodes
EC	Energy consumption
ED	Electrodialysis
EF	Electroflotation
EO	Electrochemical oxidation
ER	Electrochemical reduction
GC	Glassy carbon
ICE	Instantaneous current efficiency
MMO	Mixed-metal oxide electrode
OER	Oxygen evolution reaction
RO	Pollutant oxidation products
TCE	Total current efficiency

Rapid industrial development and population growth leads to pollution with refractory and toxic compounds, eutrophication, and fresh water depletion in water bodies. In this regard, there is a need to find new methods of water treatment because conventional processes such as biological treatment, neutralization, and particles separation methods can no longer cope with the volumes of processed water and high pollutant concentrations. Moreover, electrochemical methods do not require addition of chemicals to treat the industrial wastewater because the electrical conductivity of processed wastewater in real systems can exceed 10 mS/cm.

Electrochemical water treatment methods are physical–chemical methods. The main electrochemical water treatment processes include anodic (electrochemical) oxidation, cathodic reduction, electrocoagulation, electroflotation (EF), and electrodialysis (ED). All these methods can be used as independent treatment processes, in combination with other technologies (chemical, physical, and biological) as main treatment process or as a pretreatment or posttreatment step. Moreover, water treatment by electrochemical methods can be carried out intermittently or continuously. Electrochemical treatment is applicable for colored and turbid waters, which cannot be treated by UV and photocatalytic degradation methods, and are not sensitive to toxic compounds, which can kill microorganisms used in biological remediation processes.

2.1 ELECTROCHEMICAL OXIDATION (INCLUDING ELECTRODISINFECTION)

Electrochemical oxidation (EO) is a chemical reaction, involving the loss of one or more electrons by an atom or a molecule at the anode surface made of catalyst material during the passage of direct electric current through the electrochemical systems (anode, cathode, and an electrolyte solution).

Electrochemical Oxidation of Organic Compounds

There are two main mechanisms for EO of organic compounds in water. They are direct and indirect mechanisms (Fig. 2.1) [1−3].

In the case of indirect oxidation different electroactive species generated during the electrolysis are involved in the oxidation process. The most common process for indirect oxidation of organic compounds involves generation of highly reactive hypochlorite ions. For example, if electrolyte solution contains sodium chloride, then chlorine gas is generated at the anode (R. 2.1) with the following reaction with water molecules and formation of hypochlorous and hydrochloric acids (R. 2.2) [4,5].

$$2Cl^- - 2e^- \rightarrow Cl_2 \tag{2.1}$$

$$Cl_2 + H_2O \rightarrow HClO + HCl \tag{2.2}$$

Hypochlorous acid is partly dissociated into hypochlorite ion and hydrogen ion (R. 2.3).

$$HClO \rightarrow H^+ + ClO^- \tag{2.3}$$

In neutral and basic media hydroxide ions generated at the cathode reacts with hypochlorous acid neutralizing it and generating hypochlorite ions (R. 2.4).

$$HClO + OH^- \rightarrow H_2O + ClO^- \tag{2.4}$$

Neutralization of HClO will lead to the shifting of R. (2.2) equilibrium to the right side, which means that, in the case of good mass transfer and fast

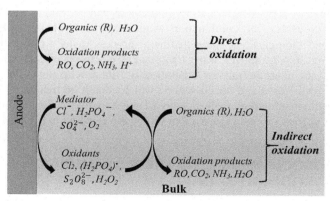

FIGURE 2.1 Direct and indirect mechanisms of electrochemical oxidation of organic compounds.

delivery of hydroxide radical from the cathode to the formed HClO with the followed acid neutralization, it can lead to complete termination of chlorine gas release at the anode. The process of hypochlorite ions formation can be interrupted by side reactions. For example, ClO^- ions can be lost in anodic and cathodic transformation reactions with the formation of chlorate and chloride ions, respectively (R. 2.5–2.6).

At anode:

$$12ClO^- + 6H_2O - 12e^- \rightarrow 4ClO_3^- + 8Cl^- + 3O_2 + 12H^+ \qquad (2.5)$$

At cathode:

$$ClO^- + H_2O + 2e^- \rightarrow Cl^- + 2OH^- \qquad (2.6)$$

The formation of hypochlorite ions is dependent on many parameters such as concentration of chloride ions in the electrolyte solution, the temperature of the electrolyte solution, applied current density, material of anode, and cathode. For example, higher concentration of chloride ions (usually higher than 3 g/L) in the electrolyte solution leads to the higher formation of hypochlorite ions; higher temperature of electrolyte solution leads to the higher evolution of oxygen, which is the waste reaction, and results to the lower formation of ClO^-. In this regard, the temperature control should be carried out during indirect EO and addition of sodium chloride salt can be required in the case of low chloride ions concentration in the treated wastewater. Anodic materials should have low overpotential toward chlorine evolution, which in turn leads to higher ClO^- production [6]. Cathodic materials should have inert properties toward ClO^- ions reduction.

One of the main disadvantages of EO of organics with electrogenerated ClO^- ions is the formation of toxic chlorinated intermediates especially while working in acidic medium [7]. Therefore, EO by other reactive species is of interest. Indirect EO of organic compounds can be enhanced by electro-Fenton reactions [8], electrogeneration of hydrogen peroxide, peroxodisulfate, peroxidiphosphate [9], and ozone [10].

$$2SO_4^{2-} \rightarrow S_2O_8^{2-} + 2e^- \qquad (2.7)$$

$$H_2PO_4^- + (H_2PO_4)^{\cdot} + e^- \qquad (2.8)$$

$$H_2PO_4^- + {}^{\cdot}OH \rightarrow (H_2PO_4)^{\cdot} + OH^- \qquad (2.9)$$

$$HPO_4^{2-} \rightarrow (HPO_4^-)^{\cdot} + e^- \qquad (2.10)$$

$$HPO_4^{2-} + HO^{\cdot} \rightarrow (HPO_4^-)^{\cdot} + OH^- \qquad (2.11)$$

$$O_2 + 2H^+ + 2e^- \rightarrow H_2O_2 \qquad (2.12)$$

$$H_2O_2 + 2H^+ + 2e^- \rightarrow 2H_2O \qquad (2.13)$$

The presence of iron catalyst in electrolyte solution and on-site cathodic generation of hydrogen peroxide initiate the reaction of hydroxyl radicals formation, which is known as electro-Fenton reaction (R. 2.14) [8].

$$H_2O_2 + Fe^{2+} \rightarrow Fe^{3+} + OH^- + {}^{\cdot}OH \tag{2.14}$$

Direct EO of organic pollutants involves direct active oxygen interaction with organic pollutants with their further oxidation. It is only possible in the case of electrodes made of catalytic materials. Active oxygen can be generated through the activation of water molecules, which is achieved in acidic media either by water dissociative adsorption below oxygen evolution reactions (OER) potential or by water electrolytic discharge above OER potential.

When bond energy of M−OH and M−H is higher than dissociative energy of water to H$^{\cdot}$ and HO$^{\cdot}$ radicals, the *dissociative adsorption of water* takes place and is described by R. (2.15−2.17).

$$H_2O + M \rightarrow M{-}OH + M{-}H \tag{2.15}$$

$$M - H \rightarrow M + H^+ + e^- \tag{2.16}$$

$$H_2O + M \rightarrow M - OH + H^+ + e^- \tag{2.17}$$

In the case when the bond energy of $M - OH$ and $M - H$ is lower than dissociative energy of water to H$^{\cdot}$ and HO$^{\cdot}$, the physical adsorption of HO$^{\cdot}$ radicals at the electrode surface takes place through the *electrolytic discharge of water* (R. 2.18).

$$H_2O + M \rightarrow M({}^{\cdot}OH) + H^+ + e^- \tag{2.18}$$

If water does not contain organic pollutants, adsorbed hydroxyl radicals are released in the form of oxygen according to R. (2.19).

$$M({}^{\cdot}OH) \rightarrow M + \tfrac{1}{2} O_2 + H^+ + e^- \tag{2.19}$$

If wastewater contains organic pollutants, adsorbed hydroxyl radicals can undergo interaction with organic pollutants through two main mechanisms [11]. In the case of strong adsorption of hydroxyl radicals on "active" anodes, higher oxides (MO) of anodic material (R. 2.20) are formed.

$$M({}^{\cdot}OH) \rightarrow MO + H^+ + e^- \tag{2.20}$$

Higher oxides can further release the free oxygen (R. 2.21) or interact with organic pollutants (R. 2.22) oxidizing them and forming pollutants oxidation products (RO) [1].

$$MO \rightarrow M + \tfrac{1}{2} O_2 \tag{2.21}$$

$$MO + R \rightarrow M + RO \tag{2.22}$$

"Nonactive" anodes adsorb weak $^{\bullet}OH$ and can either release free oxygen (R. 2.19) or provide a direct interaction of hydroxyl radicals with organic pollutants resulting in pollutants' mineralization (R. 2.23).

$$M(^{\bullet}OH) + R \rightarrow M + mCO_2 + nH_2O + H^+ + e^- \qquad (2.23)$$

Useful pollutant oxidation reactions compete with the reaction of free oxygen release occurring because of the interaction of adsorbed hydroxyl radicals with anode material as well as hydrogen (H_2) and oxygen (O_2) evolution reactions at cathode and anode, respectively, due to the presence of H^+ and OH^- ions, that are the products of water dissociation. The implementation of each specific reaction is associated to a minimum energy consumption (EC). Another undesirable side reaction occurring at the electrolysis process is the dissolution of the anode metal through R. (2.24). In this regard the use of corrosion-resistant materials is required to conduct efficiently the EO process.

$$M - ze^- \rightarrow M^{z+} \qquad (2.24)$$

According to the thermodynamics, all organic pollutants should be oxidized at potentials below the theoretical potential of OER ($E^0 = 1.23$ V). However, the complete mineralization of pollutants can only be achieved for simple C1 organic molecules while using highly reactive catalysts such as Pt. In this regard, there is a need to use electrocatalytic anodes with the high overpotential toward OER.

There is a wide variety of anodes including noble metal, carbon and graphite, PbO_2, mixed-metal oxide (MMO), and boron-doped diamond (BDD) electrodes, used for EO of organic compounds. However, neither of them are both stable and efficient. Therefore, there is a continuous search for new cost-efficient and safe electrode materials.

Noble metal electrodes have the longest history of use in laboratory scale EO applications regardless of their low efficiency toward organics mineralization. PbO_2 anodes are cost-efficient and are able to completely mineralize pollutant. However, they are potentially harmful to the environment due to the leaching Pb. Along with PbO_2 electrodes, carbon- and graphite-based anodes are cost-efficient materials. However, they have low electrochemical stability and cannot be widely used. MMO catalysts are a broad group of materials, both noble and base, able to adsorb oxygen on their structure. The commercial name for MMO anodes is dimensionally stable anodes, which was given to the materials because of their ability to preserve structural integrity, long service life lasting electrodes (up to 10 years), and low rates of degradation [12]. Most MMO electrodes are prepared on Ti substrate, which is explained by its stability, conductivity properties, and relatively low cost. Ti-based MMO electrodes have corrosion resistant properties and are persistent to dissolution. BDD electrodes consist of diamond doped with boron in the range from 10^{19} to 10^{21} atoms/cm^3, which makes them expensive. Nevertheless, the ability to completely mineralize organic pollutants on BDD anodes has a reduced efficiency toward mineralization of organics in diluted solutions [13] and at a current density higher than the limiting current, at which ˙OH radicals are consumed for the side reactions [14].

$$2BDD(\text{˙OH}) \rightarrow 2BDD + 1/2O_2 + 2H^+ + 2e^- \quad (2.25)$$

$$2BDD(\text{˙OH}) \rightarrow 2BDD + H_2O_2 \quad (2.26)$$

$$BDD(OH) + H_2O_2 \rightarrow BDD(HO_2^˙) + H_2O \quad (2.27)$$

The main advantages and disadvantages of different electrodes are listed in Table 2.1.

Anodes with low OER overpotential provide lower activity toward organic pollutants mineralization due to the strong interactions of adsorbed hydroxyl radicals with electrode material. Electrodes such as carbon-based anodes, Pt, iridium oxide, and ruthenium oxide electrodes have low overpotential of OER and are attributed to active anodes. Nonactive anodes such as PbO_2, $SnO_2-Sb_2O_5$, and BDD have high overpotential toward water oxidation and consequently can provide high mineralization efficiencies of organic compounds [11].

Fig. 2.2 shows the onset potential of OER for different electrodes in acidic media. As it can be seen, BDD anode along with $Ti/SnO_2-Sb_2O_5$ has the highest overpotential of OER followed by Ti/PbO_2 and $Ti/Ta_2O_5-SnO_2$ electrodes. This means that BDD and $Ti/Ta_2O_5-SnO_2$ anodes are the most efficient for complete mineralization of organic pollutants. Pt, $IrO_2-Ta_2O_5$, and RuO_2-TiO_2 anodes have the smallest overpotential toward OER, thus providing the low activity toward mineralization of organics.

Total current efficiency (TCE) and instantaneous current efficiency (ICE) can be calculated based on the same approach of chemical oxygen demand (COD) reduction data during electrolysis process with applied value of current [15].

$$\text{TCE, ICE} = \frac{FV[(COD_{t_0}) - (COD_t)]}{8I\Delta t} \quad (2.28)$$

where COD_{t_0} and COD_t are chemical oxygen demand at initial time (t_0) and time (t) after electrolysis process, respectively, V is the volume of electrolyte (m^3), and I is the applied current (A) [16].

One of the first works on EO of organic compound was the study on decomposition of phenolic compounds on PbO_2 anodes in the early 1970s [17]. However, the intensive studies on EO of organic compounds in wastewaters started in the late 1980s and in the beginning of 1990s [18,19]. Sharifian and Kirk investigated the phenol removal from acidic water using a packed bed reactor filled with PbO_2 pellets. They succeeded to remove $1.4 \cdot 10^{-4}$ M phenol with efficiency of 72–100% from the water in 1.5–2.5 h

TABLE 2.1 Advantages and Disadvantages of Different Group Anodes Used in Electrochemical Oxidation Applications of Organic Compounds

ElectrodeType	Advantages	Disadvantages
Noble metal electrodes (Pt, Au)	Stable in a wide range or potentials and pH Excellent repeatability properties Intensive use in laboratory scale for new process investigations	Expensive Low mineralization efficiency Low overpotential toward OER Poor use in industrial wastewater treatment application
PbO_2	Cheap Relatively high overpotential toward OER Relatively high ability to mineralize organics	Potential leaching of toxic Pb Poor performance in industrial wastewater treatment application
Carbon and graphite electrodes	Cheap Intensive use in laboratory scale for new process investigations	High electrode corrosion rates Low mineralization efficiency Low overpotential toward OER
MMO (Ti/TiO_2–RuO_2, Ti/Ta_2O_5–IrO_2, Ti/TiO_2–RuO_2–IrO_2, Ti/IrO_2–RuO_2, Ti/SnO_2–Sb_2O_5, etc.)	High stability Good conductivity properties Acceptable price Possibility to regenerate catalytic oxide coating	Sometimes it is difficult to reproduce quality of the catalyst layer Potential leaching of toxic compounds such as Sb
BDD	High overpotential toward OER High ability to mineralize organics Excellent conducting properties even at low temperatures High electrochemical stability and corrosion resistance	Expensive Reduced efficiency in diluted solutions and at increasing current density higher than a limiting current

BBD, boron-doped diamond; *MMO*, mixed-metal oxide; *OER*, oxygen evolution reactions.

while applying 1–3 A current [20]. The main intermediates of the oxidation process were benzoquinone, maleic acid, and CO_2. The similar results were obtained by Smith de Sucre and Watkinson on PbO_2 packed-bed anodes within 1.5–2 h of electrolysis while removing 93–1100 mg/L phenol from the water

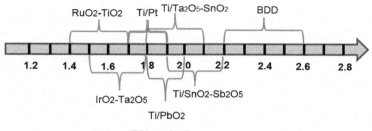

E/V vs. SHE

FIGURE 2.2 The onset potential of oxygen evolution reactions for different anode materials in acidic media. *BBD*, boron-doped diamond.

(pH 2.5) and applying 10 A current [21]. Stucki et al. studied the removal of benzoic acid at Ti/SnO$_2$—Sb$_2$O$_5$, Ti/PbO$_2$, and Ti/Pt anodes [22]. It was found that the removal of benzoic acid at Ti/SnO$_2$—Sb$_2$O$_5$ electrodes achieved 90% and at Ti/PbO$_2$ and Ti/Pt anodes achieved less than 30% while applying a current density of 30 mA/cm^2 to the electrodes. Kötz et al. could achieve a complete mineralization of 1000 mg/L phenol in alkaline media at Ti/SnO$_2$—Sb$_2$O$_5$ anodes when conducting electrolysis at 300 A/m^2 applied current density [19]. Electrolysis at PbO$_2$ electrodes could decrease total organic carbon (TOC), COD, and biological oxygen demand (BOD) values of a mixture of five monohydric phenols (phenol, p-cresol, o-cresol, 2,3-xylenol, and 3,4-xylenol) with efficiencies of 57%, 18%, and 22%, respectively, within 3 h at 526 A/m^2 applied current density [23].

Electrochemical Oxidation of Inorganic Compounds

Anodic oxidation of some inorganic compounds such as cyanides, thiocyanate, and sulfides known for their toxic properties allows complete transfer on these pollutants to a nontoxic compound. EO can be used for the treatment of wastewater from electroplating industries containing cyanides of iron, zinc, copper, and others. Anodic oxidation of cyanides leads to the formation of less toxic cyanates, ammonia, and nitrogen compounds along with CO$_2$ gas and carbonate ions and can be described through the following equations:

$$CN^- + 2OH^- - 2e^- \rightarrow CNO^- + H_2O \tag{2.29}$$

$$CNO^- + 2H_2O \rightarrow NH_4^+ + CO_3^{2-} \tag{2.30}$$

$$2CNO^- + 4OH^- - 6e^- \rightarrow 2CO_2 + N_2 + 2H_2O \tag{2.31}$$

Usually EO of cyanides is conducted in electrolyzers of continuous and bath mode using graphite, lead oxide or magnetite anodes, and steel cathode at applied current density laying in the range of 30—40 mA/cm^2 at pH of 8—9. The removal efficiency is nearly 100%.

Cyanides can be also oxidized indirectly while interacting with the products of anodic chlorine decomposition through the following reactions:

$$2Cl^- - 2e^- \rightarrow Cl_2 \tag{2.32}$$

$$CN^- + Cl_2 + 2OH^- \rightarrow CNO^- + 2Cl^- + H_2O \tag{2.33}$$

$$2CNO^- + 3Cl_2 + 4OH^- \rightarrow 2CO_2 + N_2 + 6Cl^- + 2H_2O \tag{2.34}$$

Advantage of EO of cyanides comparing to conventional chemical neutralization methods is simultaneous recovery of metals deposited at the cathode.

EO of thiocyanate leads to the formation of less toxic cyanates that can be further oxidized to carbonates, CO_2, and N_2 gas.

$$SCN^- + 10OH^- - 8e^- \rightarrow CNO^- + SO_4^{2-} + 5H_2O \tag{2.35}$$

Sulfide ions can be electrochemically oxidized to sulfates at pH 7. The formation of elemental sulfur can occur at lower pH values. Moreover, formation of polysulfides, sulfites, and tetrathionates can be observed and are described by the following reactions:

$$S^{2-} - 2e^- \rightarrow S \tag{2.36}$$

$$S + 8OH^- - 6e^- \rightarrow SO_4^{2-} + 4H_2O \tag{2.37}$$

$$S^{2-} + 8OH^- - 8e^- \rightarrow SO_4^{2-} + 4H_2O \tag{2.38}$$

$$2S^{2-} + 6OH^- - 8e^- \rightarrow S_2O_3^{2-} + 3H_2O \tag{2.39}$$

$$S^{2-} + 6OH^- - 6e^- \rightarrow SO_3^{2-} + 3H_2O \tag{2.40}$$

As it can be seen from the above reactions, there is a wide variety of intermediate products generated during the anodic oxidation of sulfides. In this regard, the use of catalytic anodes such as Pt or BDD can significantly improve the selectivity of the process toward sulfate formation [24,25].

$$Pt + HS^- + OH^- - 2e^- \rightarrow PtS + H_2O \tag{2.41}$$

$$PtS + HS_x^- \rightarrow Pt + HS_{x+1}^- \tag{2.42}$$

$$PtS + 8OH^- - 6e^- \rightarrow Pt + 4H_2O + SO_4^{2-} \tag{2.43}$$

$$BDD + S^{2-} - 2e^- \rightarrow (BDD)S \tag{2.44}$$

$$(BDD)S + S_x^{2-} \rightarrow BDD + S_{x+1}^{2-} \tag{2.45}$$

$$(BDD) S + 8OH^- - 6e^- \rightarrow BDD + 4H_2O + SO_4^{2-} \tag{2.46}$$

Electrochemical Disinfection

EO in electrodisinfection is based on the anodic generation of strong oxidants such as oxygen, ozone, or hypochlorite during water electrolysis. Similar to conventional chemical disinfection, electrodisinfection can be used for the removal and deactivation of different microorganisms from water and often it is more efficient than chemical disinfection. The main advantages of electrochemical disinfection compared with conventional chemical disinfection is to keep the working areas required for the storage and dosage of chemicals substances; compact reactors allowing to operate in situ the main line of treatment facilities and no side generation of hazardous intermediates, which are typical for chemical disinfection.

Electrodisinfection by oxygen gas. Anodic generation of oxygen, which shows some germicidal activity, is used mainly for the removal of bad odor from water in small applications where the generation of chlorine species is undesirable [26]. The most commonly used electrodes for oxygen evolution are Pt-containing anodes. Anodic generation of oxygen is shown in the following equations:

$$2H_2O - 4e^- \rightarrow O_2 + 4H^+ \quad pH \leq 7 \tag{2.47}$$

$$4OH^- - 4e^- \rightarrow O_2 + 2H_2O \quad pH > 7 \tag{2.48}$$

Electrodisinfection by chlorine gas and hypochlorite ions. Anodic generation of hypochlorite ions is described above in the section of EO organic compounds (R. 2.1−2.6). Usually the use of electrochemical generation of hypochlorites is advantageous when the concentration of chloride ions in water above 19 g/L (for example, in sea water). Anodes used for the process of electrodisinfection by hypochlorite ions should have low overpotential toward chlorine gas evolution such as Pt. However, pure Pt anodes are not used in industrial applications because of their high costs. Traditional electrodes for Cl_2 gas evolution are PbO_2 and MMO electrodes.

Active forms of chlorines are efficient for disinfection of water containing bacteria, viruses, fungi, and spores. For example, *Candida albicans* fungi die within 30 seconds while exposed to 5% NaOCl solution.

Electrodisinfection by ozone. Another oxidant, which can be electrochemically generated at the anode, is ozone. Ozone can be allocated at the anode having high overvoltage toward OERs, for example, BDD electrodes. Electrode reaction of O_3 anodic generation is described as follows:

$$3H_2O - 4e^- \rightarrow O_3 + 6H^+ \tag{2.49}$$

Bactericidal effect of ozone is associated with its high oxidation potential (2.076 V) and the ease of diffusion through the cell walls of microorganisms. For example, polio virus dies within 2 min when ozone dosage is equal to 0.45 mg/L, while chlorine kills the virus only after 3 h with a dosage of 2 mg/L.

Ozone has a devastating effect on the life of algae, leeches, mollusks, coliform bacteria, and others. However, chironomids and water mites are sensitive to neither ozone nor chlorine. Additionally, ozone improves the organoleptic properties of water.

Because both ozone and chlorine species are highly corrosive, corrosion-resistant materials should be used for the construction of equipment and pipelines.

Electrodisinfection by silver. Along with oxygen, ozone, and chlorine species, silver ions have high-antibacterial properties toward Staphylococcus, Streptococcus, typhoid, dysentery, and coliform bacteria. There are few hypotheses explaining the mechanism of silver-antibacterial properties. One of the mechanisms states that silver interacts with bacterial cell enzymes and violate cell exchange with the environment thus killing it. Another mechanism claims that silver ions penetrate the cell and bind with cell's protoplasts thus killing it. It is also suggested that silver adsorbs on the surface of cells and catalyzes reactions of cell plasma oxidation by oxygen. Anodic dissolution of silver anodes occurs by the following mechanism:

$$Ag^0 - e^- \rightarrow Ag^+ \tag{2.50}$$

The recent updates on the efficiency of EO on the removal of different pollutants are represented in Table 2.2.

2.2 ELECTROCHEMICAL REDUCTION

Electrochemical reduction (ER) is a chemical reaction, involving the gain of one or more electrons by an atom or a molecule at the cathode surface during the passage of direct electric current through the electrochemical system (anode, cathode, and an electrolyte solution).

ER can be used for organic compounds that are refractory to oxidation or for heavy metal ions such as Pb(II), Sn(II), Hg(II), Cu(II), As(III), and Cr(VI). Deposited at the cathode, metals can be further recovered. ER of different organic compounds in wastewater is expedient when the direct anodic oxidation of these compounds either does not occur or require high electric power consumption, and if products formed during the cathodic reduction of organic compounds are nontoxic (or of low toxicity) or relatively readily undergo further oxidative degradation. Increased toxicity of organic substances is explained by the presence of the halogen atoms, aldehyde, amino, nitro, and nitroso groups in their molecule structure. Thus, loss of the halogen atom or reduction of aldehydes and ketones, for example, will lead to the formation of alcohols and hydrocarbons, which are less toxic. The mechanism of direct ER of organic compounds can be shown through the following reactions:

$$RCl + H_2O + 2e^- \rightarrow RH + Cl^- + OH^- \tag{2.51}$$

TABLE 2.2 Removal Efficiency of Organic Pollutants From Water at Different Anodes

Targeted Pollutant	Anode Type	Working Parameters	Removal Efficiency	Reference
Organic Compounds				
Reactive Blue 109 (1000 mg/L)	Pt	NaCl (0.1 M) $j = 20 \, mA/cm^2$ pH 4 $t = 1.25 \, h$	96% COD	[6]
Dimethyl phthalate (0.03 mM)		Na_2SO_4 (0.2 M) $j = 20 \, mA/cm^2$ pH 7 $t = 18 \, h$	20% 10% COD	[27]
Diethyl phthalate (0.03 mM)		Na_2SO_4 (0.2 M) $j = 20 \, mA/cm^2$ pH 7 $t = 18 \, h$	50% 30% COD	
Diheptyl phthalate (0.03 mM)		Na_2SO_4 (0.2 M) $j = 20 \, mA/cm^2$ pH 7 $t = 18 \, h$	100% 100% COD	
Phenol (490 mg/L)		Na_2SO_4 (0.25 M) $t = 20 \, h$ $j = 20 \, mA/cm^2$ pH 5	100% 20% TOC	[28]

Pollutant	Electrode	Conditions	Results	Ref.
Methylene blue (40 mg/L)	Pt/MnO_2	Na_2SO_4 (0.05 M), $j = 7$ mA/cm^2, pH 8, $t = 2$ h	90%, 70% COD	[29]
Landfill leachate (COD 1414 mg/L)	Graphite carbon electrode	Na_2SO_4 (1 g/L), $j = 79$ mA/cm^2, $t = 4$ h	68% COD, 84% color	[30]
Amoxicillin (19.6 mg/L TOC)	Ti/Ti_4O_7	Na_2SO_4 (0.05 M), $j = 5$ mA/cm^2, $t = 8$ h	69% TOC	[31]
Phenol (100 mg/L)	Ti/SnO_2–Sb–Mo	Na_2SO_4 (0.25 M), $j = 10$ mA/cm^2, $t = 3.5$ h	99.6%, 82.7% TOC	[32]
Aniline (2.7 mM)	PbO_2	pH 2, $I = 2$ A, $t = 1$ h	>90%	[33]
Sanitary landfill leachate (COD 6.2 g/L)	$Ti/Pt/PbO_2$	$I = 0.3$ A, $t = 6$ h	40% COD	[34]
Sanitary landfill leachate (COD 6.2 g/L)	$Ti/Pt/SnO_2$–Sb_2O_4	$I = 0.3$ A, $t = 6$ h	40% COD	[34]
4-chlorophenol (100 mg/L)	Ti/IrO_2–RuO_2	Na_2SO_4 (0.1 mM), $j = 39$ mA/cm^2, $t = 2$ h	100%, >70% COD	[35]
Phenol (0.05 mM)	Ti/Ta_2O_5–IrO, Ti/SnO–RuO_2–IrO_2, Ti/RhO_2–IrO_2	KCl (0.01 M), $t = 8$ h, $E_{el} = 1.5$ V, pH 2, 7, 12	99%	[36]
Methylene blue (0.025 mM)	Ti/Ta_2O_5–SnO_2	Na_2SO_4 (0.1 M), $j = 9.1$ mA/cm^2, $t = 3$ h	38.6% TOC, 95% color	[37]

Continued

TABLE 2.2 Removal Efficiency of Organic Pollutants From Water at Different Anodes—cont'd

Targeted Pollutant	Anode Type	Working Parameters	Removal Efficiency	Reference
Paper mill effluent (polyphenols; 2585 mg/L COD; 101.2 m^{-1} color)	Ti/RuO$_2$–PbO$_2$	NaCl (5 mg/L) t = 2 h pH 6.3 E_{el} = 6 V	99% COD 95% color 95% polyphenols	[38]
Phthalic anhydride (2 g/L)	Ti/ RuO$_2$–IrO$_2$–SnO$_2$–TiO$_2$	Na$_2$SO$_4$ (0.1 M) j = 5 A/dm^2 pH 3 t = 4 h	88% COD	[39]
2,4-dichlorophenol (80 mg/L)	Ti/IrO$_2$/RuO$_2$/TiO$_2$	Na$_2$SO$_4$ (0.05 M) pH 9 E_{el} = 1.8 V t = 4 h	100%	[40]
Pretilachlor (60 mg/L)	Ti/SnO$_2$–Sb	Na$_2$SO$_4$ (0.1 M) j = 20 mA/cm^2 pH 7.2 t = 1 h	98.8% 43.1% TOC	[41]
Phenol (100 mg/L)		Na$_2$SO$_4$ (0.25 M) j = 10 mA/cm^2 t = 6 h	95.5% 74.5% TOC	[42]
Pyridine (100 mg/L)		Na$_2$SO$_4$ (10 g/L) j = 30 mA/cm^2 pH 3 t = 3 h	98%	[43]
Perfluorooctanoic acid (100 mg/L)		NaClO$_4$ (10 mM) j = 10 mA/cm^2 t = 1.5 h	90.3%	[44]

Perfluorooctanoic acid (100 mg/L)	Ti/SnO$_2$–Sb/PbO$_2$	NaClO$_4$ (10 mM) j = 10 mA/cm^2 t = 1.5 h	91.1%	[44]
Perfluorooctanoic acid (100 mg/L)	Ti/SnO$_2$–Sb/MnO$_2$	NaClO$_4$ (10 mM) j = 10 mA/cm^2 t = 1.5 h	37.1%	[44]
Methylene blue (30 mg/L)	PbO$_2$–ZrO$_2$	Na$_2$SO$_4$ (0.2 M) j = 50 mA/cm^2 t = 2 h	100% 72.7% COD	[45]
4-chlorophenol (8 mM)	Sn$_{0.86}$–Sb$_{0.03}$—$_{0.10}$—Pt$_{0.01}$—oxide/Ti	H$_2$SO$_4$ (0.5 M) j = 30 mA/cm^2 t = 10 h T = 23°C	100%	[46]
Dye wastewater (Color, 1565 PCU; COD, 188.94 mg/L)	Stainless Steel/SnO$_2$–CeO$_2$	E$_{el}$ = 5 V t = 2 min	83% color 48.6% COD	[47]
Olive mill wastewaters (COD 41, 000 mg/L; oil–grease 1970 mg/L)	Ti/RuO$_2$	NaCl (2M) j = 135 mA/cm^2 t = 7 h	99.6% COD 99.5% oil–grease	[48]
Phenol (490 mg/L)		Na$_2$SO$_4$ (0.25 M) t = 35 h j = 20 mA/cm^2 pH 5	99% 40% TOC	[28]
Landfill leachate (TOC 1270 mg/L; COD 1855 mg/L O$_2$)	Ti/TiO$_2$–RuO$_2$	Flow rate 2000 L/h t = 3 h j = 116 mA/cm^2	86% color 73% COD 57% TOC	[49]

Continued

TABLE 2.2 Removal Efficiency of Organic Pollutants From Water at Different Anodes—cont'd

Targeted Pollutant	Anode Type	Working Parameters	Removal Efficiency	Reference
Phenol (100 mg/L)	BDD	Na_2SO_4 (0.25 M) $j = 10$ mA/cm^2 $t = 6$ h	100% 95.4% TOC	[42]
Sanitary landfill leachate (COD 6.2 g/L)		$I = 0.3$ A $t = 6$ h	40% COD	[34]
Dimethyl phthalate (0.03 mM)		Na_2SO_4 (0.2 M) $j = 20$ mA/cm^2 pH 7 $t = 0.5$ h	100% 50% COD	[27]
Diethyl phthalate (0.03 mM)		Na_2SO_4 (0.2 M) $j = 20$ mA/cm^2 pH 7 $t = 0.5$ h	100% 50% COD	[27]
Diheptyl phthalate (0.03 mM)		Na_2SO_4 (0.2 M) $j = 20$ mA/cm^2 pH 7 $t = 0.5$ h	50% 10% COD	[27]
Amoxicillin (19.6 mg/L TOC)		Na_2SO_4 (0.05 M) $j = 5$ mA/cm^2 $t = 8$ h	36% TOC	[31]
Aniline (500 mg/L)	Ti/TiO$_x$H$_y$/Sb–SnO$_2$	Na_2SO_4 (5 wt.%) $j = 20$ mA/cm^2 $t = 5$ h	85% 71% COD	[50]

Inorganic Compounds

Na_2S (0.023 M)	Ti/RuO_2	$j = 25$ mA/cm^2 pH 13 $t = 5$ h	100%	[51]
Na_2S (60 mM)	BDD	$j = 33$ mA/cm^2 $t = 5$h NaCl 1%	100%	[52]
Na_2S (30 mM)	$Ti/IrO_2-Ta_2O_5$	$j = 13.5$ mA/cm^2 $t = 6$ h NaOH (0.25 M)	83%	[53]
Thiocyanate, SCN^-, (852.2 mg/L)	$Ti/Ru_{0.3}Ti_{0.7}O_2$	$E = 8$ V pH 9 $t = 6$ h NaCl 0.01M	90%	[54]
Cyanides (CN^-) including thiocyanate (SCN^-) (280 mg/L)	PbO_2	$j = 6.7$ mA/cm^2 pH 4 NaCl 1 g/L	99.6%	[55]
NaCN (250 mg/L)	$Ti/SnO_2-Sb-Ce$	$j = 30$ mA/cm^2 pH 13 $t = 4$ h	98.2%	[56]

Electrochemical Disinfection

Toilet wastewater *Escherichia coli Enterococcus*	Bismuth-doped TiO_2	$E = 4$ V pH 6.7–8.3 $t = 10$ min	5-log$_{10}$ reduction	[57]

Continued

TABLE 2.2 Removal Efficiency of Organic Pollutants From Water at Different Anodes—cont'd

Targeted Pollutant	Anode Type	Working Parameters	Removal Efficiency	Reference
Artemia salina	Si/BDD	$j = 255$ mA/cm^2 NaCl 3 g/L $t = 1$ h	100%	[58]
Pseudomonas aeruginosa	BDD	$j = 167$ mA/cm^2 $t = 5$ min NaCl 20 mg/L	6-log$_{10}$ reduction	[59]
Escherichia coli	PbO$_2$/graphite	$j = 253$ mA/cm^2 NaCl 20 mg/L	100% (8 min)	[60]
Enterococcus faecalis			100% (60 min)	
Artemia salina			100% (40 min)	

BBD, boron-doped diamond; COD, chemical oxygen demand; E, applied voltage at electrolysis (V); I, applied current (A); j, applied current density; t, electrolysis time (h).

$$RO + 4H^+ + 4e^- \rightarrow RH_2 + H_2O \tag{2.52}$$

$$R + 2H^+ + 2e^- \rightarrow RH_2 \tag{2.53}$$

Sometimes ER of organic compounds leads to the formation of insoluble products, which can be separated and utilized. As an example, ER of anthraquinonesulfonic acid leads to the formation of insoluble anthraquinone through the reaction of sulfonation using liquid mercury cathode. Formed sediment of anthraquinone can be utilized. To provide more efficient ER of organic compounds, electrolysis should be conducted into porous reactors.

Similar to EO, ER can occur either directly through the direct accepting of electrons by molecules and ions or indirectly through the mediated reactions with electrochemically generated species at the cathode and anode such as H_2 and iron, respectively.

At cathode:

$$2H^+ + 2e^- \rightarrow H_2, \qquad E^0 = 0.00 \text{ V} \qquad pH < 7$$
$$2H_2O + 2e^- \rightarrow H_2 + 2OH^- \quad E^0 = -0.83 \text{ V} \quad pH \geq 7$$

At anode:

$$Fe - 2e^- \rightarrow Fe^{2+}$$

Formation of molecular hydrogen from atomic hydrogen may occur by two mechanisms, either catalytic or electrochemical. Catalytic mechanism (Tafel reaction) consists of the adsorption step of hydrogen atoms on the electrode surface followed by recombination of two adsorbed atoms and formation of hydrogen molecule by the following reaction:

$$M - H + M - H \rightarrow M + H_2 \tag{2.54}$$

Electrochemical mechanism (electrochemical desorption or Heyrovsky reaction) consists of simultaneous steps of H^+ ions discharge and allocation of molecular hydrogen. This mechanism is usually favorable for metals such as iron, nickel, and copper.

$$H^+ + M - H + e^- \rightarrow M + H_2 \tag{2.55}$$

In alkaline media water is directly discharged to atomic hydrogen and hydroxide ions followed by atomic hydrogen recombination by either electrochemical or catalytic mechanism. Presence of some compounds in the solution such as sulfur or phosphorous leads to catalyst poisoning due to the enhanced adsorption of hydrogen atoms on the surface of electrodes. Extensive amount of adsorbed H⁺ atoms penetrate into the crystal lattice of cathodic metals and cause the loss of metal plasticity.

Mediated reactions of anodic material dissolution (mostly iron and aluminum) followed by reduction of different compounds such as Cr(VI) according to the reaction $Cr^{6+} + 3Fe^{2+} \rightarrow Cr^{3+} + 3Fe^{3+}$ will be discussed below in the section on electrocoagulation. An example of direct mechanism is a reduction of metals ions at the cathode surface according to the following reaction:

$$M^{z+} + ze^- \rightarrow M^0 \tag{2.56}$$

It is worth to notice that cathodes with high overpotential toward hydrogen evolution are more efficient in reduction of pollutants. Overpotential toward hydrogen evolution for different metals decreasing with ambient temperature increase, decrease of current density, and increase of the electrode surface roughness, which is directly related to the catalytic surface area increase. Analogous to EO processes, overpotential of electrodes allows cost saving due to reduction of electricity consumption for the side process and increasing selectivity of the reaction. The values of different cathodic overpotentials toward hydrogen evolution reactions are shown in Fig. 2.3. According to Fig. 2.3 electrodeposition of strongly electronegative species can be conducted while using Hg electrodes due to their high overpotential toward hydrogen evolution. ER of highly electronegative species could theoretically be conducted using Pb cathode because of their high overpotential characteristics. However, Pb electrodes interact easily with acids.

The value of overpotential strongly depends on the nature of metal needed to be discharged. Depending on the overpotential value metals can be conditionally divided into three categories. The first group consists of metals of very low overpotential not more than a thousands of a volt. They are Cd, Pb, Ag, and Hg. Second group consists of Zn, Bi, and Cu with overpotential lying in the range of a hundreds of a volt and the third group includes Co, Ni and, Fe having overpotential in the order of several tenths of a volt. The value of overpotential is also influenced by the nature of salt anion.

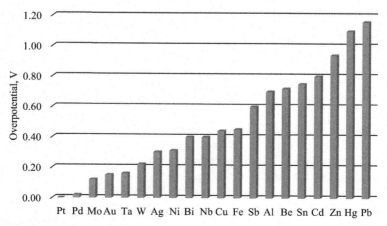

FIGURE 2.3 Overpotential of different cathode materials toward hydrogen evolution reactions.

Usually overpotential increases for different anions in the following order: $J^- > Br^- > Cl^- > ClO_4^- > NO_3^- > PO_4^{3-}$. Alkalinity of solution also affects overpotential. High alkalinity can cause additional deposition of metal hydroxides on the surface of electrodes along with metal deposition. Moreover, a number of strongly electronegative metals cannot be deposited from aqueous solutions. Also some metals, such as manganese, can undergo ER by changing only the valence electrons without the formation of solid products.

$$Mn^{3+} + e^- \rightarrow Mn^{2+} \tag{2.57}$$

$$MnO_4^- + e^- \rightarrow MnO_4^{2-} \tag{2.58}$$

$$Fe^{3+} + e^- \rightarrow Fe^{2+} \tag{2.59}$$

Electrochemical Reduction of Metals

Electrochemical treatment of wastewaters containing toxic metals is mostly conducted by electrocoagulation with soluble anodes. However, electrochemical deposition of metals from wastewater finds more applications nowadays because of the possibility of further recovery of valuable products. Moreover, there is no need for addition of chemicals usually, process is selective and operation costs are low. Along with recovery of metals, water can undergo disinfection by the formed electroactive species. As every process, ER has a disadvantage. Process efficiency is sensitive to wastewater composition and interfered easily by side reactions of hydrogen generation and oxygen reduction.

Wastewater treatment with Hg(II), Pb(II), Cd(II), and Cu(II) is usually conducted in acidic conditions at cathodes made of carbon and sulfur mixture with a different C:S ratio varying from 20:80 to 80:20. In this case reduced metals precipitate in the form of insoluble sulfides or bisulfides, which are further removed mechanically.

Direct ER methods using insoluble anodes are particularly effective for the neutralization of chromium-containing wastewater with high concentration of Cr(VI) usually above 2 g/L and allow reduction of Cr(VI) content approximately by three orders of magnitude. An optimal media for Cr(VI) reduction is considered to be acidic with pH of 2 and applied current density between 2 and 20 mA/cm^2. Reduction of Cr(VI) depending on the media conditions can be represented via the following reactions:

Acidic media:

$$Cr_2O_7^{2-} + 14H^+ + 6e^- \rightarrow 2Cr^{3+} + 7H_2O \tag{2.60}$$

Slightly alkaline media:

$$CrO_4^{2-} + 4H_2O + 3e^- \rightarrow Cr(OH)_3 + 5OH^- \tag{2.61}$$

ER of zinc at a cathode occurs with simultaneous allocation of hydrogen and deposition of zinc according to the following reactions:

$$2H^+ + 2e^- \rightarrow H_2, \quad E^0 = 0.00 \text{ V}$$
$$Zn^{2+} + 2e^- \rightarrow Zn, \quad E^0 = -0.76 \text{ V}$$

As it is seen from reduction potentials, ER of zinc should not allocate only H_2 gas; however, because of the high value of overpotential, Zn can be successfully recovered from solutions with current efficiency (CE) of 80%–90%. It is known that different electrochemical reactions occur at a particular electrode potential; therefore, when there is a mixture of different compounds in solution, reaction having more positive reduction potential thermodynamically are more favorable, thus suppressing reactions with more negative electrode potential reducing its CE and purity of deposit. In this regard, if there is a necessity to recover a metal of high purity from a solution, then solution should not contain impurities at concentrations above the maximum allowable levels.

The distance between electrodes has no significant effect on the metal recovery efficiency, whereas intensified stirring increases the deposition kinetics of metals. Studies on metal deposition kinetics on cathodes are one of the most difficult in electrochemistry. This is explained by the fact of continuous formation of deposits on electrode surface and renewal of electrodes surface along with energy distribution.

The ICE of ER can be determined by the following equation [61]:

$$\text{ICE} = \frac{zFV \cdot \dfrac{dC_i}{dt}}{I} \cdot 100\% \tag{2.62}$$

where dC_i/dt is the rate of change of metal ion concentration in solution during electrolysis, which can be determined from experimental data of $C_i = f(t)$ plot at applied current density I.

The EC required for the recovery of 1 kg metal (kWh/kg) can be calculated as follows [61]:

$$\text{EC} = \frac{\int_0^{t_{\text{tot}}} I \cdot E_{\text{cell}} \cdot dt}{3600 \cdot 1000 \cdot m_M} \tag{2.63}$$

where t_{tot} is the total electrolysis time (s), E_{cell} is the cell voltage, and m_M is the mass of recovered metal.

Electrochemical Reduction of Inorganic Nitrogen Compounds

Inorganic nitrogen in natural waters originates from ammonification process of organic nitrogen compounds contained in dead plants and animals. Moreover, some bacteria can catch the atmospheric nitrogen and reduce it to NH_4^+ forms in accordance with the following equation:

$$N_2 + 8H^+ + 6e^- \rightarrow 2NH_4^+ \tag{2.64}$$

The formed ammonia can further undergo nitrification done by bacteria forming nitrates as follows:

$$2NH_4^+ + O_2 + 2e^- \rightarrow NH_2OH + H_2O \tag{2.65}$$

$$NH_2OH + H_2O \rightarrow NO_2^- + 5H^+ + 4e^- \tag{2.66}$$

Increased agricultural and industrial activities contribute to additional wastewater and natural water pollution by nitrogen species, which cause eutrophication of water bodies and suppression of water quality and natural growth of water biota, acidify the surface water, increase the content of atmospheric ozone, etc.

The main goal of ER of inorganic nitrogen is to conduct denitrification of N-species with the formation of nontoxic N_2 gas, which is in nature conducted by anammox bacteria through the following transformations:

$$NH_4^+ + NO_2^- \rightarrow N_2 + 2H_2O. \tag{2.67}$$

However, in practice, ER of nitrogen compound is difficult to conduct because of a wide variety of stable nitrogen species such as $NO_3^-, N_2O_4,$ $NO_2^-, NO, N_2O_2^{2-}, N_2O, N_2, NH_3, NH_3OH^+, N_2H_5^+$ and N_4^+ with oxidation states varying from -3 to $+5$. In this regard, the use of electrocatalytic material is required to enhance the process of denitrification. The most favorable reaction of nitrates and nitrites is reduction to N_2 gas, which is described as follows [62]:

$$2NO_3^- + 12H^+ + 10e^- \rightarrow N_2 + 6H_2O \quad E^0 = 1.246 \text{ V} \tag{2.68}$$

$$2NO_2^- + 8H^+ + 6e^- \rightarrow N_2 + 4H_2O \quad E^0 = 1.52 \text{ V} \tag{2.69}$$

However, in practice, nitrates and nitrites are usually reduced to other more thermodynamically favorable compounds, such as NO, NO_2^-, N_2O, etc., in accordance with the following reactions:

$$NO_3^- + 2H^+ + 2e^- \rightarrow NO_2^- + H_2O \quad E^0 = 0.835 \text{ V} \tag{2.70}$$

$$NO_3^- + 4H^+ + 3e^- \rightarrow NO + 2H_2O \quad E^0 = 0.958 \text{ V} \tag{2.71}$$

$$2NO_3^- + 10H^+ + 8e^- \rightarrow N_2O + 5H_2O \quad E^0 = 1.116 \text{ V} \tag{2.72}$$

$$NO_3^- + 10H^+ + 8e^- \rightarrow NH_4^+ + 3H_2O \quad E^0 = 0.875 \text{ V} \tag{2.73}$$

$$NO_2^- + 2H^+ + e^- \rightarrow NO + H_2O \quad E^0 = 1.202 \text{ V} \tag{2.74}$$

$$2NO_2^- + 6H^+ + 4e^- \rightarrow N_2O + 3H_2O \quad E^0 = 1.396 \text{ V} \tag{2.75}$$

$$NO_2^- + 8H^+ + 6e^- \rightarrow NH_4^+ + 2H_2O \quad E^0 = 0.897 \text{ V} \tag{2.76}$$

One of the most selective catalysts toward N_2 generation from nitrates is platinum—copper electrode. So far, there is no efficient catalyst for complete conversion of nitrates and nitrites to N_2 gas. Most reduction mechanisms lead to the formation of intermediate NO gas. Therefore, there is an approach trying to find effective catalytic materials for ER of NO.

Ammonia nitrate can be reduced at graphite electrode to ammonium nitrite, which decomposes on heating to nitrogen gas.

$$NH_4NO_3 + 2H^+ + 2e^- \rightarrow NH_4NO_2 + H_2O \tag{2.77}$$

$$NH_4NO_2 \rightarrow N_2 + 2H_2O \tag{2.78}$$

Electrochemical Reduction of Organic Nitrogen Compounds

ER of organic nitrogen-containing compounds such as nitrobenzene and trinitrotoluene is conducted using lead, zinc, copper, and stainless steel cathodes in two-compartment electrolyzers where anodic and cathodic chambers are separated by a membrane. At first nitrobenzene and trinitrotoluene undergo cathodic reduction with formation of amines, which is further electrochemically oxidized in anodic compartment to nontoxic compounds. Moreover, amines are valuable intermediates for dyes and drugs production and therefore can be recycled.

It was found that diatrizoate is reduced at graphite and Pd, graphite and Pd-doped graphite felt electrodes in neutral media forming 3,5-diacetamidodiiodobenzoic acid; 3,5-diacetamidoiodobenzoic acid; and 3,5-diacetamidobenzoic acid as reaction products [63].

Electrochemical Reduction of Chlorinated Hydrocarbons

Chlorinated hydrocarbons such as polychlorinated biphenyls, dichlorodiphenyl trichloroethane, trichloroethylene, chlorinated benzenes, and others are highly toxic compounds able to cause cancer and threaten the environment biota. ER allows reducing the toxicity of chlorinated hydrocarbons through a step reduction process with a gradual abstraction of chlorine atoms. As an example of electrochemical dechlorination, the mechanisms of carbon tetrachloride cathodic reduction are shown in the following equations [64]:

$$CCl_4 + e^- \rightarrow CHCl_3 + Cl^- \tag{2.79}$$

$$CHCl_3 + e^- \rightarrow CH_2Cl_2 + Cl^- \tag{2.80}$$

$$CHCl_3 + 3H_2O + 6e^- \rightarrow CH_4 + 3Cl^- + 3OH^- \tag{2.81}$$

$$CH_2Cl_2 + e^- \rightarrow CH_3Cl + Cl^- \tag{2.82}$$

$$CH_3Cl + e^- \rightarrow CH_4 + Cl^- \tag{2.83}$$

Electrochemical Reduction of Aldehydes and Ketones

Aldehydes and ketones are organic compounds containing carbonyl group $> C = O$ in the molecule structure. Carbonyl carbon of aldehydes is connected to hydrogen atom and an organic group R (general formula, $RHC = O$), and carbonyl carbon of ketones is connected with two organic groups (general formula, $R_2C = O$). Both groups are toxic compounds having irritating and neurotoxic effects. Some of these compounds can have mutagenic and carcinogenic properties. Mechanism of reduction of carbonyl compounds consists of multistages such as hydration, dehydration, keto–enol equilibria, interactions with radical anions, etc., resulting in either formation of alcohols through a two-electron transfer process or formation of picanols as a result of one electron transfer process [65]. General mechanism of ER of aliphatic aldehydes is represented as follows [65]:

$$RHC = O + 2e^- + 2H^+ \rightarrow RCH_2OH \tag{2.84}$$

The mechanism of benzaldehyde reduction is more complex depending on pH of the medium and includes reaction of mediated radical formation.

Acidic media:

$$ArC(R) = O + H^+ \rightleftarrows ArC(R) = OH^+ \tag{2.85}$$

$$ArC(R) = OH^+ + e^- \rightarrow Ar\dot{C}(R) - OH \tag{2.86}$$

$$2\,Ar\dot{C}(R) - OH \rightarrow dimers \tag{2.87}$$

$$Ar\dot{C}(R) - OH + Hg \rightarrow organometallic\ compounds \tag{2.88}$$

$$Ar\dot{C}(R) - OH + e^- \rightarrow ArC(R)^- - OH \tag{2.89}$$

$$ArC(R)^- - OH + H^+ \rightleftarrows ArCH(R) - OH \tag{2.90}$$

Neutral media:

$$ArC(R) = O + e^- \rightarrow Ar\dot{C}(R) - O^- \tag{2.91}$$

$$ArC^{\cdot}(R) - O^- + H^+ \rightleftarrows Ar\dot{C}(R) - OH \tag{2.92}$$

Alkaline media:

$$ArC(R) = O + e^- - Ar\dot{C}(R) - O^- \tag{2.93}$$

$$Ar\dot{C}(R) - O^- + e^- \rightarrow Ar\overline{C}(R) - O^- \tag{2.94}$$

$$Ar\overline{C}(R) - O^- + 2H^+ \rightarrow ArCH(R) - OH \tag{2.95}$$

$$Ar\dot{C}(R) - O^- + Me^+ \rightarrow ArC^{\cdot}(R) - OMe \tag{2.96}$$

$$Ar\dot{C}(R) - OMe + e^- \rightarrow Ar\overline{C}(R) - OMe \tag{2.97}$$

$$Ar\overline{C}(R) - OMe + 2H^+ \rightarrow ArCH(R) - OH + Me^+ \tag{2.98}$$

Reduction efficiency of phenyl ketones in alkaline media decreases with increase of carbonyl group molecules, i.e., reduction will be easier for

$HCOH > CH_3COH > CH_3COCH_2$ or in other words the ease of ketones reducibility can be allocated in the following dependence: $C_2 > C_3 > C_4 >$ $i-C_4 > t-C_4 > Ph$, where C_2 is aceto-; C_3 is propio-; C_4 is n-butyro-; $i-C_4$ is i-butyro-; $t-C_4$ is pivalophenone; and *Ph* is benzophenone [66]. Table 2.3 contains a review on applications of ER for the removal of different compounds.

2.3 ELECTROCOAGULATION

Electrocoagulation is based on the physical—chemical process of coagulation of colloidal systems under the action of a direct electric current. While conducting electrolysis of wastewaters with steel or aluminum anodes, electrochemical dissolution of anodic metal takes place. Dissolved cations of aluminum and iron are hydrolyzed and act as coagulants, which initiate adhesion and fusion of the particles. In general, coagulation means the loss of aggregate stability in dispersed systems leading to the phase separation. A wide range of pollutants can be removed from water using electrocoagulation process. They are pathogenic microorganisms, cyanobacteria, organic pollutants, clays, and other inorganic colloids, most of which are negatively charged. In this regard, addition of positively charged cations can neutralize and destabilize the colloids forcing them to coagulate. Simplified mechanism of electrocoagulation is shown in Fig. 2.4.

Usually electrocoagulation with aluminum electrodes is used for the treatment of concentrated oil-containing wastewater (maximum concentration of oil up to 10 g/L) and organic pollutants. Wastewaters containing higher oil concentrations should be preliminarily diluted preferably by acidic wastewaters. The residual oil concentration usually does not exceed 25 mg/L. Electrocoagulation with steel electrodes is often used for the removal of Cr(IV) and other nonferrous metals such as Zn, Cu, Ni, Cd, Cr(III), etc., from wastewaters with a flow rate not exceeding 50 m^3/h; Cr(IV) concentration up to 100 mg/L; and total nonferrous metal concentration of up to 100 mg/L. The minimum total salt content of wastewater should be of 300 mg/L and the maximum suspended solids concentrations should not exceed 50 mg/L.

The Theory of Electrolyte Coagulation

The quantitative theory of electrolyte coagulation (DLVO theory) was completed by two Russian scientists Derjaguin and Landau in 1941. A similar approach to the study of the stability of colloidal systems was independently developed by two Dutch scientists Verwey and Overbeek in 1948. According to the initial letters of scientist's' family names the theory was called DLVO theory. It assumes that particles interaction energy in a liquid is a sum of Van der Waals attraction and electrostatic repulsion (double-layer repulsion) forces acting in the narrow gap separating the particles of the dispersed phase. When

TABLE 2.3 Removal Efficiency of Different Compounds From Wastewater Using Electrochemical Reduction Method

Pollutant	Reduction Products	Cathode	Working Parameters	Removal efficiency/ Current Efficiency	Reference
Chinester (424 mg/L)	Dechlorinated compounds	Graphite	$j = 6$ mA/cm^2 Flow rate 20 L/h $t = 15$ min NaOH 0.1 M Na$_2$SO$_4$ 0.1 M	80%/43%	[67]
Nitrobenzene (10 mg/L)	Nitrosobenzene, phenylhydroxylamine and aniline	TiO$_{2-x}$ SCs GC TiO$_2$	$E = -1.2$ V/SCE Flow rate 20 L/h $t = 120$ min Na$_2$SO$_4$ 0.1 M	100% 20% 80%	[68]
Diatrizoate (5 μM)	DTR-I;; DTR-2I, DTR-3I	Graphite Pd-loaded graphite	pH 7.0 $E = -1.7$ V/SCE Flow rate 6.6 L/h $t = 180$ min K$_2$HPO$_4$ 22 mM KH$_2$PO$_4$ 22 mM	87% 94%	[63]
4-t-butylcyclohexanone (0.2 M)	Corresponding alcohols	Pt Pd	$j = 59$ mA/cm^2 Pt $j = 45$ mA/cm^2 Pd	87%/87% 82%/66%	[69]
3,3,5-trimethylcyclohexanone (0.2 M)			$j = 59$ mA/cm^2 Pt $j = 90$ mA/cm^2 Pd	65%/50% 72%/60%	
Cr(VI) (100 mg/L)	Cr (III)	GC/Poly-aniline	$j = 164$ mA/cm^2 Flow velocity 0.27 m/s $t = 10$ min	99%/99%	[70]

Continued

TABLE 2.3 Removal Efficiency of Different Compounds From Wastewater Using Electrochemical Reduction Method—cont'd

Pollutant	Reduction Products	Cathode	Working Parameters	Removal efficiency/ Current Efficiency	Reference
Cr(vi) (10 mg/L)	Cr (III)	Graphite/ Polyaniline wire GC	pH < 2 $E = -0.2$ V t = 135 min H_2SO_4 0.1 M	98.7% 85.3%	[71]
Cu(ii) (10 mM) Cd(ii) (10 mM) Pb(ii) (10 mM) Zn(ii) (10 mM)	Cu(0) Cd(0) Pb(0) Zn (0)	Steel	pH = 3.5 t = 180 min $E = -0.1$ to -1.3 V	99% Cu ($E = -0.1$ V) 93% Cd ($E = -0.9$ V) 99% Pb ($E = -0.8$ V) 38% Zn ($E = -1.3$ V)	[72]
Mixture of Cu(II) Cd(II) Pb(II) Zn(II)	Cu(0) Cd(0) Pb(0) Zn (0)	Steel	pH = 3.5 t = 180 min $E = -1.3$ V	99% 92% 68% 10%	[72]
Cu(ii) (978 mg/L)	Cu (0)	Stainless steel Cu	pH = 1.4 $j = 7.4$ mA/cm^2 t = 24 h	99%	[73]
Mixture of Cu(ii) (500 mg/L) Ag(i) (500 mg/L) Pb(ii) (500 mg/L)	Cu(0) Ag(0) Pb(0)	Stainless steel Cu	pH = 1.4 $j = 7.4$ mA/cm^2 t = 24 h	11.8% 96% 93%	[73]

Nitrate-N (100 mg/L)	Ammonia-N	Fe	$j = 20$ mA/cm^2 $t = 180$ min	92.8%	[74]
Nitrate-N (50 mg/L)	N$_2$ (92%) Ammonia-N (8%) Traces of N$_2$O, NO, nitrite-N	Sn	$E = -2.5$ V $t = 90$ min K$_2$SO$_4$ 0.1 M KNO$_3$ 0.05 M	100%	[75]
Nitrate-N (50 mg/L)	Nitrite-N Ammonia-N N$_2$	Al$_2$O$_3$/Pd–Cu	pH $= 5.2$ $j = 10$ mA/cm^2 $t = 150$ min	75%	[76]
Ammonia-N	Nitrate-N N$_2$ NO$_x$	Cu/Zn Ti Fe	$j = 30$ mA/cm^2 $t = 60$ min pH 7–9 NaCl 1 g/L	100%	[77]

Chinester, ethyl-[d,1-cis, trans-2,2-dimethyl-3-|2,2' dichlorvinyl|]-1-cyclopropane carboxylate or DVCA-ethylester; *DTR-2l*, 3,5-diacetamidoiodobenzoic acid; *DTR-3l*, 3,5-diacetamidobenzoic acid; *DTR-I*, 3,5-diacetamidodiiodobenzoic acid; *GC*, glassy carbon; *TiO$_{2-x}$ SCs*, TiO$_2$ single crystals.

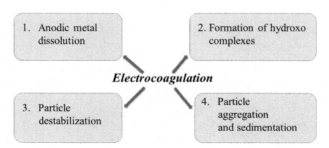

FIGURE 2.4 The main steps of electrocoagulation.

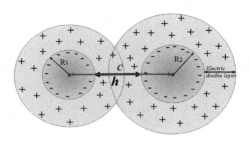

FIGURE 2.5 Overlapping of electrical double layers of two identical colloidal particles charged negatively.

considering two spherical particles of the same composition (Fig. 2.5), the London—Van der Waals forces are always attractive between those particles and can be described using Eq. (2.99) [78,79].

$$
\begin{aligned}
G_A &= \frac{\pi^2 \rho_1 \rho_2}{6} C \left[\frac{2R_1 R_2}{c^2 - (R_1 + R_2)^2} + \frac{2R_1 R_2}{c^2 - (R_1 - R_2)^2} + \ln\left(\frac{c^2 - (R_1 + R_2)^2}{c^2 - (R_1 - R_2)^2} \right) \right] \\
&= \frac{A_H}{12} \cdot \left[\frac{2R_1 R_2}{c^2 - (R_1 + R_2)^2} + \frac{2R_1 R_2}{c^2 - (R_1 - R_2)^2} + \ln\left(\frac{c^2 - (R_1 + R_2)^2}{c^2 - (R_1 - R_2)^2} \right) \right],
\end{aligned}
$$

(2.99)

Where ρ_1 and ρ_2 are electron densities of particles; C is the constant; c is the distance between particle centers; R_1 and R_2 are particle radii; $h = c - (R_1 + R_2)$ is the distance between particle surfaces; A_H is the Hamaker constant.

For two identical particles ($R_1 = R_2 = R_{12}$) at close proximity $h << R$ Eq. (2.99) can be simplified as follows:

$$
G_A = -\frac{A_H R_{12}}{12h}
$$

(2.100)

Bbecause $A_H > 0$, then $G_A < 0$ for two identical particles embedded in any fluid.

According to the second law of thermodynamics spontaneous processes proceed in the direction of decreasing free energy. In this regard, particles in water should always stick together. From another point of view, there isare many dispersed systems where particles do not adhere to each other for a long time. That means that there is some other force keeping colloids dispersed. When particles are approaching together, their double electric layers overlap, free energy of the particles increases, and as a consequence repulsive forces (G_{el}) appear, which can be described for spherical particles with a radius R ($h << R$) as follows:

$$G_{el} = 2\pi\varepsilon_r\varepsilon_0 R_{12}\phi_\delta^2 \exp(-kh), \tag{2.101}$$

where ε_r is the relative dielectric permittivity of the medium containing particles, ε_0 is the electrical permittivity of vacuum, k is the Debye length or inverse of the thickness of diffuse electrical double layer (EDL) ($1/k$), and ϕ_δ is the Stern potential. As it is seen from Eq. (2.101) the greater the thickness of the diffuse electric double layer the stronger the effect of repulsive forces.

$$\frac{1}{k} = \sqrt{\frac{\varepsilon_r\varepsilon_0 RT}{2F^2\mu}} = \sqrt{\frac{\varepsilon_r\varepsilon_0 RT}{F^2\sum_i C_i z_i^2}}, \tag{2.102}$$

where F is the Faraday constant; R is the gas constant; T is the temperature; C_i is the concentration of species i; $\mu = 1/2\sum_i C_i z_i^2$ is the ionic strength of solution; and z_i is the charge number of species i. As it is seen from Eq. (2.102) the thickness of EDL decreases with decrease in temperature and increase in electrolyte concentration and ion charge.

In summary, the energy of attraction is inversely proportional to the square of the distance between the particles. The positive electrostatic repulsion energy decreases with the distance exponentially. The action of attraction and

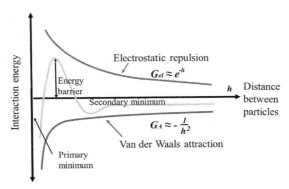

FIGURE 2.6 Attraction, repulsion, and total energy of particles interaction.

repulsion energies depends on the distance between the particles and the total energy of particles interaction $G = G_A + G_{el}$ (Fig. 2.6). The interaction forces between the particles can be distributed over distances up to hundreds of nanometers.

As it can be seen from Fig. 2.6, the attractive forces prevail at short distances between the particles ($h \rightarrow 0$, $G_A \rightarrow -\infty$), which are characterized by the primary minimum and lead to adhesion of the particles. The attraction forces also dominate at the relatively large distances between the particles because exponent decreases much faster than power function (secondary minimum at Fig. 2.5) leading to the attraction of particles through the liquid medium. Repulsive forces may prevail at medium distances (~ 100 nm), which are characterized by the maximum at the total energy curve (energy barrier) and prevent adhesion of particles. The increase of ϕ_δ contributes to the increase of the energy barrier. Usually $\phi_\delta = 20$ mV is enough for potential barrier to appear, thus providing aggregate stability of disperse systems. The potential barrier also increases with the decrease of Hamaker constant A_H.

Along with electrostatic repulsion, colloidal stability of disperse system can be explained by the presence of solvation shell of solvent molecules formed around dispersed particles and closely attached to them. The water properties of solvation shell near to the hydrophilic particles differ significantly on water properties in bulk solution. This bond water has reduced electrical permittivity and freezing temperature, increased viscosity, and boiling temperature. In this regard, when particles are approaching to each other, there is an overlapping of water boundary layers in addition to EDL overlap, which leads to additional repulsion. This is generally true for the hydrophilic particles. Despite the electrostatic energy of hydrophilic particles interaction tends to approach zero, there is still repulsion.

Let's us consider three different states of a disperse system depending on the magnitude of energy minimums and energy barrier.

1. If the magnitude of the energy barrier and the depth of the secondary minimum are insignificant, then the particles approach to the closest distance by the influence of the kinetic energy and then coagulate under the influence of the attraction forces. Such systems are unstable and coagulation is irreversible.

2. If the energy barrier is high and the depth of the secondary minimum is small, the particles cannot overcome the barrier, coagulation does not occur, and aggregate stability of the system is observed. There two ways to decrease the aggregate stability of the system. The first one is to increase the kinetic energy of particles to values higher than the energy barrier, for example, by the temperature rise of the medium or by mechanical stirring. This will lead to the increase of the number of collisions and possible

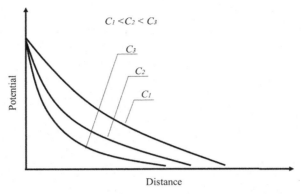

FIGURE 2.7 The dependence of the electrical double layer thickness on the concentration of the indifferent electrolyte.

particles adhesion. The second way to decrease the potential barrier is to compress the EDL of particles, which allow the particles to be in close proximity and as a result enhancement of attractive forces. This can be achieved by the addition of electrolyte to the medium. When indifferent electrolyte (nonelectroactive electrolyte, which cannot be specifically adsorbed on the surface of particles) is added to the system, the thickness of EDL decreases with the increase of supporting electrolyte concentration (Fig. 2.7). As a result the electrostatic repulsion forces and consequently the energy barrier decrease and either slow or rapid coagulation takes place.

Rapid coagulation (coagulation when all particle collisions are effective, i.e., energy barrier is absent or negligible) is described by the Schulze—Hardy rule.

According to the Schulze—Hardy rule, counterions i.e., ions having a charge opposite to the charge of the lyophobic colloidal particles, have a coagulating effect. Coagulating effect is higher for the counterions with a higher valence.

The Schulze—Hardy rule was confirmed empirically and was found to follow the inverse six power law [80]. The concept of critical coagulation concentration (CCC) was introduced. To initiate rapid coagulation the minimum concentration of counterions required is CCC, which can be estimated through the next relationship

$$CCC = \frac{C}{z^6} \tag{2.103}$$

where C is the constant and z is the valence of counterions.

In that case while comparing coagulation by monovalent (such as Na^+), divalent (such as Ca^{2+}), and trivalent (such as Al^{3+}) ions, the effectiveness of coagulation treatment by these ions will be correlated as follows:

$$Na^+ : Ca^{2+} : Al^{3+} = \frac{1}{1^6} : \frac{1}{2^6} : \frac{1}{3^6} \approx 730 : 65 : 1 \qquad (2.104)$$

This means that to initiate rapid coagulation and eliminate the energy barrier it is required to add monovalent coagulant 730 times greater than trivalent coagulant and divalent coagulant will be 65 times more effective than monovalent one.

Some cases do not obey the Schulze–Hardy rule and the exponent of z can be reduced to 2 (Eilers–Korff rule) or other values. This can be explained, for example, by an unaccounted interference with other particles or presence of significant energy barrier. This means that slow coagulation takes place and not all collisions of particles are effective and lead to particle adhesion.

Coagulating effectiveness of ions of the same charge increases with the increase in ion radii. For example Cs^+ will have the highest coagulation effect among Li^+, Na^+, K^+, Rb^+, and Cs^+.

Ions able to specifically adsorb on the surface of particles decrease the Stern potential (ϕ_δ) and electrostatic constituent of particle interaction forces, by particle charge neutralization thus destabilizing them and initiating the coagulation. For example, aluminum and iron salt are easily hydrolyzed producing species able to specifically absorb on the surface of particles. While using aluminum [$Al_2(SO_4)_3$, $AlCl_3$, or $Al_2(OH)_5Cl$] and iron [$FeSO_4$, $Fe_2(SO_4)_3$, etc.] salts as coagulants, the pH value should be carefully adjusted to maintain the consumption of coagulant at a minimum. The fact of careful pH adjustment is explained by the fact that hydrolysis products of Al and Fe salts are more effective than salts itself. Moreover, highly charged polynuclear hydroxo complexes of aluminum and iron (Fig. 2.8) formed during hydrolyses can intensify the treatment process while interacting with the negatively charged colloidal pollutants [81,82].

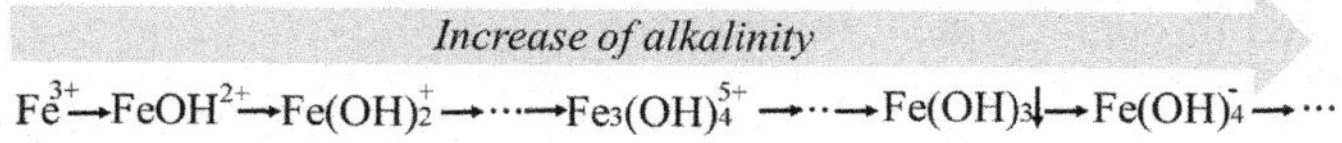

$$Al^{3+} \rightarrow AlOH^{2+} \rightarrow Al(OH)_2^+ \rightarrow \cdots \rightarrow Al_{13}(OH)_{32}^{7+} \rightarrow \cdots \rightarrow Al(OH)_3\downarrow \rightarrow Al(OH)_4^- \rightarrow \cdots$$

Increase of alkalinity

$$Fe^{3+} \rightarrow FeOH^{2+} \rightarrow Fe(OH)_2^+ \rightarrow \cdots \rightarrow Fe_3(OH)_4^{5+} \rightarrow \cdots \rightarrow Fe(OH)_3\downarrow \rightarrow Fe(OH)_4^- \rightarrow \cdots$$

FIGURE 2.8 Dependence of aluminum and iron polynuclear hydroxo complexes formation on pH.

Precipitation of $Al(OH)_3$ and $Fe(OH)_3$ occurs in the range of neutral pH and also contributes to the removal of pollutants from water initiating sweep flocculation process. Growing in size positively charged $Al(OH)_3$/$Fe(OH)_3$ aggregates electrostatically attract the negatively charged pollutants, which results to pollutants neutralization and destabilization thus precipitating them. Nevertheless, the positive effect of counterions on colloids destabilization, the dosage of coagulant should be carefully controlled to avoid the charge reversal and restabilization of colloids.

3. If the depth of the secondary minimum driving is large, the particles cannot leave the potential well and remain fixed at a distance, which corresponds to that minimum ($\sim 10^3$ nm), irrespective of the height of the energy barrier. Particles stay in an equilibrium and do not diverge and converge, continuing to exist and moving as a pair. Other particles can be attached to such pair to form more complex units based on the same "far" interaction. At the same time, each particle retains its individuality, and the system as a whole retains its dispersibility in contrast to the coagulation associated with "near" interaction and as a result in the adhesion of the particles and decrease of solid/dispersion medium interface.

Principles of Electrocoagulation

A general reaction of anodic metal dissolution of the metal leading to the formation of simple hydrated ions and complex metal ions, respectively, can be written as follows:

$$M + x\ H_2O \rightarrow Mz + x\ H_2O + ze^- \tag{2.105}$$

$$M + xH_2O + y\ A \rightarrow MA_y^{z-y} \cdot x\ H_2O + ze^- \tag{2.106}$$

The amount of metal dissolved at the electrolysis is determined by the Faraday's law and CE,

$$m = CE \cdot \frac{ItM}{Fz} \tag{2.107}$$

Theoretically while consuming 96,500 C/mol electricity, amount of dissolved aluminum and iron are 8.99 and 27.93 g, respectively.

The EC (kWh/m^3) during electrocoagulation process can be estimated using the following equation [83]:

$$EC = \frac{UIt}{V}, \tag{2.108}$$

where U is the applied voltage (V); I is the current intensity (A); t is the time of electrocoagulation (h); and V is the volume of treated solution (L).

The mechanism of electrochemical metal dissolution consists of two processes. They are anodic metal dissolution while applying electric current and chemical dissolution due to the metal interaction with the medium. The following are the most common reactions occurring at coagulation while using aluminum and iron anodes:

- Anodic metal dissolution

$$Al - 3e^- \rightarrow Al^{3+} \quad Fe - 2e^- \rightarrow Fe^{2+}$$
$$Fe - 3e^- \rightarrow Fe^{3+}$$

- Hydroxide formation

$$Al^{3+} + 3OH^- \rightarrow Al(OH)_3 \quad Fe^{2+} + 2OH^- \rightarrow Fe(OH)_2$$
$$4Fe(OH)_2 + O_2 + 2H_2O \rightarrow 4Fe(OH)_3$$

- Chemical interaction with water

$$2Al + 6H_2O \rightarrow 2Al(OH)_3 + 3H_2$$
$$2Al + 6H_2O + 2OH^- \rightarrow 2[Al(OH)_4]^- + 3H_2$$
$$2Fe^{2+} + 1/2O_2 + H_2O \rightarrow 2Fe^{3+} + 2OH^-$$

- Cathodic reduction of metals and metal films

$$Fe^{2+} + 2e^- \rightarrow Fe$$
$$Fe_3O_4 + H_2O + 2e^- \rightarrow 3FeO + 2OH^-$$

- Cathodic reduction of organic compounds

$$R + 2H^+ + 2e^- \rightarrow RH_2$$
$$RO + 4H^+ + 4e^- \rightarrow RH_2 + H_2O$$

- Cathodic water electrolysis reactions

$$2H^+ + 2e^- \rightarrow H_2, \qquad pH < 7$$
$$2H_2O + 2e^- \rightarrow H_2 + 2OH^- \quad pH \geq 7$$

The final quality of treated water during electrocoagulation is influenced by the anode materials, distance between anodes, water flow rate in the inter-electrode space, temperature, pH, anionic and cationic composition of the medium, the frequency of polarity change, the current density, etc. The influence of these parameters on electrocoagulation is given below.

Anode Material

To improve the mechanical properties metals are often doped with different elements. The presence of a doping element in anode material alloys can either enhance or suppress the anodic dissolution of metal. For example, the main doping elements for aluminum alloys are copper, manganese, silicon, magnesium, and zinc. It is known that Al alloy containing as little as 0.5%−1% of Mn, Cu, or Si increase the corrosion potential of aluminum in acidic and neutral media and thus have the highest rate of Al electrochemical dissolution. The intensification of metal dissolution in this case can be explained by formation of corrosion initiation sites in the presence of doping elements. On the contrary, Al alloys doped with as little as 0.5% of Mg or Zn decrease the corrosion potential of Al and suppress the rate of its electrochemical dissolution in basic media. The latter can be explained by the formation of film on the metal surface containing insoluble hydroxides under these pH conditions.

Ambient Temperature

Increase on water temperature prevents the formation of sediments on the surface of electrodes and facilitates the rate of passivating film dissolution. Active aluminum anode dissolution occurs at water temperatures between 2 and 60°C, and a current density of 1−4 mA/cm^2. Higher temperature leads to decrease in the yield of aluminum and the current increases in the voltage across the electrodes. The maximum CE during steel electrode dissolution is achieved in a temperature range of 10°−25°C and an applied current density of about 6 mA/cm^2 [84].

Anionic Composition of Electrolyte

The presence of anions in the water has a significant impact on the corrosion and consequently dissolution of the anode. Among different anions chloride ions have the greatest impact on the aluminum anode activity because of their small dimensions and ease of penetration through the oxide film. A subsequent oxygen displacement by chloride results in the inhibition of oxide film formation and enhanced Al corrosion. Chloride ions cause 1000 times more severe corrosion than the equivalent amount of sulfates. The effect of different anions on the enhancement of Al corrosion reduces in the following order:

$$Cl^- > Br^- > J^- > F^- > SO_4^{2-} > NO_3^- > PO_4^{3-}$$

On the other hand, the presence of chloride ions in solution can significantly reduce the dissolution of iron anodes that is explained by the chlorine evolution (R. 2.109) at potentials higher than a certain value. Moreover, while comparing chloride and sulfate solution, the increase of pH after electrocoagulation treatment was lower in chlorine solution than in sulfate one, which is explained by R. (2.110–2.111).

$$2Cl^- -2e^- \rightarrow Cl_2 \qquad (2.109)$$

$$Cl_2 + 2H_2O \rightarrow 2HOCl + H_2 \qquad (2.110)$$

$$2Fe(OH)_2 + HOCl + OH^- \rightarrow 2Fe(OH)_3 + Cl^- \qquad (2.111)$$

Applied Current Density

Current density plays an important role in electrocoagulation process. The higher the applied current density, the better the electrode surface used and the faster the rate of electrochemical metal dissolution. Higher current densities increase the CE of anodic metal dissolution and enhanced production of hydrogen bubbles at cathode, which facilitate the flotation of hydroxide flocs. From another side, the higher the applied current density, the higher the polarization and passivation of electrodes, which lead to the increase of potentials and loss of energy for the side processes. Iron electrodes are more prone to passivation than aluminum ones. In this regard, there are no recommendations on the optimal current densities for electrocoagulation treatment because they depend on system conditions and desired treatment effect. In general the most commonly used applied current densities are in the range of $1-50$ mA/cm^2. However, there are studies reporting the use of higher values over 100 mA/cm^2 for the treatment of sewage wastewater [85, 86]. Usually higher current densities are used for significant wastewater volumes with a high concentration of COD, BOD, and suspended solid values, whereas lower current densities are used in the processes where electrocoagulation is incorporated with other conventional water treatment methods. In the case of iron anodes, increase of current densities leads to the rise of solution pH, reduction of dissolved iron, and increase of $Fe(OH)_2$ deposition.

Formation of oxide films on the electrode surface (anode passivation) significantly hinders electrocoagulation. To prevent the sediments deposition on the electrodes, i.e., to prevent the passivation of electrodes, it is recommended to reverse the electrode's polarity with a periodicity of minutes, hours, or weeks. The higher the current density, the more often polarity of electrodes should be changed. Despite the positive effect of polarity reversal, it is not always effective for the removal of deposits because the CE drops significantly at the moment of polarity change.

Distance and Water Flow Rate Between Electrodes

To reduce the EC during the electrolysis process, the distance between electrodes should be reduced, because the greater the distance between electrodes, the higher the EC. In this regard the optimal distance between electrodes is in the range of 10–20 mm.

The higher the water flow rate between electrodes, the faster the anodic metal dissolution, which prevents the metal oxide films formation. Consequently, the higher the water flow rate, the lower the EC and passivation of electrodes.

Effect of pH

The effect of pH is more complicated in electrocoagulation process compared to conventional coagulation. While conducting electrolysis process with aluminum anodes, a significant change of pH values is observed after the treatment process. Increase of pH is observed in acidic media (pH up to 4) and decrease of pH occurs in the case of alkaline solutions (pH above 8–9). In the case of neutral solutions insignificant shift of pH to more positive values can be observed. This is explained by buffering capacity at these pH and production of hydroxyl ions at cathode. The decrease of pH in alkaline conditions ions can be explained by the formation of $Al(OH)_4^-$ complexes. Acidic (pH < 4) and alkaline (pH > 8) conditions lead also to the enhanced aluminum anode dissolution. This is explained by the solubility of the oxide films at these conditions and as a consequence elimination of passivation film, which prevents electrochemical and chemical metal dissolution. While using iron anodes for electrocoaguation, significant increase of pH is observed analogous to aluminum anodes [87].

Practical Applications of Electrocoagulation

Electrocoagulation can be effectively used for a variety of organic and inorganic pollutants as well as microorganisms. Examples of removal efficiencies of different pollutants are given in Table 2.4 and the mechanism of their removal is considered below.

Chromium and Other Metals Removal

Nonferrous metal companies are the main source of chromium-containing wastewaters, where electrolytes containing Cr(VI) are used for etching and passivation of rolling metal. The concentration of Cr(VI) in wastewaters of nonferrous metal companies is within 5–80 mg/L. Conventional methods for the removal of Cr(VI) from wastewaters include chemical reduction of Cr(VI) to Cr(III) by, for example, sulfur dioxide, sodium sulfite and bisulfite, sodium sulfide, etc., and further precipitation of Cr(III) in the form of hydroxide by alkali reagent (base, sodium bicarbonate, or lime). The main disadvantage of chemical reduction of Cr(VI) containing wastewaters is the increase of total salt content in the treated effluent, which limits the reuse of treated water. Another

TABLE 2.4 Removal Efficiency of Different Pollutants From Water Using Electrocoagulation Treatment

Pollutant (Initial Concentration)	Anode Type	Working Parameters	Removal Efficiency	EC, kWh/m³	Reference
Cu (45 mg/L) Cr (44.5 mg/L) Ni (394 mg/L)	Al–Fe	pH 3 j = 10 mA/cm² t = 20 min	100% 100% 100%	10.07	[83]
Cr(vi) (100 mg/L)	Fe	pH 1 j = 1 A/L t = 14 min NaCl 1 g/L	50%		[94]
S²⁻ (7.5—9.5 mg/L) Phosphorous (1.38 mg/L)	Fe	pH 6.5—7.1 j = 17.9 mA/cm² Q = 60 C/L	88% 40%	8 C/mg	[95]
Fluoride (5—25 mg/L)	Al	pH 4—8 j = 12—50 A/m² t = 23—53 min Flow rate: 150—400 mL/min	90%—95%		[96]
Fluoride (4.84 mg/L)	Al	pH 7 E = 30 V t = 5 min	90%		[97]
As(v) (150 µg/L)	Fe Al	pH 6.5 j = 7.5 mA/cm² t = 5 min	93% 97%	0.007 0.011	[91]
As(v) (100 µg/L)	Fe	pH 7 j = 4.5 mA/cm² t = 5 min	95%	0.15	[92]

Sample	Electrode	Conditions	Removal	Value	Ref.
Tannery wastewater: COD 2400 mg/L TOC 1000 mg/L Sulfide 110 mg/L NH_3–N 220 mg/L	Al	pH 7.9 j = 7.9 mA/cm^2 t = 20 min	50% COD 50% TOC 30% ammonia 9% sulfide		[98]
	Mild steel		48% COD 35% TOC 25% ammonia 95% sulfide		
Olive mill wastewater COD 20 000 mg/L	Al	pH 4.2 j = 25 mA/cm^2 t = 15 min	88% color 80% COD		[99]
TSS 300 mg/L Turbidity 150 NTU (Nephelometric Turbidity Unit)	304 stainless steel	pH 7 j = 5.9 mA/cm^2 t = 6 min	99% TSS 98% turbidity		[100]
Nitrate (100 mg/L)	Al	j = 1.1 mA/cm^2 t = 60 min NaCl 1 g/L	45%		[101]
Landfill leachate: NH_3–N 386 mg/L COD 2566 mg/L	Fe	pH 6.4 j = 2.98 mA/cm^2 t = 90 min	45% COD 44% NH_3–N		[102]
NO_3–N (500 mg/L)	Al	pH 7 j = 2.5 mA/cm^2 t = 60 min	95.9%	6.26	[103]
Microcystis aeruginosa $1.2 \cdot 10^9$ $-1.4 \cdot 10^9$ cells/L	Fe Al	pH 7 j = 1 mA/cm^2 t = 45 min	79% 100%	0.4	[104]
Escherichia coli	Al	pH 7.5 j = 5.5 mA/cm^2 t = 35 min	97%		[105]

COD, Chemical oxygen demand; EC, energy consumption; Q (C/L), is the applied electric charge density; TSS, total suspended solids.

common methods used for the removal of Cr(VI) ions from wastewaters are ion exchange and ED; however, these methods are more expensive than chemical removal or electrocoagulation. Electrocoagulation allows Cr(VI) removal in a wide range of pH. The mechanism of Cr(VI) removal by electrocoagulation using iron anodes is based on the chemical reduction by the produced Fe(II) ions and $Fe(OH)_2$ depending on the solution pH as indicated in the following reactions (R. 2.112–2.114).

Acidic media:

$$6Fe^{2+} + Cr_2O_7^{2-} + 14H^+ \rightarrow 6Fe^{3+} + 2Cr^{3+} + 7H_2O \qquad (2.112)$$

Slightly acidic and neutral media:

$$6Fe(OH)_2 + Cr_2O_7^{2-} + 7H_2O \rightarrow 6Fe(OH)_3 + Cr(OH)_3 + 2OH^- \qquad (2.113)$$

Slightly alkaline media:

$$3Fe(OH)_2 + CrO_4^{2-} + 4H_2O \rightarrow 3Fe(OH)_3 + Cr(OH)_3 + 2OH^- \qquad (2.114)$$

In addition, the mechanism of cathodic Cr(VI) reduction is taking place in accordance with R. (2.115, 2.116).

Acidic media:

$$Cr_2O_7^{2-} + 14H^+ + 6e^- \rightarrow Cr^{3+} + 7H_2O \qquad (2.115)$$

Slightly alkaline media:

$$CrO_4^{2-} + 4H_2O + 3e^- \rightarrow Cr(OH)_3 + 5OH^- \qquad (2.116)$$

Removal of other toxic metals such as Zn, Cu, or Ni is achieved by their sedimentation in the form of metal hydroxides then proper pH is established according to the generalized R. (2.117).

$$M^{z+} + zOH^- \rightarrow M(OH)_z \qquad (2.117)$$

Sulfide Removal

Leather tanning industry is one of the main sources of sulfide-containing wastewater with the pollutant concentration of up to 300 mg/L [88]. Sodium sulfide is used for the hair removal from animal skin in tannery. Simplified removal mechanism of sulfides is based on the reaction of dissolved Fe(II) and produced Fe(III) with sulfide ions and formation of insoluble precipitates (R. 2.118–2.120). The use of aluminum anodes is not effective for sulfide removal because of the formation unstable Al_2S_3 (R. 2.121), which decomposes to $Al(OH)_3$ with the release of sulfide [89].

$$Fe^{2+} + HS^- \rightarrow FeS\downarrow + H^+ \qquad (2.118)$$

$$2Fe^{3+} + HS^- \rightarrow 2Fe^{2+} + S\downarrow + H^+ \qquad (2.119)$$

$$2Fe(OH)_3 + 3H_2S \rightarrow 2FeS + S + 6H_2O \tag{2.120}$$

$$2Al^{3+} + 3S^{2-} \rightarrow Al_2S_3\downarrow \tag{2.121}$$

Fluoride Removal

The most common mechanism of fluoride removal from wastewaters is based on adsorption (R. 2.122) and coprecipitation (R. 2.123) of fluoride ions with aluminum polynuclear hydroxo complexes ($Al_n(OH)_{3n}$, Fig. 2.3) in the range of solution pH of 6−8 [85].

$$Al_n(OH)_{3n} + mF^- \rightarrow Al_nF_m(OH)_{3n-m} + mOH^- \tag{2.122}$$

$$nAl^{3+} + (3n-m)OH^- + mF^- \rightarrow Al_nF_m(OH)_{3n-m} \tag{2.123}$$

Higher pH values of treated solution can contribute to the removal of fluoride from wastewaters through the mechanism indicated in (R. 2.124). However, this mechanism is often avoided because of the high residual pH of the treated effluent.

$$Al(OH)_4^- + F^- \rightarrow Al(OH)_3F + OH^- \tag{2.124}$$

It was reported that coexistence of different anions and cations in electrolyte solution can both facilitate and suppress the rate of defluoridation. Such presence of increased amount of SO_4^{2-} ions in electrolyte solution can compete with F^- ions for the place inside the hydroxo-fluoro complexes of aluminum according to the following reaction [90]:

$$Al(OH)_{3-x}F_x + ySO_4^{2-} \rightarrow Al(OH)_{3-x}F_{x-2y}(SO_4)y + 2yF^- \tag{2.125}$$

The presence of Ca^{2+} ions can enhance defluoridation of wastewaters through the formation of insoluble CaF_2 (R. 2.126) precipitates and coprecipitation with hydroxo complexes of aluminum (R. 2.127) [90].

$$Ca^{2+} + 2F^- \rightarrow CaF_2 \tag{2.126}$$

$$mAl^{3+} + nCa^{2+} + (3m+2n)H_2O$$
$$\rightarrow Al_mCa_n(OH)_{3m+2n} + (3m+2n)H^+ \tag{2.127}$$

Arsenic Removal

Arsenic exists in a variety of forms such as arsenious (H_3AsO_3) and arsenic (H_3AsO_4) acids in wastewaters. Both acids dissociate with the formation of following ions: $H_2AsO_3^-$, $HAsO_3^{2-}$ AsO_3^{3-} and $H_2AsO_4^{2-}$ and $HAsO_4^{2-}$. High removal efficiencies of arsenic are achieved by both iron and aluminum anodes through the mechanism of adsorption on the polynuclear hydroxo complexes of aluminum and iron (R. 2.128−2.132) [91,92].

$$2FeO(OH) + H_2AsO_4^- \rightarrow (FeO)_2HAsO_4 + H_2O + OH^- \qquad (2.128)$$

$$3FeO(OH) + HAsO_4^{2-} \rightarrow (FeO)_3AsO_4 + H_2O + 2OH^- \qquad (2.129)$$

$$Fe(OH)_3 + AsO_4^{3-} \rightarrow [Fe(OH)_3 \cdot AsO_4^{3-}] \qquad (2.130)$$

$$\equiv Al - OH + HAsO_4^{2-} \rightarrow \equiv Al - OAs(O)_2OH^- + OH \qquad (2.131)$$

$$mAl^{3+} + (3m - n)OH^- + nHAsO_4^{2-} \rightarrow Al_m(OH)_{3m-n}(HAsO_4) \qquad (2.132)$$

Color, Organic Matter, and Turbidity

Wastewaters of dye, leather tanning, pharmaceutical, and other industries as well as surface waters can be polluted with different organic compounds, suspended solids, colloidal particles of both organic and inorganic origin, etc., which contribute to waters turbidity, color, COD, TOC, and organic matter content. The main mechanism of colloids removal during electrocoagulation is adsorption onto formed hydroxides of aluminum and iron as well as trapping of pollutants into flocs. The higher the amount of formed hydroxide content, the higher the efficiency of organic matter and turbidity removal, which is achieved by the increased applied current densities or time of electro-coagulation. Moreover, higher current densities provide higher production of H_2 gas bubbles, which also accelerate the removal of pollutants through the flotation. The studies on organic matter removal depending on pH of treated solution in general did not reveal a significant impact of pH on pollutants removal efficiency. On the other hand, increasing the initial concentration of organic matter in the treated solution reduces the rate of contaminants removal.

Cyanobacteria and Bacteria Removal

The presence of pathogenic bacteria in water directly influences the human health. During algae blooming, amount of cyanobacteria in surface and wastewaters significantly increases, which can threaten the environment due to the release of cyanotoxins. Therefore, waters containing bacteria should be disinfected. One of the methods for algae and bacteria removal from water is electrocoagulation. Because cyanobacteria can be considered as colloids, it is possible to remove them by applying electrical field and major mechanisms of coagulation. Moreover, along with coagulation, EF by hydrogen gas bubbles as well as disinfection by in situ generated electrochemical species contribute to the removal of cyanobacteria from water [93]. The main mechanism of bacteria removal includes the sorption of bacteria species on formed hydroxide as well as the disinfection by the generated electroactive species.

Ammonia and Nitrate Nitrogen Removal

Ammonia is a widely common pollutant in wastewaters of agricultural and tannery industries, livestock farming, domestic wastewater, etc. Nitrogen compounds are the main pollutants causing eutrophication of water bodies, which in turn leads to depletion of fish resources. There are two ways by which ammonia nitrogen can be removed from wastewater during electrocoagulation. The first way is the volatilization at alkaline conditions (pH > 9.5). The second way is the decomposition of ammonia to nitrogen gas by hypochlorous acid and hypochlorite ion, generated in situ from chlorine gas according to reaction (R. 2.133−2.136) [89].

$$2Cl^- - 2e^- \rightarrow Cl_2 \tag{2.133}$$

$$Cl_2 + H_2O \rightarrow HOCl + H^+ + Cl^- \tag{2.134}$$

$$2NH_3 + 3HOCl \rightarrow N_2 + 3H^+ + 3Cl^- + 3H_2O \tag{2.135}$$

$$2NH_3 + 2OCl^- \rightarrow N_2 + 2H^+ + 2Cl^- + 2H_2O \tag{2.136}$$

The mechanism of nitrates removal is the interaction with anodic aluminum through the reduction of nitrates to nitrites, nitrites to ammonia, and ammonia to nitrogen gas (R. 2.137−2.139) [85].

$$3NO_3^- + 2Al + 3H_2O \rightarrow 3NO_2^- + 2Al(OH)_3 \tag{2.137}$$

$$NO_2^- + 2Al + 5H_2O \rightarrow NH_3 + 2Al(OH)_3 + OH^- \tag{2.138}$$

$$2NO_2^- + 2Al + 4H_2O \rightarrow N_2 + 2Al(OH)_3 + 2OH^- \tag{2.139}$$

2.4 ELECTROFLOTATION

In 1904, Elmore patented and proposed EF as one of the first flotation processes of ore enrichment for the activation of oil flotation. However, practical application of the method in the flotation of minerals started to be used at the end of the 1960s.

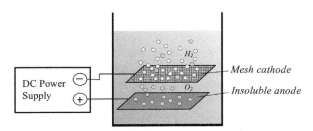

FIGURE 2.9 Schematic diagram of the electroflotation process.

EF is a physical chemical method of water treatment from insoluble (dispersed) substances. The method is based on conducting the water electrolysis on insoluble electrodes and flotation effect (Fig. 2.9). While carrying out the water electrolysis, gas bubbles of oxygen and hydrogen are generated at the anode and cathode, respectively. When gas bubbles rise up to the liquid surface due to the density difference between gas and liquid phases, they are faced with dispersed particles of pollution, adhere to them and float them to the water surface to form a stable flotosludge layer, which is later removed by skimmers or other mechanical devices. Nevertheless, the formation of both hydrogen and oxygen gas bubbles hydrogen bubbles produced at the cathode play the main role in the flotation process. Usually high removal efficiency from dissolved compounds can be achieved with the optimum concentration of dissolved pollutants in wastewater in the range between 10 and 100 mg/L and the maximum concentration below 200 mg/L [106].

At anode:

$$2H_2O - 4e^- \rightarrow O_2 + 4H^+ \quad E^0{}_a = -1.23 \text{ V} \quad pH \leq 7$$
$$4OH^- - 4e^- \rightarrow O_2 + 2H_2O \quad E^0{}_a = -0.40 \text{ V} \quad pH > 7$$

At cathode:

$$2H^+ + 2e^- \rightarrow H_2, \quad E^0{}_c = 0.00 \text{ V} \quad pH < 7$$
$$2H_2O + 2e^- \rightarrow H_2 + 2OH^- \quad E^0{}_c = -0.83 \text{ V} \quad pH \geq 7$$

To remove dissolved compounds from wastewater by EF, they should be converted into insoluble form. Often the preliminary neutralization of acidic and alkaline solutions as well as the transfer of metal ions into solid phase by coagulation and/or flocculation is conducted to intensify the process of EF and enhance the treatment efficiency. This can also be achieved by in situ electrogeneration of hydroxides, which react with metal ions forming insoluble metal hydroxide particles.

EF is used mainly for the removal of metal ions; fine-dispersion metal hydroxides such as iron, copper, nickel, cadmium, chromium, magnesium, etc.; suspended solids; phosphates; suspensions; resinous and emulsified substances; mineral and industrial oils; and grease and surfactants from industrial wastewaters.

EF allows achieving high-pollutant removal efficiencies reaching up to 99% within only 5—10 min of process operation. The efficiency of the EF (α, %) depends on many parameters such as applied current density, material of anodes, acidity and composition of the medium, size of bubbles and dispersed particles and their surface charge, arrangement of electrodes inside electroflotator (equipment used for carrying out of the EF process), solution flow rate, etc. It can be estimated through the following equation:

$$\alpha \, (\%) = \frac{C_0 - C_t}{C_0} \cdot 100\%, \tag{2.140}$$

where C_0 and C_t are the initial and final concentration of the dispersed phase in aqueous solution after time t of the treatment, respectively.

Fluid saturation degree (β) by the hydrogen bubbles is directly proportional to the applied current density (j) and inversely proportional to the rise rate (u_{H_2}) of hydrogen bubbles

$$\beta = k \cdot CE \cdot \frac{j}{\rho_g \cdot u_{H_2}}, \qquad (2.141)$$

where k is the electrochemical equivalent of hydrogen (g/C); ρ_g is the density of gas bubbles.

The cell voltage of electroflotator (E_{EF}) consists of potential drop in all parts of the circuit and can be determined as follows:

$$E_{EF} = E^0 + \Delta E_a + \Delta E_c + \Delta E_{sol} + \Delta E_{contacts} \qquad (2.142)$$

where E^0 is the standard potential of water decomposition ($E^0 = 1.23$ V); ΔE_a and ΔE_c are anodic and cathodic overpotentials, respectively; $\Delta E_{sol} = j \cdot \rho \cdot l \cdot k_{sol}$ is the potential drop in electrolyte solution and $\Delta E_{contacts}$ is the potential drop in terminals; ρ is the resistivity of electrolyte solution ($\Omega^{-1} cm^{-1}$); j is the applied current density (A/cm^2); l is the distance between electrodes; and k_{sol} is the co-efficient taking into account the increase in the electrolyte resistance due to the gas supply (usually $k_{sol} = 1.15 - 1.35$).

The volume of hydrogen gas produced during electrolysis can be calculated as follows:

$$V_{H_2} = k \cdot I \cdot t \cdot CE, \qquad (2.143)$$

where k is electrochemical equivalent of hydrogen ($k_{H_2} = 4.56 \cdot 10^{-4} \, m^3 / A \cdot h$); t is the electrolysis time (h); and I is the current (A).

The EC ($W \cdot h/m^3$) required for the allocation of 1 m^3 hydrogen under the normal operation can be calculated using the following equation:

$$EC = \frac{EzF}{V_{mH_2}} \qquad (2.144)$$

where V_{mH_2} is the molar volume of 1 mol hydrogen $\left(V_{mH_2} = 0.0224 \, m^3 \right)$.

EF process is a complex process, which can be described through the following steps:

- Electrochemical generation of gas bubbles
- Interaction of dispersed particles with gas bubbles and formation "particle—gas bubbles" flotocomplexes
- Flotation of formed complexes to the water—air interface and formation of the froth layer (flotosludge)

The size of gas bubbles depends on the conductivity of the wastewater, i.e., the lower the electrical conductivity of water, the higher the electric field strength and the bubbles are smaller. In general the size of gas bubbles formed during EF is smaller than those in conventional flotation methods. The size of hydrogen bubbles is about twice smaller than oxygen bubbles evolved at the anode. The diameter of hydrogen bubbles in EF varies from 20 to 40 μm. The properties of the electrode surface, the shape of electrodes, pH and temperature of the medium, the surface tension at the electrode—solution interface, and the current density influence the size of the generated gas bubbles.

The main advantage of EF is the ability to conduct the process without additional chemicals and high dispersion of bubbles in the range of microns and tens of microns, which is up to two orders of magnitude smaller than in the conventional froth flotation. However, reagents may only be used for the formation of precipitation and their flocculation, which allows separation of fine particles including ions (ion flotation). The absence of organic reagents causes no side pollution of waters, which favors the establishment of production facilities for extraction of some components of marine and thermal waters by EF. Moreover, simultaneous aeration of wastewater during EF allows simultaneous decomposition of easily oxidizable substances and microorganisms. Electroflotators are compact and are easy to operate and automate. Another advantage of EF is the ability to significantly accelerate the process of sedimentation and separation of precipitates, which takes 2—6 h in conventional chemical processes. Moreover, flotosludge formed during EF is of high density comparable to traditional froth flotation where water content in the froth layer can be up to 10%. This allows reducing the water loss.

EF process can be conducted using both soluble and insoluble electrodes. While using soluble iron or aluminum electrodes, along with gas generation, there is a dissolution of iron and aluminum cations forming metal hydroxides, which trap dispersed particles of pollutants and coagulate. Limited distance between electrodes and significant sizes of coagulating flocs allow good adhesion of gas bubbles to the flocs. When using insoluble anodes, electrophoresis (movement of dispersed particles under the influence of external electric field), discharge of particles at the electrode surface, and presence of compounds able to deplete particle solvation layer affect strongly the efficiency of EF. The use of insoluble electrodes is usually favorable at low particles concentration, whereas soluble electrodes show better performance in the case of high concentrations of stable pollutants resistant to aggregation.

The Theory of Flotation

Flotation phenomenon is associated with the phenomena of wetting and nonwetting of particles by water. The better the wetting of particles by water is, the worse the adhesion of particles to gas bubbles. The contact angle (θ_c) is a measure of wettability of solid body by liquid. The contact angle is measured

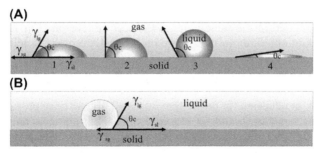

FIGURE 2.10 Illustration of contact angles formed by liquid (A) and gas (B) bubbles on a solid surface. γ_{sg}, γlg, and γsl are surface tension at the interface of solid—gas, liquid—gas, and solid—liquid phases, respectively; θ_c is the contact angle.

at a tangent to liquid surface (Fig. 2.10). Complete wetting of a solid surface with water is characterized by $\theta_c = 0$ degrees (for example, quartz) and complete nonwetting (perfect adhesion to a gas bubble) is described by $\theta_c = 180$ degrees. If contact angle is in the range between 0 and 90 degrees (cases 1 and 4 of Fig. 2.4A), then wettability of solid surface by water is favorable (for example, wettability of graphite is characterized by $\theta_c = 55$ degrees). On the contrary, if $\theta_c = 90-180$ degrees (case 3 of Fig. 2.4A), then solid surface has low wettability ($\theta_c = 106$ degrees for paraffin) and liquid droplets tend to minimize their surface energy thus reducing the contact area between liquid droplets and solid particles.

During the interaction of two insoluble phases at equilibrium, each phase tends to reduce excess of its potential energy at the interface with the other phase. That is how the surface tension is formed and it can be generally referred to the surface energy per unit area or force per unit length. All constituents of the surface energy between different phases can be expressed in the vector form (Fig. 2.10). The equilibrium between surface energies in the system of water droplet on a smooth homogeneous solid surface (Fig. 2.10A) and for the system where gas bubble interact with a solid surface (Fig. 2.10B) can be described through the vectors of surface tension using Young's equation as follows:

$$\gamma_{sg} = \gamma_{sl} + \gamma_{lg} \cdot \cos \theta_c \qquad (2.145)$$

The efficiency of EF can be characterized through the work of adhesion. Adhesion is a spontaneous process and according to the second law of thermodynamics it can take place only with the decrease in the free energy of the system. This, for example, means that flotation of solid particles by gas bubbles (adhesion of solid particles to gas bubbles) is higher for more prominent decrease in free energy (ΔW). The strength of adhesive bonds is characterized by the work of adhesion (decrease of system's free energy, ΔW), which is expressed in the same units as the surface tension (J/m^2).

The work of adhesion for the system on Fig. 2.4A can be calculated via the Young–Dupré equation:

$$\Delta W_{sl} = \gamma_{lg}(1 + \cos \theta_c) \tag{2.146}$$

The work of adhesion for the system on Fig. 2.4B can be expressed through the following equation (valid for the case of a small particles and large gas bubble):

$$\Delta W_{sg} = \gamma_{lg}(1 - \cos \theta_c) \tag{2.147}$$

The contact angle and wettability are directly related to the hydrophilicity and hydrophobicity of solid particles. Let' us consider three cases of θ_c boundary conditions in the system of gas bubble-solid particle interaction in the liquid medium, which affect flotation.

1. If $\theta_c = 0$ degrees, then particles are completely hydrophilic ($\cos \theta_c = 1$ and $W_A = 0$) and flotation is not observed.
2. If $\theta_c = 90$ degrees, then $\cos \theta_c = 0$ and $W_A = \gamma_{lg}$.
3. If $\theta_c = 180$ degrees, then particles are completely hydrophobic and $\cos \theta_c = -1$ and the strength of adhesion is maximum ($W_A = 2\gamma_{lg}$).

From the boundary conditions it is seen that flotation is possible at the values of $\theta_c > 0$ degrees. However, it can be also concluded that the greater the contact angle, i.e., the smaller the wettability of particles by water (particles are more hydrophobic), the higher the decrease of the system free energy. This means that flotation is more prominent and hence more efficient for hydrophobic particles.

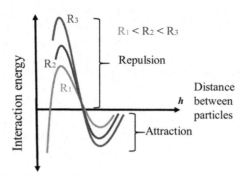

FIGURE 2.11 Dependence of total interaction energy between a gas bubble and a solid particle on the size of solid particle. R_1, R_2, and R_3 are radii of solid particles.

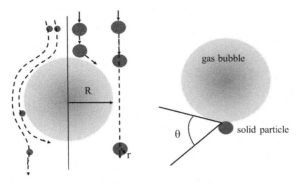

FIGURE 2.12 Illustration of gas bubble (radius R) interaction with solid particles (radius r).

The ratio between particle and gas bubble size influences significantly the efficiency of flotation because it determines the efficiency of the collision and the subsequent attraction. There is a critical particle size for a particular gas bubble size below which collision between gas bubbles and particles do not occur. This is explained by the fact that the depth of secondary energy minimum (Fig. 2.11) increases with the increase of particle radius at relatively great distances. Moreover, larger particles have greater weight, which allows linear movement of solid particles under the influence of inertial forces. This condition corresponds to the case when the distance between the centers of gas bubble and solid particle is less than the sum of particle diameter (d) and bubble radius (R) (Fig. 2.12).

On the contrary, very small particles have low weight and move along with water current, thus flowing around the popping up gas bubbles (Fig. 2.12). The

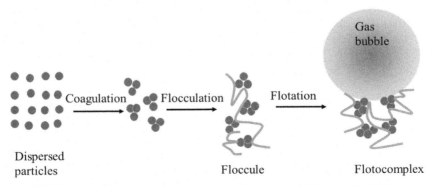

FIGURE 2.13 The sequence of water treatment steps often combined with flotation.

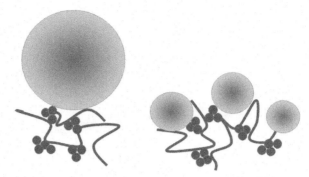

FIGURE 2.14 Different ways of pollutants flotation by gas bubbles of different size.

probability of formation of the "bubble—particle" complex can be determined using the following equation.

$$\omega = \frac{\left[n\frac{4}{3}\pi(R + r)^3 - n\frac{4}{3}\pi R^3\right]}{V} = C_g\left[\left(1 + \frac{r}{R}\right)^3 - 1\right], \qquad (2.148)$$

where n is the number of gas bubbles with radius R; r is the particle radius; V is the volume of fluid, and $C_g = n\frac{4}{3}\pi R^3/V$ is the volume concentration of the gas phase. As it is seen in the equation above, the larger particles increase the probability of the "bubble—particle" complex formation. In this regard in real applications flotation is often preceded by coagulation (destabilization of colloids) and flocculation (aggregation of destabilized particles under the influence of high-molecular compounds or flocculants) processes (Fig. 2.13). In general dispersed particles and flocs can be floated by either one larger or few smaller gas bubbles (Fig. 2.14).

The mechanism of flocculant action is that large flocs having good sedimentation properties are formed between flocculant and dispersed particles. The formation of flocs occurs because of the adsorption of flocculant molecule on the surface of particles, which can take place by three different

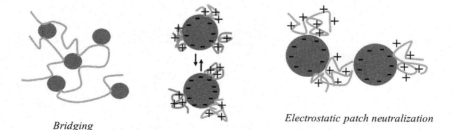

Bridging

Charge neutralization

Electrostatic patch neutralization

FIGURE 2.15 Mechanism of flocs formation.

mechanisms. They are charge neutralization, electrostatic patch, and bridging flocculation (Fig. 2.15) [107]. The mechanism of charge neutralization leads to adhesion of particles because of the decrease of electrostatic repulsion that takes place when oppositely charged molecules of flocculants are adsorbed on particle surface, thus neutralizing its charge. This mechanism is typical for flocculants of low molecular weight (around 10^4). The bridging mechanism of floc formation is typical for high-molecular-weight nonionic flocculants, occurs most frequently and includes the simultaneous adsorption of flocculant molecule on the surface of few particles leading to the formation of polymer bridges between them. Mechanism of electrostatic patch neutralization takes place in the case of particles with low-density charge and flocculants of high-density charge. In that case polymer flocculant can electrostatically bond to and neutralize few oppositely -charged particles. The bridging mechanism of flocculation is considered to have the strongest bonds followed by the electrostatic patch neutralization and charge neutralization mechanism having the weakest flocculants/particles bond. There are few classifications of flocculants. They can be organic (for example, polyethylene oxide and polyacrylamide) and inorganic (for example, active silicic acid). Organic flocculants can be divided into natural, such as starch, or natural resins or synthetic ones. Synthetic flocculants in turn can be classified into nonionic such as polyethylene oxide, polyelectrolyte [either anionic ($-COOH$, $-PO(OH)_2$, $-SO_3H$, etc.) or cationic ($-NH$, NH_2, etc.) electrolyte depending on the functional groups], and copolymers (which can comprise both cationic and/or anionic functional groups). The rate of floccules sedimentation is hundreds and thousands times higher than sedimentation of separate particles. The dose of flocculants is usually 10-fold lower than the dose of coagulants.

The balance of forces affecting the bubble−particle complex in liquid can be described through the Archimedes buoyancy ($F_A = \rho_l \cdot g \cdot (V_g + V_s)$) of the formed complex, gravity force of a particle ($F_g = \rho_s \cdot g \cdot V_s$), and drag force of liquid medium ($F_d = \frac{1}{2} \rho_l \cdot u^2 \cdot A \cdot c_d$) as follows:

$$F_A = F_g + F_d, \tag{2.149}$$

where ρ_s and ρ_l are the solid particle and fluid densities, respectively; V_g and V_s are the volumes of gas bubble and particle, respectively; g is the gravitational acceleration; c_d is the drag coefficient; $A = \pi \cdot d_{gs}^2/4$ is the cross-sectional area of gas bubble−particle complex in the direction of its motion; d_{gs} is the equivalent diameter of gas bubble−solid particle complex; and u is the rising velocity of gas bubble−particle complex. According to the balance of forces the rising velocity can be calculated as follows:

$$u = \sqrt{\frac{2g[\rho_l(V_g + V_s) - \rho_s V_s]}{c_d A \rho_l}} \tag{2.150}$$

If the movement of a gas bubble—particle complex is laminar, then the rising velocity is equal to

$$u = \frac{g[\rho_l V_g - V_s(\rho_s - \rho_l)]}{3\pi\eta d_{gs}}$$ (2.151)

where η is the dynamic viscosity of the liquid medium.

If the drag force, $F_d = 0$, then the ratio between the volume of a gas bubble (V_g) and the volume of a solid particle (V_s) can be calculated as follows:

$$\frac{V_g}{V_s} = \frac{\rho_s - \rho_l}{\rho_l}$$ (2.152)

The velocity of the particles (u_s) and bubbles (u_g) movement relative to the medium is determined by the following equations:

$$u_s = -\frac{2}{9}\left(\frac{g\rho_l r^2}{\eta}\right)\left[(1 - C_s)\left(\frac{\rho_s}{\rho_l} - 1\right) + C_g\right]$$ (2.153)

$$u_g = \frac{1}{9}\left(\frac{g\rho_l R^2}{\eta}\right)\left[1 + C_s\left(\frac{\rho_s}{\rho_l} - 1\right) - C_g\right]$$ (2.154)

where C_s and C_g are the volume concentration of solid particles and gas bubbles in water, respectively. The rate of particles flotation is described by the first-order kinetics model as follows:

$$\frac{dC_s}{dt} = -k_f C_s$$ (2.155)

where k_f is the coefficient of flotation rate.

The ratio between gas and solid phases in the flotation medium determines particle separation efficiency. When the weight ratio between gas and solid phases $m_g/m_s = 0.01 \div 0.1$, the highest efficiency of particles separation can be achieved:

$$\frac{m_g}{m_s} = 1.3\frac{b(\beta P - 1)Q_l}{C_s Q}$$ (2.156)

where m_g and m_s are weight of air and solid particles, respectively (g); b is the air solubility in water at atmospheric pressure and a given temperature (cm^3/L); β is the degree of saturation; P is the absolute pressure at which water is saturated with the air; Q_l is the amount of water-saturated air (m^3/h); Q is the water flow (m^3/h); and C_s is the concentration of suspended particles.

Nevertheless, the abovementioned conclusions about the preferable flotation of larger particles, flotation of fine hydrophilic particles is also observed in practice. The explanation of this phenomenon was described by Russian scientists Derjaguin and Dukhin. They developed the flotation theory of small- and medium-sized particles (<50 μm in radius) and considered the process of interaction of a gas bubble and solid particle as a process of heterocoagulation (aggregation of dissimilar in chemical nature particles) [108]. It was explained that regardless the size and weight, small particles can approach the surface of a gas bubble from time to time due to the turbulent currents of water at the bubble tail and close proximity of water flows to the surface of the bubble in its upper half-part. When solid particle approaches the bubble surface, adhesion is possible in the case of energy barrier elimination (Fig. 2.15). This can be achieved either in the case when surface charge of gas bubble is equal to zero or when the surface of solid particle has the charge opposite to the surface charge of gas bubble.

Influence of Current Density on Electroflotation Efficiency and Time of Eectroflotation

Current density plays an important role in the EF process. It directly affects the amount of produced gas and consequently the efficiency of pollutants removal by EF. The higher the applied current density, the more gas bubbles are produced. In this regard it can be concluded that efficiency of particles removal increases with the current density due to higher degree of saturation by gas, which lead to higher amount of successful collisions between bubbles and particles. The size of produced bubbles is also dependent on the applied current density. In general, higher current densities should lead to smaller gas bubbles generation, which is explained by the fact of intensive electrode reactions at higher densities, accumulation of dissolved gases at the electrode's surface, shorter time of gas bubbles nucleation, and faster growth and detachment of grown bubble from the surface of electrode [109,110]. It was reported that diameter of H_2 bubbles in alkaline media of pH 10 decreased from 35 μm at applied current density of 15 mA/cm^2 to around 31 μm at 80 mA/cm^2 for 316 mesh stainless steel electrode and from 47 μm at 15 mA/cm^2 to 34 μm at 80 mA/cm^2 for mirrored stainless steel electrode [109]. The larger the generated bubbles are, the greater the minimum size of particles, which can adhere to these bubbles. This means that there is a limited adhesion of smaller particles, which leads to reduced efficiency of EF. Moreover, an excess of gas bubbles can accumulate under the flotosludge layer, thus thickening it and mixing with the clarified water. Therefore, there is an optimum current density for particular wastewater, which should be determined experimentally.

EF time is usually determined empirically and is closely related to the applied current density or potential difference and concentration of suspended particles.

Influence of Electrode Material on Electroflotation Efficiency

A significant part of the electric energy losses occurs while overcoming resistance of the electrolyte and overpotentials at the electrodes, which depends on the electrode material, their surface state, current density, solution temperature, and other factors. The material of electrodes used in EF process influences the amount of produced gas bubbles. Such anodes with high overvoltage toward OER and cathodes with a high overvoltage toward hydrogen evolution reactions produce fewer bubbles at the applied potential difference or applied current compared to the electrodes with a lower overvoltage toward hydrogen and OERs. As mentioned above, the efficiency of EF process is mostly determined by the amount and quality of formed hydrogen gas at the cathode. Therefore, cathode material is of great importance while planning the EF treatment. There are different cathode materials, which are used in EF. They are, for example, graphite- and carbon-based electrodes, MMO electrodes, stainless steel cathodes, copper, nickel, etc. However, probably the most commonly used cathodes are made of stainless steel because of its low cost, stability, and high performance toward hydrogen gas allocation. The lower the cathode overvoltage toward hydrogen evolution is, the smaller bubbles of hydrogen (range of nanometers) are formed. Consequently higher overvoltage leads to the formation of larger hydrogen bubbles. Anodes can be either insoluble like MMO electrodes or soluble made of aluminum and iron in the case if EF is enhanced by electrocoagulation process.

Influence of Particle Charge on Electroflotation Efficiency

All particles dispersed in water are charged. The charge of particles is greatly dependent on the pH of solution. Interaction of similar charged particles and bubbles contributes to the electrostatic repulsion and increase of energy barrier. Higher efficiency of EF is obtained when countercharged bubbles and particles adhere, thus providing better formation of flotosludge [106]. Such, for example, metal ions are charged positively. Bubbles of hydrogen and oxygen produced at the cathode and anode, respectively, have a charge at the moment of their detachment from the electrode surface. Oxygen bubbles are charged positively and hydrogen bubbles are charged negatively. Therefore, EF of metals by hydrogen bubbles provides higher metal removal efficiencies than the same process with the prevalence of oxygen bubbles. Consequently, efficiency of EF is higher in the case of countercharged or zero-charged gas bubbles and particles.

Liquid Medium Composition

The higher the concentration of particles in the flotation medium is, the higher the possibility of particles and bubbles collision, adhesion, and flotation.

The presence of anions in the treated solution influences the service life of the electrodes and consequently the frequency of electrode change. High concentration of Cl^-, SO_4^{2-}, NO_3^-, Br^-, and F^- anions starts to facilitate the corrosion of electrodes (especially steel electrode). The trace amount of chlorides and sulfates at concentrations as low as 1 mg/L is enough to initiate the electrodes intensive corrosion, while the concentration of the anions above 100 mg/L to the electrodes aggressive dissolution requiring the use of more expensive stable electrodes. On the contrary, the presence of OH^- or SiO_3^{2-} anions can form the protective layer on the electrode surface. Similar to the electrocoagulation process it is recommended to change the polarity of electrodes occasionally to prevent the sedimentation of carbonates on the electrode surface. If the polarity change is not possible, then the deposition of carbonates can be prevented by addition of conditioning reagents such as lime or soda.

Addition of flocculants into the treated solution usually enhances the efficiency of EF by reducing the time required for the achievement of the same removal values. The mechanism of EF process enhancement is similar to that observed in traditional flotation and includes the aggregation of particles into large floccules and resulting into higher probability of successful adhesion between aggregated particles and gas bubbles. In the case of countercharged flocculants and destabilized particles adsorption takes place because of electrostatic attraction forces while the adsorption of flocculants and particles of the same charge occurs through the ion-exchange mechanism. Moreover, decreased time of EF in the presence of flocculants leads to lower EC. The higher the molecular weight of flocculants, the better enhancement of EF is achieved, which is explained by stronger trapping of flocculants to particle due to the bridging mechanism of flocs formation.

The presence of surfactants in the treated solution stabilizes the gas bubbles preventing their coalescence and thus keeping their size in a smaller range. The smaller the size of gas bubbles, the higher their surface area, the slower their rising speed, and as a result the greater the possibility of successful adhesion to suspended particles.

pH of the Medium

The information on the influence of medium pH on the size of electrochemically generated bubbles is contradictory. Some studies reported that smaller hydrogen bubbles are produced in neutral and acidic condition, while the influence of alkaline conditions on hydrogen bubble size is not obvious and the size of oxygen bubble increases at higher pH [110,111]. Such that it was reported that small hydrogen bubbles of 16 μm diameter was obtained at pH 3−4 while recovering sphalerite [111] and hydrogen bubbles with the size below 30 μm can be generated at pH of 2−6 at iron and carbon cathodes [112]. Another study reported the opposite dependence of the smaller hydrogen bubbles generation in alkaline media (~33 μm at pH 9−12) compared to acidic one (~35−36 μm at pH 2−4) at the 316 stainless steel cathode [109]. In turn, the size of oxygen bubbles generated at ruthenium-coated titanium

TABLE 2.5 Removal Efficiency of Different Pollutants From Water Using Electroflotation (EF) and Combined Electrocoagulation—EF Processes

Pollutant (Initial Concentration)	Anode/Cathode	Working Parameters	Removal Efficiency	EC, kWh/m³	Reference
Pb (15 mg/L) Ba (15 mg/L) Zn (15 mg/L)	Stainless steel/Stainless steel	pH 8 $j = 35$ mA/cm² $t = 20$ min	97% 97% 45%	14	[113]
Oil (50 mg/L)	Al/Al	pH 4.7 $E = 2.5$ V $t = 30$ min NaCl 4 mg/L	90%	0.67	[114]
Chlorella vulgaris algae (10⁸ cell/L)	Al/Graphite Cu/Graphite	pH 7 $j = 2$ mA/cm² $t = 30$ min	96% 94.5%		[115]
Desmodesmus subspicatus microalgae (1.84·10⁷ cell/mL)	Al/Al Fe/Fe	pH 5.6 $j = 5.6$ mA/cm² $t = 20$ min	95.4% 99.7% turbidity 64.7% 99.6% turbidity		[116]
Cu (100 mg/L) Ni (100 mg/L)	Ti/RuO₂/Stainless steel	pH 5 (for Cu) pH 6 (for Ni) $j = 2.3$ mA/cm² $t = 120$ min Na₂SO₄ (1 g/L)	98% 25%		[117]
Cu (37.8 mg/L) CN⁻ (114 mg/L)	Stainless steel/Cu	pH 1.6 $j = 240$ mA/cm² $t = 75$ min	99.3% 99.9%	5.48 5.33	[118]

Textile wastewater (COD 340 mgO$_2$/L SS 300 mg/L Turbidity 300 NTU)	Al/Al	pH 8.7 $j = 11.55$ mA/cm^2 $t = 10$ min	79.7 % COD 85.5% SS 76.2% turbidity		[119]
Pond water: Cl$^-$ (10.4 mg/L) Nitrate (7 mg/L) Phosphate (0.6 mg/L) TSS ($1.1 \cdot 10^{-3}$ mg/L)	Al/Al	pH 6.94 $j = 26$ mA/cm^2 $t = 30$ min	0% 62% 99% 46%		[120]
Wastewater from *Panaeus latisulcatus* prawn culturing tanks: COD (36.5 mgO$_2$/L) TAN (2.83 mg/L) Nitrite (2.98 mg/L) SS (46.2 mg/L)	Ti/IrO$_2$–SnO$_2$–Sb$_2$O$_5$/Ti	pH 7.5 Flow rate 0.9 L/h $j = 2.5$ mA/cm^2	79% COD 91% TAN 92% Nitrite 91% SS	1.75	[121]
Diclofenac (0.1 mM) Ibuprofen (0.1 mM) Ketoprofen (0.1 mM)	Al/Al	NaCl 0.01 M $j = 8.3$ mA/cm^2 $t = 20$ min	14% 44% 10%		[122]
Ankistrodesmus falcatus (2.88 g/L) *Scenedesmus obliquus* (2.76 g/L)	Carbon/Carbon	$j = 8.3$ mA/cm^2 $t = 30$ min	91% 68%	1.76	[123]
Ni (100 mg/L) Pb (100 mg/L) Cd (100 mg/L) Zn (100 mg/L)	Al/Al	pH 8 $E = 20$ V $A_{el} = 4.6$ cm^2 $t = 20$ min	99.2% 100% 98.5% 98.7%		[124]
Oil-bearing waste water from hot-rolling mills (114 mg/L petroleum products)	Graphite/Steel mesh	pH 7.5 $j = 360$ mA/cm^2 $t = 10$ min	82.4%		[125]

COD, chemical oxygen demand; *SS*, suspended solids; *TAN*, total ammonia nitrogen.

electrode increased with the pH change from acidic media with O_2 bubble size of 33 μm at pH 2 to alkaline media with O_2 bubble size of 45 μm at pH 8–12 [109].

Similar to conventional flotation process, EF is usually combined with electrocoagulation to intensify the removal efficiency of dispersed particles. In this case, the process will look the same as in Fig. 2.9, only instead of an insoluble anode, sacrificial/soluble electrode, such as aluminum or iron, will be used. Table 2.5 summarizes efficiencies of EF as well as combined EF and electrocoagulation processes for the removal of different pollutants.

2.5 ELECTRODIALYSIS

ED is a method for the removal of electrolytes from water by transporting electrolyte ions through ion-exchange membranes to other solution under the influence of constant DC (direct current) directed perpendicular to the membrane plane. The driving force of the process is the electrical potential gradient [126]. ED is based on the phenomena of electrolytic dissociation of salts, directed movement of ions in an electric field, and selective transfer of ions through ion-exchange membranes.

The first mention of the use on ED for solution desalination is dated back to 1890. Maigrot and Sabates tried to desalinate sugar syrup in an electrochemical cell divided into compartments by parchment membrane, which was a prototype of porous nonactive membrane. However, the intensive studies on ED for desalination purposes began in 1950s from the synthesis of ion-exchange membranes by Juda and McRae and construction of first electrolyzers [127,128]. In 1952 the process was successfully implemented for the purification of coolant water from the pressurized water reactor giving demineralization efficiency in the range of 66%–72% [129]. In 1954 desalination unit with ion-exchange membranes was installed for Saudi Arabian Oil Company.

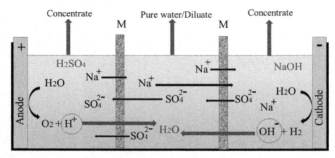

FIGURE 2.16 Schematic illustration of a simple electrodialyzer with nonactive porous membranes. *M*, membrane.

The simplest ED cell (electrodialyzer, Fig. 2.16) consists of an anode and a cathode separated by two membranes, which divide the reactor volume into three compartments.

Let us consider the case when electrolyte solution (for example, Na_2SO_4) is fed to all compartments (it can also be fed to the middle compartment along with distilled water pumped into anodic and cathodic compartments) of a simple electrodialyzer (Fig. 2.13). When applying the potential difference to electrodes cations start to move toward the cathode (negative electrode) and anions move toward the anode (positively charged electrode). Electrolyte concentration of the middle compartment (dilute or retentate compartment) will gradually become desalinated and filled with clean water. At the same time electrode reactions will lead to the oxygen gas generation and formation of acid (H_2SO_4) at the anode (R. 2.157) and hydrogen gas evolution as well as alkali (NaOH) formation at the cathode (R. 2.158). Thus, near-electrode compartments become enriched with alkaline or acid electrolyte (concentrate or permeate).

$$2H_2O - 4e^- \rightarrow O_2 + 4H^+ \text{ or}$$
$$2SO_4^{2-} + 2H_2O - 4e^- \rightarrow O_2 + 2H_2SO_4 \tag{2.157}$$

$$2H_2O + 2e^- \rightarrow H_2 + 2OH^- \text{ or}$$
$$2Na^+ + 2H_2O + 2e^- \rightarrow H_2 + 2NaOH \tag{2.158}$$

If electrolyte solution contains chloride ions, chlorine gas would be allocated in the anodic chamber along with the oxygen in accordance with the following reaction:

$$2Cl^- - 2e^- \rightarrow Cl_2 \tag{2.159}$$

In the case of nonactive porous membrane use (Fig. 2.16), the increase of acidity and alkalinity in the near-electrode chambers will force the excess OH^- and H^+ ions migrate through the membranes toward oppositely charged electrodes. While passing to the middle chamber OH^- and H^+ ions are neutralized forming water. The process of OH^- and H^+ ions transport to the middle chamber will interfere with the transport of electrolyte ions thus causing the loss of useful current. In this regard the use of selective ion-exchange membranes allows to enhance the CE and to decrease EC for the process of ED due to preferable transport of ions with either positive or negative charge. While using porous membranes, the CE of ED does not usually exceed 20%. An opposite use of ion-exchange membranes can provide around 100% CE. The CE of the ED process can be calculated similar to other electrochemical processes based on the Faraday's law using the following equation:

$$CE = \frac{(C_0 - C_t)VF}{It}, \tag{2.160}$$

where C_0 and C_t are initial and final concentration of electrolyte, respectively, after time t of the treatment and V is the volume of treated water. The higher the CE, the more efficient the ED process.

To avoid anodic dissolution of metal and metal ions transfer to the middle compartment through the anodic membrane a special attention should be paid to the selection of anodic material. Usually anodes are made of Pt deposited on titanium, lead oxide, or graphite. Cathode can be made of different metals and often stainless steel or titanium electrodes are used.

Classification of membranes used in ED. In general membranes used in ED processes can be divided into nonactive (porous), active (ion exchange), and ideally active (ion exchange). In turn, active membranes can be divided into cation exchange and anion exchange depending on the prevalent ions passing through the membranes.

1. Nonactive membranes contain pores of a certain size that allow mechanical passing of smaller compounds through the membranes. These membranes do not change the ion transport number (t_i) that means that cations and anions transport number in the solutions $(t^0_c$ and $t^0_a)$ is equal to the transport number of cations (t_c) and anions (t_a) in the membrane, respectively. It is worth to notice that the ion transport number is the fraction of the total electricity carried by a specific ion i. Pore radius of nonactive membranes is significantly larger than the thickness of EDL. Nonactive membrane can be used for the separation of nonelectrolytes from electrolytes.

2. Active/selective membranes change the ion transport number of either cations or anions depending on the membrane charge. For example, if a membrane increases the cation transport number $(t_c > t^0_c)$, which is typical for negatively charged membranes (cation-exchange membrane carrying a negative charge of anions fixed in the membrane matrix), the anion transport number will be decreased $(t_a < t^0_a)$ by the membrane. In contrast when positively charged membrane (anion-exchange membrane carrying a positive charge of cations fixed in the matrix) increases anion transport number $(t_a < t^0_a)$, then $(t_c > t^0_c)$. Pore size of these membranes is smaller than those of nonactive membranes. The smaller the pore size the more prominent the change in the ion transport number.

3. Ideally active (ideally selective) membranes can let pass only cations or anions (if $t_c = 1$ then $t_a = 0$ and in contrast if $t_a = 1$ then $t_c = 0$). All electricity is transferred through the membrane by counterions. It is worth to notice that the higher the cation and transport numbers in cathodic and anodic membrane, respectively, the higher the CE of ED.

Transport of ions through ion-exchange membranes in ED is controlled by diffusion, electromigration, and convection and the total flux J_i (mol/m2s) of ions through the membrane can be found as follows:

$$J_i = \vartheta \cdot C_i - D_i \frac{dC_i}{dx} - \frac{z_i \cdot F \cdot C_i \cdot D_i}{R \cdot T} \frac{\partial \phi}{\partial x}, \qquad (2.161)$$

where ϑ is the velocity of ion in solution due to convection (m/s); C_i is the concentration of ion i (mol/m^3); D_i is the diffusion coefficient (m^2/s); x is the coordinate of direction (m); z_i is the charge number of species; ϕ is the electric potential (V); R is the gas constant; and T is the temperature.

Ion transport number (t_i) is the ratio of a part of electricity, which is transferred through the membrane by ions countercharged to ions fixed in the membrane matrix related to the total electricity.

$$t_i = \frac{z_i \cdot J_i}{\sum_{i=1}^{n} J_i} \tag{2.162}$$

While conducting experimental measurements, it is difficult to make a division between diffusion migration and convection fluxes while determining the total flux of ions. Therefore, for the cases when concentration gradient is negligible and the membrane is ideally active, i.e., has only one type of counterions moving, ion transport number can be found as follows [130]:

$$t_i = \frac{z_i \cdot J_i \cdot F}{j} \tag{2.163}$$

where $J_i = J_{mig}$ and $j = F \cdot \sum_{i=1}^{n} z_i \cdot J_i$ is the current density passing through a membrane (A/m^2).

Ion-exchange membranes used for ED are produced from ion-exchange resins and thermoplastic polymer material such as polyethylene, polyethylene cross-linked with divinylbenzene, or polypropylene usually in the form of flexible sheet of rectangular shape. Depending on type, ion-exchange membranes can have acidic or basic ion exchange groups such as $SO_3{}^-$, COO^-, $SeO_2{}^-$, $-N^+\equiv$, $-N^+(CH_3)_3$, $-N^+(R)_3$, etc., in their matrix. The nature of fixed charges and counterions significantly affects the selectivity of the membranes and electrical conductivity. Most of commercial cation-exchange membranes contain $SO_3{}^-$ and anion-exchange membranes contain groups of quaternary ammonium bases, for example, $[-(CH_3)_3N^+]$. Membranes should have high mechanical and chemical strength, high conductivity, and high permeability for ions along with a high selectivity and low electrical resistance ($2-10$ Ω/cm^2 area of the ion exchange membrane) and thickness. To improve the mechanical strength of membranes, it is possible to reinforce them by attaching to the porous or mesh substrate made of polymer, graphite, metal materials, etc. Use of homogeneous membranes allows achieving more even distribution of charges and better electrochemical separation properties in the membranes. Homogeneous membranes are produced by the direct introduction of ionic groups into the polymer film, for example, by sulfonation or amination of polyethylene film. There is a continuous search for new polymer matrixes for production of homogeneous membranes and methods for the surface modification of membranes to obtain charge selectivity and resistance to poisoning by organic components.

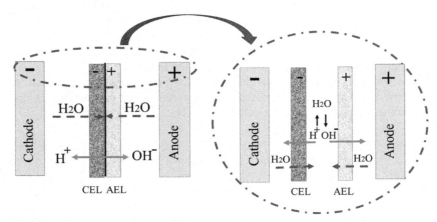

FIGURE 2.17 Working principle of bipolar membrane. *AEL*, anion-exchange layer; *CEL*, cation-exchange layer.

Ion-exchange membranes can be either monopolar (either cation exchange or anion exchange), which are described above, or bipolar. The matrix of a monopolar anion-exchange membrane contains the fixed cationic groups, which are neutralized by the charge of mobile anions located in the membrane pores. The anions of the electrolyte solution can enter the membrane matrix and replace the initially present anions. At the same time cations from the electrolyte solution cannot penetrate into the anion-exchange membrane matrix because of electrostatic repulsive forces caused by the rigidly fixed matrix cations. Cation-exchange membranes have the similar mechanism of work and contain the fixed anionic groups in the membrane matrix. Bipolar membrane (Fig. 2.17) consists of anion- and cation-exchange layers tightly connected to each other.

Anion side of bipolar membrane is directed to anode and cation side is directed to cathode in electrodialyzer. When membrane is placed into solution, water molecules start to diffuse into the membrane to the "junction" between ion-exchange layers of bipolar membrane. Under standard conditions a small amount of water dissociates into hydrogen ions (protons) and hydroxide ions as follows: $H_2O \leftrightarrow H^+ + OH^-$ and H^+ and OH^- ions are always present in water. If the temperature of water is constant, then the amount of dissociated ions is also constant and $[H^+] = [OH^-] = 10^{-7}$ mol/L for deionized water. To diminish high electric resistivity of deionized water due to small amount of charge carriers and hence low conductivity, the distance between anion- and cation-exchange layers is minimized. Under the influence of electric field, H^+ ions start to move from the intermediate layer of bipolar membrane toward cathode and OH^- ions move toward anode. Decrease of dissociated ions in the junction of membrane is replenished by new dissociation of water in accordance with dissociation equilibrium and decrease of water in intermediate layer is compensated by diffusion of water molecules from both sides of membrane. Electric field established over the membrane enhances splitting of water to hydrogen and hydroxide ions in addition to electrolytic water dissociation. The mechanism of water splitting intensification is still unclear and there are two theories explaining this effect. According to the first theory, water molecules interact with the fixed charged groups of membranes and produce H^+ and OH^- ions through the protonation—deprotonation reactions. The second theory considers the Wien effect (the effect of significant increase of electric conductivity in solutions of weak electrolytes at high electric field strength). Under normal conditions weak electrolyte ions are surrounded by an ionic atmosphere containing an excess of countercharged ions. When strong electric field is applied ion atmosphere is destroyed and equilibrium of water dissociation shifts to the right, which leads to the increase of dissociation degree [131].

Ion-exchange membranes are used mainly for water decontamination and desalination purposes by means of decrease/increase of electrolyte concentration. ED units are widely used for the desalination of sea water in the production of kitchen and other salts, pretreatment of water for thermal power plants, and preparation of drinking and industrial water. It can be used, for example, in the treatment of wastewaters from electroplating industry. ED can be conducted at elevated temperatures with highly acidic and alkaline solutions. ED using conventional electrodialyzers becomes cost-efficient compared to reverse osmosis and distillation if the concentration of dissolved salts is below or equal to 10 g/L. The salt content in the treated solution directly affects the EC required for desalination. In general, water desalination with the salt content of 2.5—3 g/L requires about 2 Wh/L of energy, which increases to 1—5 Wh/L when salt content is in the range of 5—6 g/L. Another approximation says that about 1 kWh of power is required for the removal of 1 kg of salts.

Principles of Electrodialysis

Use of three-compartment electrodialyzers is a cost-consuming process due to the loss of energy to the electrode side reactions. In this regard a significant improvement of cost efficiency can be achieved by using a large number of membranes. Usually electrodialyzers can contain up to 2000 of ion-exchange membranes pairs. There are three ways of electrodialyzer operation depending on the type of used membranes.

1. The principle of work of conventional multichamber electrodialyzers with membranes of different charge (Fig. 2.18) is similar to that of the simple electrodialyzer shown on Fig. 2.16. Membranes are placed in between two electrodes. Electrolyte solution (wastewater) is usually fed to all compartments; however, a separate feed of dilute, concentrate, and electrode

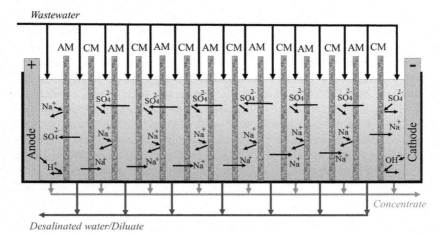

FIGURE 2.18 The working principle of multi-compartment electrodialyzer. *AM*, anion-exchange membrane; *CM*, cation-exchange membrane.

compartments is also used in some applications. While applying the direct current, cations move toward the cathode through the cation-exchange membrane placed from the cathode side. Anion membrane is placed on the anode side. The cation-exchange and anion-exchange membranes are alternated. The anions move in the opposite direction toward the anode through the anion exchange membrane. While passing the anion-exchange membrane, anions are retained in the concentrate stream by a cationic membrane located from the side of anode. On the contrary, cations are retained by the anionic membrane located on the cathode side. Thus, the overall process results into the increase of ions concentration in each odd chamber and decrease of ions concentration in each even compartment. Multicompartment electrodialyzer allows reducing the EC and as a result minimizing costs of the process.

2. *ED with bipolar membranes* is an effective replacement to conventional water electrolysis and allows direct production of bases and acids through the water splitting to hydrogen and hydroxide ions without forming oxygen and hydrogen gas. The process gives significant cost savings especially when using multicompartment electrodialyzers. Generation of O_2 and H_2 gases during anodic and cathodic reactions, respectively, consumes up to half of useful energy in the process of conventional water electrolysis [131]. A simple scheme of acid and base production during ED is shown in Fig. 2.19. Electrolyte solution is fed to the compartments between anion-exchange and cation-exchange membranes and water is fed to electrode and near bipolar membrane compartments. When electric field is applied, compartments on

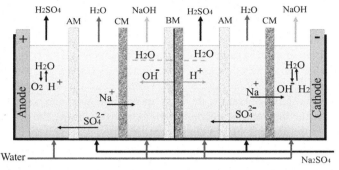

FIGURE 2.19 Working principle of electrodialyzer with bipolar membranes for base and acid production. *AM*, anion-exchange membrane; *CM*, cation-exchange membrane.

both sides of bipolar membrane start to be filled with electrolyte ions migrating from near compartments and with H^+ (from cathode side) and OH^- (from anode side) ions migrating from intermediate part of bipolar membrane thus producing acid and base. The more detailed mechanism of ED process with bipolar membrane is represented in Fig. 2.19.

3. *Substitutional ED* with membranes of the same type i.e., only cation-exchange or only anion-exchange membranes packed between two electrodes are used for special application such as for obtaining organic acids from their salts or reducing the acidity of citrus juices and rarely used in water treatment applications. For example, to reduce the acidity of orange juice, electrodialyzer should entirely consists of anion-exchange membranes. When applying potential difference between the electrodes, citrate ions $(C_3H_5O(COO)_3^{3-})$ present in the juice start to move toward anode passing through the anion-exchange membrane to the compartment containing alkali solution. Then an equivalent amount of hydroxide ions move out from the alkali compartment through the next anion-exchange membrane to the next juice compartment where they are neutralized by hydrogen ions of the juice, thus deacidifying it.

Depending on the direction of dilute flow, ED systems can be divided into the following systems:

- one -pass flow systems (continuous process) when the desired degree of desalination is achieved in a single pass of dilute through the electrodialyzer;
- batch systems (discontinuous process) or circulating ones when dilute and feeding water of concentrated compartments circulate through the electrodialyzer several times until the desired value of desalination is achieved;
- partially circulating process, where a part of desalinated dilute is circulated through the electrodialyzer along with the feeding dilute.

Also there are a mixed type of systems where multiple circulation of the dilute through the ED unit is carried out the required amount of times until the desired desalination degree is achieved. The washing solution of the concentrate chambers in such systems is discharged after one pass.

Electrodialyzers like other electrochemical cells can be operated either in galvanostatic (at a constant current) or potentiostatic (at a constant voltage) mode. Potentiostatic operation mode is more favorable and safe, which is explained by the fact that the uncontrolled voltage rise during the galvanostatic mode can cause damage of ED equipment.

Pros and Cons of Electrodialysis Compared to Other Membrane Technologies

ED has the following advantages compared to other membrane technologies and particularly reverse osmosis [132]:

- High water recovery in the range of 80%−90%.
- Ability to obtain highly concentrated brines, which facilitates their further processing and allows recovery of valuable components.
- Long service life of membranes, which can last up to 7−10 years.
- Lower requirements for water quality entering to treatment by ED compared to water treated by reverse osmosis. Water silt density index (SDI) for the process of ED should be usually below 12 while for reverse osmosis SDI should not exceed 3.
- Resistance of membrane to elevated temperatures, drying up, bacteria decontamination, and free chlorine content up to 1 mg/L. However, there is a possibility of membranes fouling with microorganisms.
- Mechanical strength of membranes and of manual cleaning of membranes.
- Membranes contribute to about 50% of ED equipment costs.
- Compact sizes of ED equipment.
- Simplicity of operation and ease of automation.
- Low operating pressures (0.3−0.4 atm) of electrodialyzers.
- No chemicals addition to the process except the occasional use of softeners.

The main disadvantages and parameters reducing the efficiency of ED are as follows:

1. Membrane passivation and polarization due to mostly concentration polarization.

    Let us consider the phenomenon of concentration polarization on the surface of the cation-exchange membrane. When applying the potential difference to the electrodes, cations start to migrate toward the cathode, which leads to the increase of cation transport number in membranes and decrease the cation transport number in the solution. As a result, the concentration of cations near to the membrane surface on the cathode side will increase and decrease on the anode side. This will occur as long as concentration gradients are established on the side of cathode and anode. When concentration gradients are established, the flow of cations in solution caused by diffusion and potential difference becomes equal to the flow of cations in membrane. Thus, the concentration of cations increases at the surface of membrane located on the side of concentrated solution and decreases on the side of diluted solution. The presence of carbonates and sulfates in water along with calcium and manganese can lead to the precipitation of $CaSO_4$ and $CaCO_3$ on the surface of membranes in the concentrate compartment. Decrease of cation concentration on the side of diluted (becoming desalinated) solution inhibits the current density, which can be used for the process of ED. The formation of hydroxides ions will cause the precipitation of $Mg(OH)_2$ on the surface of cation-exchange membrane on the cathode side. Membrane deposits cause the increase of membrane resistivity, decrease of its conductivity, increase of the current flux through the free surface of membrane, and consequently increase of current density passing through the membrane. At some point of passivation, the values of passing current significantly exceed the limiting diffusion current density for this membrane. This leads to the extensive membrane passivation due to the enhanced water dissociation at the membrane/solution interface and precipitation of metal hydroxides and significant inhibition of ED [133]. From another side, formation of OH^- and H^+ ions at the membrane surface can improve the conductivity of solution, thus improving ions mass transport. Moreover, the performance of membranes can be decreased because of the presence of gas bubbles in solution, which occupy the active area of membranes.

To decrease the passivation and polarization of membranes the practice of periodical change of electrodes polarity is used during ED (several times per hour), which entails a change of direction of ion movement in solution. However, such setup requires more complex control system. Removal of deposits from membranes can be achieved also by washing them in acidic solution of HCl, which can dissolve precipitates. Pretreatment from hardness can be used as another step to prevent the precipitation of carbonates and hydroxides on the membrane surface. Removal of calcium and magnesium ions can be achieved by addition of alkaline solution to dilute and allowing formation of magnesium and calcium hydroxides, which can be precipitated and filtrated from the dilute. Both alkaline and acidic solutions can be obtained during ED with the use of bipolar membranes [133].

2. Necessity in the solution pretreatment from suspended solid and hardness before the process (up to seven stages). The allowed concentration of suspended solids for ED is 3 mg/L, COD is below 5 mg O_2/L, boron and iron content below 0.1 mg/L.

3. Frequent membrane replacement (often after 6−12 months).

4. Impossibility to remove uncharged molecules and organic compounds.

    However, ED can be used for the desalination of noncharge compounds such as pharmaceuticals, sugar, and vine.

5. Concentrate solution requires additional treatment or recycling.

    To improve the efficiency of the process and reduce the amount of formed wastewater to be recycled or utilized, recirculation of concentrate solution is often used.

6. Ion transport number in membranes usually differs from the ion transport number in the electrolyte solution.

7. Electroosmotic transfer of solvent molecules captured by migrating ions through the membrane can be observed during ED and change the composition and volume of electrolyte solution in electrodialyzer compartments.

8. Electrophoretic movement of particles can cause additional precipitates on the surface of membranes.

    The principles and measures of elimination of electrophoretic deposits are similar to the principles and measures conducted to avoid passivation of membranes.

9. Dialysis of dissolved solute is opposite to the direction of migrating ions in ED. Dialysis is a diffusion process of solute transfer through the membrane due to the concentration gradient built on the opposite sides of the separation membrane. Movement of the solvent in dialysis in the opposite direction of migrating ions reduces the mass transfer of ions and becomes prominent along with electroosmosis and electrophoresis when the composition of concentrate solution exceeds the allowable values for a particular wastewater.

The Main Calculations of Electrodialysis Process

The EC (kWh/m^3) during ED process can be estimated using the following equation:

$$EC = \frac{UIt}{V},$$ \hfill (2.164)

where U is the voltage of electrodialyzer (V); I is the current (A); t is the duration of ED (h); and V is the volume of treated solution (L).

It was also reported that EC of significantly salty ocean water does not obey the Faraday's law and can be found from the following empirical equation [132]:

$$EC = \frac{3.5 - 0.5 \cdot C_{TDS}}{C_{TDS}},$$ \hfill (2.165)

where C_{TDS} is the salt content in water (g/L).

Desalination ratio (α) and CE in ED can be determined as follows [134,135]:

$$\alpha = 1 - \frac{C_t}{C_0} = 1 - \frac{C_p}{C_f}, \qquad (2.166)$$

where C_t and C_0 are final outlet and initial inlet concentrations of electrolyte solution, respectively. For the batch process C_t and C_0 can be represented through the product and feed concentration, C_p and C_f, respectively.

$$CE = \frac{F \cdot J_i}{j} = \frac{z \cdot F \cdot Q_d \cdot (C_0 - C_t)}{N \cdot I} = \left(\overline{t^{CEM}} + \overline{t^{AEM}} - 1 \right), \qquad (2.167)$$

where Q_d is the flow rate of dilute (m^3/s); N is the number of cell pairs; I is the current (A); and $\overline{t^{CEM}}$ and $\overline{t^{AEM}}$ are the transport number of cation- and anion-exchange membranes, respectively. If assume that $\overline{t^{CEM}} = \overline{t^{CEM}} = \overline{t^{IEM}}$ then,

$$\overline{t^{IEM}} = \frac{z \cdot F \cdot Q_d \cdot (C_0 - C_i) + NI}{2NI} \qquad (2.168)$$

The volume of oxygen or hydrogen gas (V_i) allocated at electrodes during ED can be determined based on the Faraday's law as follows:

$$V_i = \frac{\delta_i \cdot I \cdot t \cdot R \cdot T}{z \cdot F \cdot p}, \qquad (2.169)$$

where δ_i is the stoichiometric coefficient of i gas and p is the gas storage pressure (Pa).

Parameters Influencing the Efficiency of Electrodialysis

The efficiency of electrodialysis depends on such parameters as the type and material of membranes, distance between membranes, applied current density, etc. As it was mentioned before, active membranes provide higher CE and smaller EC. Moreover, membranes have to possess low electrical resistance, high selectivity, moderate swelling, and sufficient mechanical strength. The optimal distance between membranes is about 1−2 mm. ED allows obtaining water with the quality higher than distilled water quality obtained by the reverse osmosis.

Current Density and Cell Voltage

There is a general dependence of demineralization efficiency on the applied current density. The higher the current density of ED units operation, the higher the demineralization efficiency and faster the process. However, too very high current densities increase the concentration gradient leading to significant decrease of ion concentration at the membrane surface. When the concentration of ions at the membranes is equal to zero, the process is characterized by the limiting current (j_{lim}), above which the efficiency of ED is becoming suppressed [136].

$$j_{\lim} = \frac{F \cdot D_i \cdot C_i^b}{\left(\overline{t_i} - t_i\right) \cdot \delta},$$
(2.170)

where C^b_i is the concentration of electrolyte solution in the bulk; $\overline{t_i}$ is the ion transport number in the membrane; t_i is the ion transport number in solution; and δ is the thickness of diffusion layer.

The influence of cell voltage on CE of ED process is similar to the effect of current density. Higher cell voltage increases the CE of ED in the range of small electrolyte concentrations. When high voltage is applied to ED cell containing high electrolyte concentrations (>20 g/L), the effect of concentration polarization becomes more prominent and CE of ED significantly decreases.

Electrolyte Composition and Concentration

Efficiency of ED process depends on the dilute composition and concentration. The lower the concentration of electrolyte, the higher the electrochemical activity of membrane and consequently the higher the efficiency of desalination. Presence of gas bubbles, suspended solids, phosphates, microorganisms, and other compounds in the amount exceeding the maximum allowable concentrations for particular ED process significantly depresses the efficiency of desalination and membrane performance.

Flow Rate of Dilute and Concentrate Feed

Flow rate of water in dilute and concentrate compartments affects membrane polarization and hence the efficiency of desalination. If the feed rate of the solutions is below the critical level, polarization of membranes and decomposition of water into hydrogen ions and hydroxide ions start to be prominent. From another side generated hydrogen and hydroxide ions are involved in the transfer of current, which leads to the decrease of CE and transfer of desalinated ions. At some high value of water flow, CE can decrease to zero and suppress the efficiency of ED. In this regard, there is an optimal value of flow rate, which should be determined empirically for each particular process.

2.6 SUMMARY

Because the need for clean water constantly increases when the volumes of wastewater become greater and pollution of water bodies with toxic refractory compounds is growing, conventional water treatment methods do not provide anymore sufficient degree of purification. Electrochemical water treatment technology is an effective alternative and complement to traditional methods.

Electrochemical cleaning methods have a number of significant advantages over conventional methods:

- mineralization of water stay at the same level, which is important for circulating water supply systems;
- smaller amount of precipitates is formed, for example, when comparing coagulation and electrocoagulation methods;
- no need to organize working areas for chemical storage and no need for chemical transportation system;
- easy automatable system;
- electrochemical treatment plants require a small working area compared to conventional wastewater treatment methods;
- usually requires no chemicals to conduct the process of water treatment.

EO and ER, electrocoagulation, EF, and ED are the most developed and studied methods of electrochemical water treatment technology. EO and ER are the methods for the wastewater treatment, which take place at the anode and cathode, respectively. EO is successfully used for the degradation of organic, such as pharmaceuticals, dyes, and pesticides, and inorganic, such as cyanides, thiocyanate, and sulphides compounds. Organic compounds are either mineralized to simple inorganic compounds, such as H_2O and CO_2, during the oxidation, or transformed to other organic intermediates. The mechanism of destruction of compounds during oxidation depends largely on the materials of the anodes used. Oxidation of organic pollutants at noble metal and metal alloy, carbon and graphite anodes usually occur with low mineralization efficiency, while MMO, BDD, and PbO_2 electrodes provide relatively high degrees of pollutant mineralization. EO can be direct as a result of direct loss of electrons by a pollutant molecule on the surface of the anode or indirect associated with the interaction of pollutants with products of electrolysis of electrolyte solution. Indirect oxidation is used also for water disinfection. The most common mechanism of water disinfection is electrochemical generation of hypochlorite ions from water and chlorine gas, which is in turn formed as the main product of sodium chloride salt electrolysis. ER can be used for reduction of organic compounds, to transform them to less toxic intermediates, inorganic nitrogen compounds, to conduct denitrification to nontoxic N_2 gas final products, and metals, to remove them from the treated solution by depositing at the cathode. Deposition of metals simultaneously makes the recovery of valuable components from wastewater and technological solutions·possible.

The mechanism of EC process is very similar to that of conventional electrolyte coagulation. The main difference is that the coagulant is generated during the dissolution of the anodes when conducting electrolysis. Iron and aluminum anodes are used as sacrificial anodes. When dissolved, metal ions are hydrolyzed forming various polynuclear hydroxo complexes, which act as coagulants. Coagulants adhere to pollutant particles, thus destabilizing and precipitating them. EC usually provides a higher degree of treatment compared to electrolyte coagulation due to additional electrochemical effects, such as EF, electrophoresis, and cathodic reduction, occurring during electrolysis. EC can remove a wide range on pollutants. They are, for example, bacteria, ammonia and nitrate hydrogen, sulfide, arsenic, color, organics, and turbidity.

EF removes insoluble contaminants from water by floating them to the surface of solution by electrogenerated gas bubbles, mostly H_2 gas bubbles. EF can successfully remove oils and greases, metal hydroxides, emulsified substances, and others. The efficiency of pollutant removal during EF strongly depends on the size of gas bubbles and particles to be removed, but in general, the smaller the size of gas bubbles, the better separation efficiency is achieved during EF.

ED is a water demineralization technology, providing the separation of ions by means of electrical potential gradient and ion-exchange membranes assembled in an electrochemical reactor. Electrolyzer should contain at least two membranes that divides reactor into three compartments. The separation occurs due to the movement of ions through the membranes, and as a result, their concentration in some chambers and dilution in others. ED has advantages over the reverse osmosis process, which is commonly used for water desalination. They are low operation pressures, lower requirements for water quality entering to the treatment, no additional chemical for the exception of softeners, long service life of membranes, and others. To save cost for the reactor operation, it is recommended to install up to 2000 membrane pairs between electrodes.

Because efficiency of electrochemical water treatment methods usually depends on the applied current density, electrode material and shape, acidity of the treated electrolyte solution and its composition, selection of optimal electrode materials, membranes, operating parameters, as well as control of the electrolyte composition are the main conditions for effective treatment.

SELF-CONTROL QUESTIONS

1. What are the main interactions between the particles according to the DLVO theory? Please define the following concepts: primary and secondary minimums and energy barrier.
2. Please give a definition for the Schulze—Hardy rule and explain its practical application for coagulation processes.
3. What are the main parameters affecting coagulation? Why coagulant dosage should be carefully controlled?
4. What are the main chemical reactions occurring during electro-coagulation? What are parameters affecting the dissolution of anode?
5. What are pollutants, which can be successfully removed by electro-coagulation? What is the mechanism of fluoride, arsenic, and organic matter removal by electrocoagulation?

6. What are the main parts of a simple electrodialyzer? Please describe the mechanism of electrodialysis (ED). How ED can be performed depending on membrane types and flow of feed solutions?

7. What kind of membranes can be used in electrodialysis? Please specify their differences.

8. What are the mechanisms and steps of electroflotation (EF)? Which pollutants can be removed by EF method?

9. How does wettability influence the adhesion between gas bubbles to dispersed particles? How does the contact angle influence the flotation efficiency?

10. What kind of forces acts to a formed flotocomplex? How they are equilibrated?

11. What are the main parameters influencing the efficiency of elctroflotation process?

12. What are the main advantages and disadvantages of electrodialysis? How the polarization and decontamination of membranes can be reduced?

13. What is the flocculation? Why flocculation and coagulation are often combined with flotation?

14. What are the mechanisms of flocculation? How flocculants are usually classified?

15. What are the pros and cons of ED compared to other membrane technologies?

16. What is the difference between electrochemical oxidation and electrochemical reduction (ER) water treatment process? Which contaminants can be removed by ER methods?

17. What are the main mechanisms of electrochemical removal of inorganic and organic nitrogen-containing compounds, metals aldehydes, and ketones from water?

18. What are the main application areas or electrochemical disinfection and reduction processes?

REFERENCES

[1] C. Martínez-Huitle, S. Ferro, Electrochemical oxidation of organic pollutants for the wastewater treatment: direct and indirect processes, Chem. Soc. Rev. 35 (2006) 1324–1340.

[2] O.J. Murphy, G. Hitchens, L. Kaba, C. Verostko, Direct electrochemical oxidation of organics for wastewater treatment, Water Res. 26 (4) (1992) 443–451.

[3] D. Valero, V. García-García, E. Expósito, A. Aldaz, V. Montiel, Electrochemical treatment of wastewater from almond industry using DSA-type anodes: direct connection to a PV generator, Sep. Purif. Technol. 123 (2014) 15–22.

[4] M. Deborde, U. von Gunten, Reactions of chlorine with inorganic and organic compounds during water treatment — kinetics and mechanisms: a critical review, Water Res. 42 (2008) 13–51.

[5] J. March, M. Gual, Studies on chlorination of greywater, Desalination 249 (2009) 317–322.

[6] N. Nordin, F. Amir, Riyanto, M. Othman, Textile industries wastewater treatment by electrochemical oxidation technique using metal plate, Int. J. Electrochem. Sci. 8 (2013) 11403–11415.

[7] S. Aquino Neto, A. de Andrad, Electrooxidation of glyphosate herbicide at different DSA® compositions: pH, concentration and supporting electrolyte effect, Electrochim. Acta 54 (7) (2009) 2039–2045.

[8] M. Çelebi, N. Oturan, H. Zazou, M. Hamdani, M. Oturan, Electrochemical oxidation of carbaryl on platinum and boron-doped diamond anodes using electro-Fenton technology, Sep. Purif. Technol. 156 (3) (2015) 996–1002.

[9] A. Sánchez, J. Llanos, C. Sáez, P. Canizares, M. Rodrigo, On the applications of peroxodiphosphate produced by BDD-electrolyses, Chem. Eng. J. 233 (2013) 8–13.

[10] K. Juttner, U. Galla, H. Schmieder, Electrochemical approches to environmental problems in the process industry, Electrochim. Acta 45 (15–16) (2000) 2575/2594.

[11] C. Comninellis, Electrocatalysis in the electrochemical conversion/combustion of organic pollutants for waste water treatment, Electrochim. Acta 39 (11) (1994) 1857–1862.

[12] S. Trasatti, Electrocatalysis: understanding the success of DSA®, Electrochim. Acta 45 (2000) 2377–2385.

[13] P. Patel, N. Bandre, A. Saraf, J.P. Ruparelia, Electro-catalytic materials (electrode materials) in electrochemical wastewater treatment, Procedia Eng. 51 (2013) 430–435.

[14] M. Panizza, G. Cerisola, Direct and mediated anodic oxidation of organic pollutants, Chem. Rev. 109 (12) (2009) 6541–6569.

[15] C.P.C. Comninellis, Anodic oxidation of phenol for waste water treatment, J. Appl. Electrochem. 21 (1991) 703–708.

[16] A. Fabianska, A. Ofiarska, A. Fiszka-Borzyszkowska, P. Stepnowski, E. Siedlecka, Electrodegradation of ifosfamide and cyclophosphamide at BDD electrode: decomposition pathway and its kinetics, Chem. Eng. J. 276 (2015) 274–282.

[17] A. Nilsson, A. Ronlán, V. Parker, Anodic oxidation of phenolic compounds. Part III. Anodic hydroxylation of phenols. A simple general synthesis of 4-alkyl-4-hydroxycyclohexa-2,5-dienones from 4-alkylphenols, J. Chem. Soc., Perkin Trans. 1 (1973) 2337–2345.

[18] C. Comninellis, E. Plattner, Electrochemical wastewater treatment, Chimia 42 (1988) 250–252.

[19] R. Kötz, S. Stucki, B. Carcer, Electrochemical waste water treatment using high overvoltage anodes. Part I: physical and electrochemical properties of SnO_2 anodes, J. Appl. Electrochem. 21 (1991) 14–20.

[20] H. Sharifian, D. Kirk, Electrochemical oxidation of phenol, J. Electrochem. Soc. 133 (1986) 921–924.

[21] V. Smith de Sucre, A. Watkinson, Anodic oxidation of phenol for waste water treatment, Can. J. Chem. Eng. 59 (1981) 52–59.

[22] S. Stucki, R. Kötz, B. Carcer, W. Suter, Electrochemical wastewater tretament using high overvoltage anodes. Part II: anode performance and applications, J. Appl. Electrochem. 21 (1991) 99–104.

[23] M. Chettiar, A. Watkinson, Anodic oxidation of phenolics found in coal conversion effluents, Can. J. Chem. Eng. 61 (1983) 568–574.

[24] F. Al Kharafi, A. Saad, B. Ateya, I. Ghayad, Electrochemical oxidation of sulfide ions on platinum electrodes, Mod. Appl. Sci. 4 (3) (2010) 2–11.

[25] P. Caliari, M. Pacheco, L.L.A. Ciríaco, Anodic oxidation of sulfide to sulfate: effect of the current density on the process kinetics, J. Braz. Chem. Soc. 0 (2016) 1–10.

[26] A. Kraft, Electrochemical water disinfection: a short review, Platinum Metals Rev. 52 (3) (2008) 177–185.

[27] H. Li, X. Zhu, Y. Jiang, J. Ni, Comparative electrochemical degradation of phthalic acid esters using boron-doped diamond and Pt anodes, Chemosphere 80 (2010) 845–851.

[28] X.-Y. Li, Y.-H. Cui, Y.-J. Feng, Z.-M. Xie, J.-D. Gu, Reaction pathways and mechanisms of the electrochemical degradation of phenol on different electrodes, Water Res. 39 (10) (2005) 1972–1981.

[29] A. Alaoui, K. El Kacemi, K. El Ass, S. Kitane, S. El Bouzidi, Activity of Pt/MnO$_2$ electrode in the electrochemical degradation of methylene blue in aqueous solution, Sep. Purif. Technol. 154 (2015) 281–289.

[30] M. Bashir, M. Isa, S. Kutty, Z. Awang, H. Aziz, S. Mohajeri, I. Farooqi, Landfill leachate treatment by electrochemical oxidation, Waste Manage. 29 (9) (2009) 2534–2541.

[31] S. Ganiyu, N. Oturan, S. Raffy, M. Cretin, R. Esmilaire, E. van Hullebusch, G. Esposito, M. Oturan, Sub-stoichiometric titanium oxide (Ti$_4$O$_7$) as a suitable ceramic anode for electrooxidation of organic pollutants: a case study of kinetics, mineralization and toxicity assessment of amoxicillin, Water Res. 106 (2016) 171–182.

[32] J. Liang, C. Geng, D. Li, L. Cui, X. Wang, Preparation and degradation phenol characterization of Ti/SnO$_2$-Sb–Mo electrode doped with different contents of molybdenum, J. Mater. Sci. Technol. 31 (2015) 473–478.

[33] D. Kirk, H. Sharifian, F. Foulkes, Anodic oxidation of aniline for waste water treatment, J. Appl. Electrochem. 15 (2) (1985) 285–292.

[34] A. Fernandes, D. Santos, M. Pacheco, L. Ciríaco, A. Lopes, Nitrogen and organic load removal from sanitary landfill leachates by anodic oxidation at Ti/Pt/PbO$_2$, Ti/Pt/SnO$_2$-Sb$_2$O$_4$ and Si/BDD, Appl. Catal., B 148–149 (2014) 288–294.

[35] H. Wang, J. Wang, Electrochemical degradation of 4-chlorophenol using a novel Pd/C gas-diffusion electrode, Appl. Catal., B 77 (2007) 58–65.

[36] M. Makgae, M. Klink, A. Crouch, Performance of sol–gel titanium mixed metal oxide electrodes for electro-catalytic oxidation of phenol, Appl. Catal., B 84 (2008) 659–666.

[37] M. Shestakova, M. Vinatoru, T. Mason, M. Sillanpää, Sonoelectrocatalytic decomposition of methylene blue using Ti/Ta$_2$O$_5$-SnO$_2$ electrodes, Ultrason. Sonochem. 23 (2015) 135–141.

[38] T. Zayas, M. Picazo, L. Salgado, Removal of organic matter from paper mill effluent by electrochemical oxidation, J. Water Resour. Prot. 3 (2011) 32–40.

[39] S. Chellammal, P. Kalaiselvi, P. Ganapathy, G. Subramanian, Anodic incineration of phthalic anhydride using RuO$_2$–IrO$_2$-SnO$_2$–TiO$_2$ coated on Ti anode, Arabian J. Chem 9 (2016) S1690–S1699.

[40] Y.-Y. Chu, W.-J. Wang, M. Wang, Anodic oxidation process for the degradation of 2, 4-dichlorophenol in aqueous solution and the enhancement of biodegradability, J. Hazard. Mater. 180 (2010) 247–252.

[41] J. Wei, Y. Feng, X.L.J. sun, L. Zhu, Effectiveness and pathways of electrochemical degradation of pretilachlor herbicides, J. Hazard. Mater. 189 (2011) 84–91.

[42] J. Lv, Y. Feng, J. Liu, Y. Qu, F. Cui, Comparison of electrocatalytic characterization of boron-doped diamond and SnO$_2$ electrodes, Appl. Surf. Sci. 283 (2013) 900–905.

[43] D. Li, J. Tang, X. Zhou, J. Li, X. Sun, J. Shen, L. Wang, W. Han, Electrochemical degradation of pyridine by Ti/SnO$_2$-Sb tubular porous electrode, Chemosphere 149 (2016) 49–56.

[44] H. Lin, J. Niu, S. Ding, L. Zhang, Electrochemical degradation of perfluorooctanoic acid (PFOA) by Ti/SnO_2-Sb, Ti/SnO_2-Sb/PbO_2 and Ti/SnO_2-Sb/MnO_2 anodes, Water Res. 46 (2012) 2281–2289.

[45] Y. Yao, C. Zhao, M. Zhao, X. Wang, Electrocatalytic degradation of methylene blue on PbO_2-ZrO_2 nanocomposite electrodes prepared by pulse electrodeposition, J. Hazard. Mater. 263 (2013) 726–734.

[46] K. Meaney, S. Omanovic, Sn0.86–Sb0.03–Mn0.10–Pt0.01-oxide/Ti anode for the electro-oxidation of aqueous organic wastes, Mater. Chem. Phys. 105 (2007) 143–147.

[47] J. Zhang, Y. Wei, G. Jin, G. Wei, Active stainless steel/SnO_2-CeO_2 anodes for pollutants oxidation prepared by thermal decomposition, J. Mater. Sci. Technol. 26 (2) (2010) 187–192.

[48] U. Un, U. Altay, A. Koparal, U. Ogutveren, Complete treatment of olive mill wastewaters by electrooxidation, Chem. Eng. J. 139 (3) (2008) 445–452.

[49] P. Moraes, R. Bertazzoli, Electrodegradation of landfill leachate in a flow electrochemical reactor, Chemosphere 58 (1) (2005) 41–46.

[50] X. Li, D. Shao, H. Xu, W. Lv, W. Yan, Fabrication of a stable Ti/TiOxHy/Sb-SnO_2 anode for aniline degradation in different electrolytes, Chem. Eng. J. 285 (2016) 1–10.

[51] L. Wang, P. Lu, C. Liu, L. Wang, Electro-oxidation of sulfide on Ti/RuO_2 electrode in an aqueous alkaline solution, Int. J. Electrochem. Sci. 10 (2015) 8374–8384.

[52] K. Waterson, D. Bejan, N. Bunge, Electrochemical oxidation of sulfide ion at a boron-doped diamond anode, J. Appl. Electrochem. 37 (2007) 367–373.

[53] J. Haner, D. Bejan, N. Bunce, Electrochemical oxidation of sulfide ion at a Ti/IrO_2–Ta_2O_5 anode in the presence and absence of naphthenic acids, J. Appl. Electrochem. 39 (2009) 1733–1738.

[54] Q. Gu, J. Zheng, Study on the electrochemical oxidation treatment of coking wastewater by DSA anode, in: Asia-Pacific Energy Equipment Engineering Research Conference (AP3ER 2015), Zhuhai, 2015.

[55] I. Sasidharan Pillai, A. Gupta, Anodic oxidation of coke oven wastewater: multiparameter optimization for simultaneous removal of cyanide, COD and phenol, J. Environ. Manag. 176 (2016) 45–53.

[56] H. Xu, A. Li, L. Feng, X. Cheng, S. Ding, Destruction of cyanide in aqueous solution by electrochemical oxidation method, Int. J. Electrochem. Sci. 7 (2012) 7516–7525.

[57] X. Huang, Y. Qu, C. Cid, C. Finke, M.R. Hoffmann, K. Lim, S. Jiang, Electrochemical disinfection of toilet wastewater using wastewater electrolysis cell, Water Res. 92 (2016) 164–172.

[58] E. Lacasa, E. Tsolaki, Z. Sbokou, M. Rodrigo, D. Mantzavinos, E. Diamadopoulos, Electrochemical disinfection of simulated ballast water on conductive diamond electrodes, Chem. Eng. J. 223 (2013) 516–523.

[59] M. Rajab, C. Heim, T. Letzel, J. Drewes, B. Helmreich, Electrochemical disinfection using boron-doped diamond electrode – the synergetic effects of in situ ozone and free chlorine generation, Chemosphere 121 (2015) 47–53.

[60] S. Chen, W. hu, J. Hong, S. Sandoe, Electrochemical disinfection of simulated ballast water on PbO_2/graphite felt electrode, Marine Pollut. Bull. 105 (1) (2016) 319–323.

[61] D.-T. Chin, Metal recovery from wastewater with an electrochemical method, Chem. Eng. Edu. 36 (2) (2002) 149–155.

[62] M. Figueiredo, Electrocatalytic Reduction of Nitrogen Conaining Compounds on Platinum Surface, University of Alicante, Alicante, 2012.

[63] J. Radjenovic, V. Flexer, B. Donose, D. Sedlak, J. Keller, Removal of the x-ray contrast media diatrizoate by electrochemical reduction and oxidation, Environ. Sci. Technol. 47 (2013) 13686−13694.

[64] J. Gui, Method for Destruction of Chlorinated Hydrocarbons. USA Patent US 5569809 A, 29 October 1996.

[65] P. Zuman, Aspects of electrochemical behavior of aldehydes and ketones in protic media, Electroanalysis 18 (2) (2005) 131−140.

[66] J. Philip, J. Leone, Mechanism of the electrochemical reduction of phenyl ketones, J. Am. Chem. Soc. 80 (5) (1958) 1021−1029.

[67] E. Szebényi-Győri, E. Gagyi-Pálffy, G. Bajnóczy, E. Prépostffy, Dechlorination of chlorinated hydrocarbons in a monopolar packed bed electrochemical reactor, Periodica Polytechnica Ser. Chem. Eng. 43 (2) (1999) 65−76.

[68] C. Liu, A.-Y. Zhang, D.-N. Pei, H.-Q. Yu, Efficient electrochemical reduction of nitrobenzene by defect-engineered TiO_2-x single crystals, Environ. Sci. Technol. 50 (2016) 5234−5242.

[69] J. Coleman, R. Kobylecki, J. Utley, Stereoselective electrochemical reduction of cyclic ketones, Chem. Commun. (1972) 104−105.

[70] L. Ruotolo, J. Gubulin, Optimization of Cr(VI) electroreduction from synthetic industrial wastewater using reticulated vitreous carbon electrodes modified with conducting polymers, Chem. Eng. J. 149 (1−3) (2009) 334−339.

[71] X. Fang, G. Zhang, J. Chen, D. Wang, F. Yang, Electrochemical reduction of hexavalent chromium on two-step electrosynthesized one-dimensional polyaniline nanowire, Int. J. Electrochem. Sci. 7 (2012) 11847−11858.

[72] C. Ascensão, L. Ciríaco, M. Pacheco, A. Lopes, Metal recovery from aqueous solutions, Portugaliae Electrochim. Acta 29 (5) (2011) 349−359.

[73] J. Paul Chen, L. Lim, Recovery of precious metals by an electrochemical deposition method, Chemosphere 60 (10) (2005) 1384−1392.

[74] M. Li, C. feng, Z. Zhang, S. Yang, N. Sugiura, Treatment of nitrate contaminated water using an electrochemical method, Bioresour. Technol. 101 (2010) 6553−6557.

[75] I.I.D. Katsounaros, C. Polatides, G. Kyriacou, Efficient electrochemical reduction of nitrate to nitrogen on tin cathode at very high cathodic potentials, Electrochimi. Acta 52 (3) (2006) 1329−1338.

[76] Z. Zhang, Y. Xu, W. Shi, W. Wang, R. Zhang, X. Bao, B. Zhang, L. Li, F. Cui, Electrochemical-catalytic reduction of nitrate over $Pd-Cu/\gamma Al_2O_3$ catalyst in cathode chamber: enhanced removal efficiency and N_2 selectivity, Chem. Eng. J. 290 (2016) 201−208.

[77] Y. Wang, X. Guo, J. Li, Y. Yang, Z. lei, Z. zhang, Efficient electrochemical removal of ammonia with various cathodes and Ti/RuO_2-Pt anode, Open J. Appl. Sci. 2 (2012) 241−247.

[78] H. Hamaker, The London − van der Waals attraction between spherical particles, Physica 4 (10) (1937) 1058−1072.

[79] J. Polte, Fundamental growth principles of colloidal metal nanoparticles − a new perspective, Cryst. Eng. Commun. 17 (2015) 6809−6830.

[80] J.-P. Hsu, B.-T. Liu, Critical coagulation concentration of a colloidal suspension at high particle concentrations, J. Phys. Chem. B 102 (1998) 334−337.

[81] G. Domazetis, M. Raoarun, B. James, Studies of mono- and polynuclear iron hydroxy complexes in brown coal, Energy Fuels 19 (2005) 1047−1055.

[82]   A. Sarpola, J. Saukkoriipi, V. Hietapelto, J. Jalonen, J. Jokela, P. Joensuu, Identification of hydrolysis products of $AlCl_3$ •$6H_2O$ in the presence of sulfate by electrospray ionization time-of-flight massspectrometry and computational methods, Phys. Chem. Chem. Phys. 9 (2007) 377−388.

[83]   F. Akbal, S. Camci, Copper, chromium and nickel removal from metal plating wastewater by electrocoagulation, Desalination 269 (2011) 214−222.

[84]   W.-L. Chou, Y.-H. Huang, Electrochemical removal of indium ions from aqueous solution using iron electrodes, J. Hazard. Mater. 172 (2009) 46−53.

[85]   I. Kabdaslı, I. Arslan-Alaton, T. Ölmez-Hancı, O. Tünay, Electrocoagulation applications for industrial wastewaters: a critical review, Environ. Technol. Rev. 1 (1) (2012) 2−45.

[86]   M. Nasrullah, M.N.I. Siddique, A.W. Zularisam, Effect of high current density in electrocoagulation process for sewage treatment, Asian J. Chem. 26 (14) (2014) 4281−4285.

[87]   A. Ciblak, X. Mao, I. Padilla, D. Vesper, I. Alshawabkeh, A. Alshawabkeh, Electrode effects on temporal changes in electrolyte pH and redox potential for water treatment, J. Environ. Sci. Health. Tox. Hazard. Subst. Environ. Eng. 47 (5) (2012) 718−726.

[88]   O. Tünay, I. Kabdslı, D. Orhon, E. Ates, Characterization and pollution profile of leather tanning industry in Turkey, Water Sci. Technol. 32 (1995) 1−9.

[89]   J. Feng, Y. Sun, Z. Zheng, J. Zhang, S. Li, Y. Tian, Treatment of tannery wastewater by electrocoagulation, J. Environ. Sci. 19 (2007) 1409−1415.

[90]   Q. Zuo, X. Chen, W. Li, G. Chen, Combined electrocoagulation and electroflotation for removal of fluoride from drinking water, J. Hazard. Mater. 159 (2−3) (2008) 452−457.

[91]   M. Kobya, U. Gebologlu, F. Ulu, S. Oncel, E. Demirbas, Removal of arsenic from drinking water by the electrocoagulation using Fe and Al electrodes, Electrochim. Acta 56 (2011) 5060−5070.

[92]   C. Ucar, M. Baskan, A. Pala, Arsenic removal from drinking water by electrocoagulation using iron electrodes, Korean J. Chem. Eng 30 (10) (2013) 1889−1895.

[93]   S. Monasterio, F. Dessi, M. Mascia, A. Vacca, S. Palmas, Electrochemical removal of Microcystis Aeruginosa in a fixed bed reactor, Chem. Eng. Trans. 41 (2014) 163−168.

[94]   Y.A. Taweel, E. Nassef, I. Elkherriany, D. Sayed, Removal of Cr(VI) ions from waste water by electrocoagulation using iron electrode, Egypt. J. Pet. 24 (2015) 183−192.

[95]   M. Vepsäläinen, J. Selin, P. Rantala, M. Pulliainen, H. Särkkä, K. Kuhmonen, A. Bhatnagar, M. Sillanpää, Precipitation of dissolved sulphide in pulp and paper mill wastewater by electrocoagulation, Environ. Technol. 32 (12) (2011) 1393−1400.

[96]   M. Emamjomeh, M. Sivakumar, Fluoride removal by a continuous flow electrocoagulation reactor, J. Environ. Manag. 90 (2) (2009) 1204−1212.

[97]   V. Khatibikamal, A. Torabian, F. Janpoor, G. Hoshyaripour, Fluoride removal from industrial wastewater using electrocoagulation and its adsorption kinetics, J. Hazard. Mater. 179 (1−3) (2010) 276−280.

[98]   J.-W. Feng, Y.-B. Sun, Z. Zheng, J.-B. Zhang, S. Li, Y.-C. Tian, Treatment of tannery wastewater by electrocoagulation, Journal of Environmental Sciences 19 (2007) 1409−1415.

[99]   F. Hanafi, O. Assobhei, M. Mountadar, Detoxification and discoloration of Moroccan olive mill wastewater by electrocoagulation, J. Hazard. Mater. 174 (2010) 807−812.

[100] K. Sadeddin, A. Naser, A. Firas, Removal of turbidity and suspended solids by electrocoagulation to improve feed water quality of reverse osmosis plant, Desalination 268 (1−3) (2011) 204−207.

[101] H. Hossini, A. Rezaee, Optimization of nitrate reduction by electrocoagulation using response surface methodology, Health Scope 3 (3) (2014) e17795.

[102] X. Li, J. Song, J. Guo, Z. Wang, Q. Feng, Landfill leachate treatment using electro-coagulation, Procedia Environ. Sci. 10 (2011) 1159−1164.

[103] J. Lakshmi, G. Sozhan, S. Vasudevan, Recovery of hydrogen and removal of nitrate from water by electrocoagulation process, Environ. Sci. Pollut. Res. 20 (2013) 2184−2192.

[104] S. Gao, J. Yang, J. Tian, F.T.G. Ma, M. Du, Electro-coagulation−flotation process for algae removal, J. Hazard. Mater. 177 (2010) 336−343.

[105] C. Ricordel, C. Miramon, D. Hadjiev, A. Darchen, Investigations of the mechanism and efficiency of bacteria abatement during electrocoagulation using aluminum electrode, Desalination Water Treat. 52 (2014) 5380−5389.

[106] V. Kolesnikov, V. Kudryavtsev, Theoretical & applied aspects of using electroflotation method for wastewater treatment in the surface finishing industry, in: Proceedings of the Aesf Annual Technical Conference, Baltimore, 1995.

[107] Z. Zhu, T. Li, J. Lu, D. Wang, C. Yao, Characterization of kaolin flocs formed by poly-acrylamide as flocculation aids, Int. J. Miner. Process 91 (2009) 94−99.

[108] B. Derjaguin, S. Dukhin, Theory of flotation of small and medium-size particles, Prog. Surf. Sci. 43 (1−4) (1993) 241−266.

[109] S.G. da Cruz, A.J. Dutra, M. Monte, The influence of some parameters on bubble average diameter in an electroflotation cell by laser diffraction method, J. Environ. Chem. Eng. 4 (3) (2016) 3681−3687.

[110] C. Jimenez, B. Talavera, C. Saez, P. Canizares, M. Rodrigo, Study of the production of hydrogen bubbles at low current densities for electroflotation processes, J. Chem. Technol. Biotechnol. 85 (2010) 1368−1373.

[111] C. Llerena, J. Ho, D. Piron, Effect of pH on electroflotation of sphalerite, Chem. Eng. Commun. 155 (1) (1996) 217−228.

[112] G. Bshaskar, P. Khangaonkar, Electroflotation - a critical review, Trans. Indian Inst. Met 37 (1) (1984) 59−66.

[113] I. de Oliveira da Mota, J. de Castro, R. de Góes Casqueira, A. de Oliveira Junior, Study of electroflotation method for treatment of wastewater from washing soil contaminated by heavy metals, J. Mater. Res. Technol. 4 (2) (2015) 109−113.

[114] R. Bande, B. Prasad, I. Mishra, K. Wasewar, Oil field effluent water treatment for safe disposal by electroflotation, Chem. Eng. J. 137 (3) (2008) 503−509.

[115] K. Tumsri, O. Chavalparit, 2011 2nd international conference on environmental science and technology, in: Optimizing Electrocoagulation-electroflotation Process for Algae Removal, Singapore, 2011.

[116] F. Baierle, D. John, M. Souza, T. Bjerk, M. Moraes, M. Hoeltz, A. Rohlfes, M. Camargo, V. Corbellini, R. Schneider, Biomass from microalgae separation by electroflotation with iron and aluminum spiral electrodes, Chem. Eng. J. 267 (2015) 274−281.

[117] A. Khelifa, S. Moulay, A. Naceur, Treatment of metal finishing effluents by the electro-flotation technique, Desalination 181 (2005) 27−33.

[118] C. Poon, Electroflotation for groundwater decontamination, J. Hazard. Mater. 55 (1997) 159−170.

[119] B. Merzouk, K. Madani, A. Sekki, Using electrocoagulation−electroflotation technology to treat synthetic solution and textile wastewater, two case studies, Desalination 250 (2) (2010) 573−577.

[120] C. Ricordel, A. Darchen, D. Hadjiev, Electrocoagulation−electroflotation as a surface water treatment for industrial uses, Sep. Purif. Technol. 74 (3) (2010) 342−347.

[121] X. Yunqing, L. Jianwei, Application of electrochemical treatment for the effluent from marine recirculating Aquaculture systems, Procedia Environ. Sci. 10 (2011) 2329−2335.

[122] Y.-J. Liu, S.-L. Lo, Y.-H. Liou, C.-Y. Hu, Removal of nonsteroidal anti-inflammatory drugs (NSAIDs) by electrocoagulation–flotation with a cationic surfactant, Sep. Purif. Technol. 152 (25) (2015) 148–154.

[123] A. Guldhe, R. Misra, P. Singh, I. Rawat, F. Bux, An innovative electrochemical process to alleviate the challenges for harvesting of small size microalgae by using non-sacrificial carbon electrodes, Algal Research 19 (2016) 292–298.

[124] M. Belkavem, M. Khodir, S. Abdelkrim, Treatment characteristics of textile wastewater and removal of heavy metals using the electroflotation technique, Desalination 228 (2008) 245–254.

[125] E. Maksimov, A. Ostsemin, Intensifying the cleaning of emulsion- and oil-bearing waste water from rolled-product manufacturing by electroflotation, Metallurgist 58 (11–12) (2015) 945–949.

[126] K. Mohammadi, M. Malayeri, Curved electrodialysis membranes: an innovative approach to enhance ion separation in EDMEM stacks, Desalination Water Treat. 57 (29) (2016) 13367–13376.

[127] W. Juda, W. McRae, Coherent ion-exchange gels and membranes, Am. Chem. Soc. J. 72 (2) (1950) 1044.

[128] W. Juda, Construction of cells for electrodialysis. USA Patent US 2741595 A, 10 April 1956.

[129] P. Cohen, Membrane electrodialysis of simulated pressurized water reactor coolant, Ind. Eng. Chem. 51 (1) (1959) 66–67.

[130] K. Kontturi, A. Ekman, P. Forssell, A method for determination of transport numbers in ion exchange membranes, Acta Chemica Scandinavica A 39 (1985) 273–277.

[131] T. Franken, Bipolar membrane technology and its applications, Membr. Technol. 125 (2000) 8–11.

[132] B. Pilat, Practice of water desalination by electrodialysis, Desalination 139 (2001) 385–392.

[133] V. Shaposhnik, N. Zubets, I. Strygina, B. Mill, Depassivation of ion-exchange membranes in electrodialysis, Russian J. Appl. Chem. 74 (10) (2001) 1653–1657.

[134] Y. Tanaka, A computer simulation of continuous ion exchange membrane electrodialysis for desalination of saline water, Desalination 249 (2) (2009) 809–821.

[135] M. Sadrzadeh, T. Mohammadi, Treatment of sea water using electrodialysis: current efficincy evaluation, Desalination 249 (2009) 279–285.

[136] J. Krol, Monopolar and Bipolar Ion Exchange Membranes. Mass Transport Limitations, Grootegast, Print Partners Ipskamp, Enschede, 1997.

Chapter 3

Emerging and Combined Electrochemical Methods

Mika Sillanpää, Marina Shestakova

Lappeenranta University of Technology, Lappeenranta, Finland

NOMENCLATURE

Latin Alphabet

x_i	Exchange capacities of i-ion	mol/g
K	Concentration exchange constant	
a_{ei}	Activity of i-ion in electrolyte solution at equilibrium	mol/L
K_d	Weight distribution coefficient	mL/g
K_M	Michaelis–Menten constant	mol/L
α	Separation factor	
U_e	Mean interstitial velocity	m/h
A	Cross-sectional/surface area	m^2
U	Superficial speed of the fluid	m/h
U_V	Voltage	V
U_T	Thermal voltage	V
U_{cell}	Cell voltage	V
$\varepsilon_{V,avg}$	Average void fraction of the packed ion-exchange resin	
ε_V	Void fraction of resin in diluate compartment	
ε	Relative dielectric permittivity	
ε_0	Electrical permittivity of vacuum	$8.854 \cdot 10^{-12}$ F/m
V_t	Total volume of the diluate chamber	m^3
V_s	Total volume taken by the ion-exchange resin in diluate compartment	m^3
V	Volume	m^3
V_{MFC}	Working volume of MFC reactor	m^3
ρ	Density	kg/m^3
A_{sp}	Specific surface area	cm^2/g
$x(r)$	Fraction of resin beads with radius r	
r	Radius	m
λ_i	Ionic conductivity of i substance	1/(Ohm cm)
λ	Light wavelength	m

Electrochemical Water Treatment Methods. http://dx.doi.org/10.1016/B978-0-12-811462-9.00003-7

d	Diameter of ultrasonic transducer	m
D_i	Diffusion coefficient	m^2/s
$C_{i,v}$	Volumetric concentration of current carriers	mmol/mL
C_b	Bulk concentration	
E	Photon energy	eV
E_F	Fermi energy	eV
E_{gap}	Band gap energy	eV
E_{redox}	Reduction potential	V
E^0	Standard redox reaction potential	V
E_{cell}	Cell potential	V
$\overline{E_{cell}}$	Mean cell potential	V
$\Delta\overline{E}$	Change of potential through ion exchanger bed	V
I_{stack}	EDI module ("stack") current	A
$R_{solution}$	Resistance of interstitial solution	Ohm
$\Delta R_{solution}$	Changes in resistance of interstitial solution	Ohm
R_{resins}	Resistance of ion-exchange resins	Ohm
ΔR_{resins}	Changes in resistance of ion-exchange resins	Ohm
$\phi_{water\ splitting}$	Potential of water splitting reactions	V
$\Delta\phi_{water\ splitting}$	Changes in potential of water splitting reactions	V
c	Speed of sound	m/s
c_l	Speed of light	$c_l = 3.00 \cdot 10^8$ m/s
C_0	Initial/inlet concentration of substances	mol/L, mg/L
C_t	Final/outlet concentration of substances	mol/L, mg/L
C_i	Concentration of ions	mol/m^3
$C_{mi,l}$	Concentration of ions in micropores	mol/m^3
C_{mA}	Concentration of ions in macropores	mol/m^3
$C_{ions,mi}^{ch}$, $C_{ions,mi}^{disch}$	Ion concentration in micropores at the end of charge and discharge steps in the CDI process	mol/m^3
C_{out}, C_{inflow}	Output and inflow salt concentration in the spacer channel	mol/m^3
C_{ox}, C_{red}	Concentrations of oxidized and reduced forms	
$\overline{C_{sp}}$	Average concentration of ions in the spacer channel	mol/m^3
$C_{T,m}$	Total concentration of counter- and co-ions	mol/m^3
C	Capacitance	F
C_T	Total electrical capacity of the interface	$\mu F/cm^2$
C_{E-H}	Electrical capacity of the compact double layer	$\mu F/cm^2$
C_{H-S}	Electrical capacity of the diffuse layer	$\mu F/cm^2$

$C_{E-H,vol}$	Volumetric capacitance of the Stern layer	F/m^3
Q	Flow rate	L/min
Q_r	Reaction quotient	
q	Charge	C
F	Faraday constant	96485.33289(59) C/mol
ΔG	Gibbs free energy	J/mol
M	Molar mass	g/mol
I	Current	A
I_a	Acoustic intensity	W/m^2
z_i	Valence of i-species	
R	Gas constant	8.314 J/K mol
T	Temperature	K
T_a	Sound wave period	1/Hz
ϕ	Electric potential	V
ϕ_δ	Stern potential/potential at OHP	V
dx	Distance	m
μ	Chemical potential	J/mol
$\overline{u_i}$	Ion mobility in the bed of ion exchanger	$m^2/V\ s$
t	Time	h
t_{HRT}	Hydraulic retention time	h
W_{max}	Maximum energy, which can be stored in capacitors	J
d	Thickness/distance	m
γ	Interfacial surface tension of solution	J/m^2
m	Mass of electrode material	kg
ν	Scan rate	mV/s
N	Near field distance	m
Z''	Imaginary part of impedance	Ohm
ω	Angular frequency of the applied AC signal	rad/s
f	Sound wave frequency	Hz
μ_{att}	Excess attractive chemical potential	
μ	Specific growth rate of microorganisms	h^{-1}
μ_{max}	Maximum specific growth rate of the microorganisms	h^{-1}
$\Delta\phi_D$	Donnan potential	
$\Delta\phi_m$	Potential difference across the membrane	V
ϕ_{mA}	Dimensionless potential in the macropores	
ϕ_1	Potential in the electrode matrix	V
$\Delta\phi_{sp}$	Potential difference across a half of a spacer channel	V
σ_{mi}	Micropore charge density	mol/m^3

σ_{mi}^{ch}, σ_{mi}^{disch}	Micropore charge densities at the end of charge and discharge steps respectively in the CDI process	mol/m^3
Γ_{salt}	Adsorption	mg/g or mol/g
$\sum F$	Charge density	C/g
V_{mi}	Volume of micropores	mL/g
Λ	Theoretical charge efficiency	
J_i	Total flux of ions	$mol\ s/m^2$
p_{mA}, p_{meso}, p_{mi}	Macro-, meso-, and microporosity of the electrode	vol.%
p_m	Membrane porosity	vol.%
R_{EE}	External electronic resistance	$\Omega\ m^2$
R_{sp}	Resistance in the spacer channel	$\Omega\ m^2$
R_m	Ionic resistance of membranes	$\Omega \cdot m^2$
Φ_v	Feeding water flow rate through the cell	m^3/h
ν_{sp}	Transport coefficient of salt from outer volume into the spacer channel	mol/s
ν_l	Frequency of light	Hz
L_{sp}	Thickness of the spacer channel	m
θ	Wetting angle	
ϑ	Rate of product formation	mol/min
ϑ_{max}	Maximum rate of product formation	mol/min
$[S]$	Substrate concentration	mol/L
S	Synergetic effect	
η_c, η_a, η_{ohm}	Concentration, activation and ohmic resistance losses	V
P	Power	W
h	Plank constant	$4.13 \cdot 10^{-15}$ eV s
k	Boltzmann constant	$1.38 \cdot 10^{-23}$ J/K
k_{EO}, k_{US}, k_{SEC}	Rate constants in EO, US and SEC degradation processes	1/min
p	Sound pressure	Pa
p_g, p_v	Pressure of gas and vapor inside the cavitation bubble, respectively	Pa
p_{hs}	Hydrostatic pressure	Pa
ν	Sound particle velocity	m/s

Abbreviations

AM	Anion-exchange membrane
AR	Anion resin
BDD	Boron-doped diamond electrode
CB	Conduction band
CC	Constant current
CE	Current efficiency
CEDI	Continuous electrodeionization
CV	Constant voltage
CM	Cation-exchange membrane
CR	Cation resin
CDI	Capacitive deionization
COD	Chemical oxygen demand

DC	Direct current
DR	Desalination ratio
EDI	Electrodeionization
EFC	Enzymatic fuel cells
EC	Energy consumption
EDLs	Electrical double layers
EO	Electrochemical oxidation
EF	Electro-Fenton
FB	Flow between or flow by process
FTECDI	Flow-through electrode capacitive deionization
GDE	Gas diffusion electrode
M/E	Membrane/electrode interface
MFC	Microbial fuel cell
MCDI	Membrane capacitive deionization
OHP	Outer Helmholtz plane
PEC	Photoelectrocatalysis
RVD	Reverse-voltage desorption
S/M	Membrane/separator interface
SCL	Space charge layer
SEC	Sonoelectrocatalysis
SHE	Standard hydrogen electrode
TDS	Total dissolved substances
TOC	Total organic carbon
US	Ultrasound
VB	Valence band
ZVD	Zero-voltage desorption

3.1 ELECTRODEIONIZATION

Electrodeionization (EDI) is an electrical water treatment process, which combines the properties of ion-exchange resins and ion-selective (ion-exchange) membranes to deionize water. In fact, the process is conducted into an "electrodialyzer," compartments of each are filled with ion-exchange resins, thus providing a synergic effect of electrodialysis and ion-exchange processes for even deeper water demineralization. As it was mentioned in the second chapter, electrodialysis is not effective for demineralization of water with low salt content, since the high electrical resistance of water leads to a significant increase of power consumption and reduces the efficiency of the process. Ion-exchange resins incorporated into compartments of EDI unit improve ion mobility and provide additional electrical path for ions due to the enhanced electrical conductivity, thus enhancing the current efficiency (CE) of the process. This allows to obtain large amounts of high-purity water (up to 0.1 μS/cm) without significant expenditure of reagents required for the regeneration of ion-exchange material. EDI finds applications in pharmaceuticals, electronics, power, automotive, and other sectors. The efficiency of the method depends mainly on the initial concentration of contaminants in water, flow rate of permeate in the system, preliminary pretreatment steps, working pressure in the system (usually in the range between 1.5 and 4 atm), and the value of the applied current density or potential difference. Nevertheless, the rate of ion transfer in the cross section of the layer and the speed of continuous electroregeneration of resin are the crucial parameters, which affect the efficiency of EDI unit. The rate of ion mass transfer is limited by diffusion of the counterions from water to the surface of the resin and co-ions from the resin in the water. The influence of diffusion processes can be decreased by reducing the applied current density. To keep the efficiency of EDI process high, the surface area of resin should be high. In practice, this is achieved by limiting the thickness of the diluate compartments. EDI allows removal of weakly ionized species such as silica, CO_2, boron, and ammonia from water. For example, efficiency of the removal of total dissolved substances (TDS) from water can reach up to 99% depending on the composition of organic substances in water and pretreatment steps. In average, EDI consumes 0.05—0.08 kWh for cubic meter of product water depending on the feed water quality [1].

The main advantages of EDI process are as follows:

- No chemicals required for resin regeneration, which is opposite for example to a single ion-exchange process using caustic soda, sulfuric acid, and hydrochloric acid for resin regeneration.
- High-quality water is achievable.
- Cost-effective process in terms of maintenance cost because there is no need to change the resin.

Costs of production of 1 m³ of deionized water by EDI process is about 20%—25% lower compared with the ion-exchange process.

- Efficient removal of >90% is achievable for TDS, silica, CO_2, etc.
- Easy automated process.
- A nonstop process due to the simultaneous and continuous treatment of water and regeneration of resin.
- Constant quality of product water in a contrast to the ion-exchange process where quality of treated water is getting worse by the end of the filtration cycle.

Along with the advantages of EDI a number of disadvantages can be identified. Because EDI is a combination of ion exchange and electrodialysis, the process combines the disadvantages of these two methods, which are as follows:

- Formation of concentrate solution, which requires regeneration.

Composition of the concentrate solution is directly related to the original composition of feed water. The main problem of concentrate recycling is related to defluorination of the solution, which is especially prominent when artesian water from the deep aquifers is used as the feed solution. Other contaminants are usually removed efficiently by conventional water treatment methods or their concentration is not exceeding the maximum allowable concentrations.

- High fluctuations of pH in compartments, which do not contain resin.
- Pretreatment of feed solution is required.
- Limitation and process efficiency dependence on the feed water quality.

Operation temperature should be in the range between 5 and 45°C; electrical conductivity below 60 $\mu S/cm$; CO_2 content should not exceed 5 mg/L; chlorine, iron (III), and calcium content should be below 10, 50, and 500 $\mu g/L$, respectively; and inlet pressure below 5.5 atm. Moreover, the presence of microorganism in feed water can lead to resin fouling. Therefore, water should be pretreated and rinsing of the system with peracetic acid, sodium hydroxide, and others is required from time to time.

Usually, only 10%−20% of the applied power is consumed on the transport of salt ions and the rest goes to split the water. Such a low efficiency of power consumption contributes to the fact that EDI process becomes very efficient when the total salt content of feed water does not exceed 100 mg/L. In addition, the probability of calcium carbonate deposits on the surface of anion-exchange membranes should be considered when calcium ions are present in the feed water. Typically, EDI systems are operated with 95% efficiency of feed water use, i.e., 95% of the initial water is a product (permeate) and only 5% is discharged in the drainage (concentrate). If the content of free carbonic acid in the feed water is greater than 5 mg/L and calcium is above 0.5 mg/L, the ratio of permeate/concentrate is lowered so as to prevent concentration polarization effect and the formation of calcium carbonate scale. This allows lowering the calcium content in the concentrate and, as a result, the amount of calcium that enters the anion-exchange membrane from the concentrate (selectivity of anion-exchange membranes to reject cations never reaches 100% and some part of calcium ions present in the water enter the membrane). The presence of other multivalent metal ions such as magnesium, iron, and aluminum also contributes to the clogging of the EDI chamber [2].

- Impossibility to remove uncharged molecules and organic compounds
- Membrane passivation and polarization
- Periodical membrane replacement

Simple EDI unit consists of three compartments Fig. 3.1 where the middle compartment is separated by the ion-exchange membranes. Similar to the electrodialysis process, anion-exchange membrane is placed on the side of anode, thus forming anodic concentrate compartment, and cation-exchange membrane is placed on the side of cathode, thus forming cathodic concentrate/permeate compartment. The space formed between two membranes is diluate compartment, where water purification takes place. Depending on the

FIGURE 3.1 Working principle of electrodeionization unit. *AM*, anion-exchange membrane; *AR*, anion resin; *CM*, cation-exchange membrane; *CR*, cation resin.

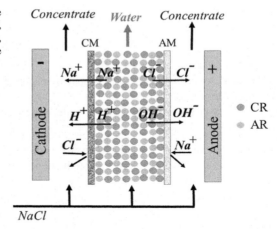

water composition and desired water quality to be obtained, either all three compartments or only the middle compartment is filled with a mixture of cation and anion resins. Contaminated water can be supplied either to all compartments or only to the middle one. In that case near-electrode compartments are rinsed with a brine solution. Let us consider the working principle of water deionization on an example of sodium and chlorine ions removal by simple EDI unit where only diluate compartment is filled with resin (Fig. 3.1).

Contaminated water is supplied to all compartments, two flows of water pass through the concentrate compartments, and one is supplied to the purification compartment. Mixed anion- and cation-exchange resin adsorbs dissolved ions in the diluate compartment. When potential difference is applied between electrodes, it enables a continuous movement of ions through the ion-exchange resin to the electrodes toward the appropriate electrode. Anions move through the anion-exchange membrane toward the anode and cations pass the cation-exchange membrane and move toward the cathode. As a result, water in the middle compartment becomes clean and ions are concentrated in the near-electrode compartments producing two flows of high-salinity water, which should be utilized. Simultaneously to the movement of ions in EDI unit under DC, water splitting onto hydrogen and hydroxide ions takes place in the resin compartment on the surface of resin beads [3]. Produced H^+ cations and OH^- anions due to their high mobility are exchanged with the adsorbed ions, thus allowing continuous regeneration of the resin. Hydrogen and hydroxyl ions, which are not entered into the exchange reaction with resins, are transported to the concentrate flow together with the dissolved salts and undergo recombination into water molecules. The process of resin regeneration can be intensified by either intensifying the hydrolysis of water molecules or increasing the conductivity in the concentrate compartments. The rate of water hydrolysis is controlled by resin and ion-exchange membrane properties as well as by the thickness of the resin layer. Conductivity of solution in the concentrate compartments is achieved when filling these compartments with the resin or dosing strong electrolyte solution in them. Because the resin is regenerated continuously when electric current is applied, EDI process can be conducted in a nonstop operation mode. The process performed in such mode is called continuous electrodeionization or CEDI.

The mechanism of EDI can be divided into three stages:

- Ion exchange between ion-exchange resin and solution
- Transfer of ions through the resin under applied electric field
- Electrolytic regeneration of resin

The Principles of Ion-Exchange: Ion-Exchange Resins

Soil scientists were the first who drew attention to the ion-exchange process. When electrolyte solution is passed through the soil, a part of ions from solution are absorbed by the soil, whereas other ions are leached into solution. Soils are mostly exchanged with cations. Ion exchange takes place in a strictly equivalent amount, that is, the amount of absorbed gram equivalents of elements is equal to the amount of gram equivalents of elements washed out/desorbed from the soil. Adsorbability of different ions is different. The higher the charge of the ions, the better they are adsorbed due to the greater attraction. Moreover, adsorbability also depends on the solvation shell of ions, and the smaller the solvation shell, the better the ion adsorbed by the resin.

The theory of ion-exchange was developed by Russian physical chemist Boris Nikolsky in 1930. The following relation was obtained for adsorption of two ions from solution:

$$\frac{x_1^{\frac{1}{z_1}}}{x_2^{\frac{1}{z_2}}} = K \frac{a_{e1}^{\frac{1}{z_1}}}{a_{e2}^{\frac{1}{z_2}}} \tag{3.1}$$

where x_1 and x_2 are exchange capacities of the first and second ions (equilibrium amounts of ions adsorbed on a unit of mass of a resin; g eq^{-1}/g or mol/g); z_1 and z_2 are charge (valence) of these ions; K is the concentration exchange constant; and a_{e1} and a_{e2} are activities of ions in electrolyte solution at equilibrium (mol/L). Activities can be replaced by concentrations. Exchange constant K depends on many parameters such as temperature of the process, size of ions, and their hydration. The higher the charge of the ion and the higher the atomic number of the element, the higher the exchange constant.

Ion-exchange resins are solid particulate materials substantially insoluble in water and common solvents, which contain active (ionic) group of a basic or acidic nature with mobile ions. In general, ion-exchange resins are used in the processes of water softening and desalination, isolation of rare metal, regeneration of metal wastes, purification and separation of different substances, and as a catalyst for organic synthesis. Ion-exchange allows changing the ionic composition of the treated liquid without changing the total number of charges present in the liquid before the exchange process.

Ion-exchange resins can be classified according to the following types:

- Ion charge
  - Cation-exchange resins having formula of R—C, where R is a polymer resin matrix with a rigidly fixed ion on its structure. In the case of cation-exchange membranes, permanently fixed ions carry negative charge (Fig. 3.2). C is a countercation (for example, Na^+, H^+), which is mobile and able to be exchanged with other cations found in water. Depending on the degree of ionization, cation-exchange resins can be divided into strong and weak acid resins. Strong acid cation-exchange resins behave as strong acids and can contain the following functional (ionic) groups in their structure such as sulfonic $-SO_3H$ or phosphorous $-PO(OH)_2$ acid groups. Weak acid resins are for example resins with carboxylic $-COOH$ and phenolic $-C_6H_5OH$ groups. Strong cationic resins are capable to exchange ions (dissociate) at any pH (0—14), while weak acid cation resins dissociate at a pH > 7. Weakly acidic resins are slightly ionized and do not interact with neutral salt such as NaCl.
  - Anion-exchange resins having formula of R—A, where R is a polymer resin with a rigidly attached cations in its structure, which give a positive net charge to the resin and are stabilized with counteranions, A (Fig. 3.2). Anion-exchange resins also can be divided into strong and weak base anion-exchange resins. Strong base anion-exchange resins contain usually hydroxide $-OH$ group, a quaternary ammonium base $-NR_3OH$, and weak basic primary $-NH_2$ and secondary amino group $-NHR$. Strong base anion-exchange resins can dissociate over the entire

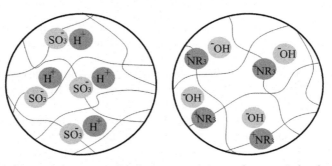

Cation-exchange resin bead   *Anion-exchange resin bead*

FIGURE 3.2 Schematic representation of cation-exchange and anion-exchange resin beads.

pH range (0−14), while weak base resins only at pH < 7. Weakly basic resins are not ionized and do not interact with neutral salts. However, ion-exchange reaction has become possible when weakly basic resins are protonized with strong acid.

- Structure of the resin matrix
  - Gel-type structure. Gel-type resins contain no true pores and can exchange ion only when swelled. To achieve this state, resins are placed in water for a while. The pore size in this type of resins is about 1 nm.
  - Macroporous structure. Ion-exchange resins with macroporous structure have large number of pores on the surface of the resin, which contribute to the ion-exchange process. The pore size of such resins is about 100 nm.
  - Intermediate structure. Resins of this group have intermediate properties of gel and porous structure type of resins.

    A distinctive feature of ion-exchange resins with different structures is that the structure of the gel-type resins has higher exchange capacity than macroporous resin structure. In turn, ion-exchange resins with a porous structure have better osmotic, chemical, and thermal stabilities compared to the gel-type resins. This means that porous resins can remove impurities at a virtually any feed water temperature.

- Type of the resin matrix

  Industrial ion-exchange resins are made of polymer materials. Approximately 90% of commodity synthetic ion-exchange resins are obtained by reaction of copolymerization of polystyrene or polyacrylate with divinyl benzene. Therefore, the majority of ion-exchange resins usually have either polystyrene or polyacrylic matrix.

- Bead size

  The shape and size of the resin beads should ensure effective contact with the treated water in the absence of excessive pressure drop. Resin beads can have either polydispersed (beads size in the range of 0.3−1.2 mm) or monodispersed (0.5−0.6 mm) particle size distribution. In general, polydispersed resins are more preferable to use. The presence of too small beads in the resin can cause their carryover from a resin bed while conducting backwashing. This results in a decrease of exchange capacity of the resin bed. Moreover, the absence of large beads provides greater surface area to volume ratio of resin and, as a result, better exchange kinetics and more efficient backwashing.

  To describe the equilibria of ion-exchange process, two main parameters are used. They are weight distribution coefficient (K_d) and separation factor (α).

$$K_d = \frac{x}{C_e} \tag{3.2}$$

where C_e is the concentration of metal ion remained in the electrolyte solution after equilibrium was established.

$$\alpha = \frac{K_{d1}}{K_{d2}} \qquad (3.3)$$

where K_{d1} and K_{d2} are distribution coefficients for two different ions.

The Main Features of Electrodeionization Process

Usually EDI is conducted in a unit consisted of many compartments separated by alternating anion- and cation-exchange membranes similar to electrodialyzers. Each even compartment of EDI unit is the diluate (purifying) compartment and every odd compartment is a concentrate one. Diluate compartment are always filled with a mixture of anion- and cation-exchange resin, whereas concentrate compartment can be both filled with and free of resin. Feed solution is usually fed to all compartments. However, when concentrate compartments are not filled with the resin, feed solution can be separately supplied to diluate and electrode compartments and brine solution is injected to concentrate compartments and plays the role of additional transport solvent assisting removal of weaker electrolytes. Fig. 3.3 shows a typical EDI unit used for water demineralization. Let us consider the similar case of sodium chloride removal from water. All sections of the unit are filled with resin and permeate is supplied to all compartments. The ion-exchange resin on both sides of the membrane enhances the transport of anions and cations through the membrane. This is explained by a large concentration gradient due to the adsorption of transported ions by resin. Moreover, this eliminates the step of brine supply to the concentrate and electrode compartment and simplifies the

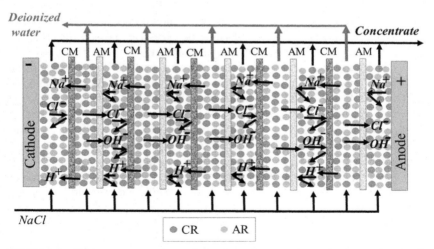

FIGURE 3.3 Multicompartment electrodeionization unit. *AM*, anion-exchange membrane; *AR*, anion resin; *CM*, cation-exchange membrane; *CR*, cation resin.

process. When a potential difference is applied between electrodes, Na^+ cations from a diluate compartment move toward the cathode through a cation-exchange membrane passing into the next concentrate compartment. Similarly, Cl^- anions passing through anion-exchange membranes move toward the anode and enter the concentrate compartment next to the left section. The further movement of ions is limited because the cation-exchange membrane prevents the flow of anions toward the anode and anion-permeable membrane prevents the flow of cations toward the cathode. As a result, the ions are concentrated in concentrate compartments, from which they are discharged, and high-quality deionized water is obtained in diluate compartments.

In the case when concentrate compartments are not filled with the resin, the intensity of electroregeneration is limited by the conductivity of the concentrate solution. The conductivity of solution is directly connected with the amount of salt in the concentrate solution, which should be maintained at a certain level. From another side, mobility of ions is directly related to the value of applied current. The higher the current generated in EDI cell, the faster the velocity of transported ions. Because velocity of ions inside the membrane is greater than that in solution, enhanced electric current (limiting current, I_{lim}) can lead to the depletion of charge carriers at the membrane/solution interface. This will result in dissociation of water to H^+ and OH^-, which start to compensate the lacking charge carriers and also result in decreased efficiency of the process. In this regard, EDI unit is usually operated at the current value equal to 80% of I_{lim} [4]. Moreover, to improve conductivity, concentrate solution is often partially recirculated through the system (Fig. 3.4).

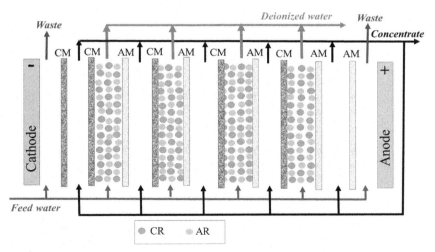

FIGURE 3.4 Multicompartment electrodeionization unit with a partial concentrate recirculation. *AM*, anion-exchange membrane; *AR*, anion resin; *CM*, cation-exchange membrane; *CR*, cation resin.

Recirculation of the concentrate also assists the better removal of the weaker electrolytes, allowing reaching the desired quality of product water. Taking into account that cathodic compartment becomes enriched with a small amount of hydrogen gas and anodic compartment becomes contaminated with oxygen and chlorine gas during electrode reactions (3.4–3.6), solution formed in the near-electrode compartments is often separated from the overall concentrate flow and collected separately to prevent membrane deterioration caused by the electrode gases.

Anode:

$$2H_2O \rightarrow O_2 + 4H^+ + 4e^- \tag{3.4}$$

$$2Cl^- \rightarrow Cl_2 + 2e^- \tag{3.5}$$

Cathode:

$$2H_2O + 2e^- \rightarrow H_2 + 2OH^- \tag{3.6}$$

The rate of concentrate solution recirculation and the share of its discharge to the drainage should be at a value preventing sedimentation in the concentrate compartment.

As mentioned earlier, water splitting during EDI process occurs in the places where beads of ion-exchange resin are in contact with each other and with the membranes, i.e., in places with the highest concentration overpotential. Under the action of electric current, hydrogen ions do not react with the resin, move toward the cathode, and react with bicarbonates anions, resulting in a carbon dioxide formation.

$$H^+ + HCO_3^- \rightarrow CO_2 + H_2O \tag{3.7}$$

Excessive hydroxyl ions react with bicarbonate anions resulting in the formation of carbonate ions. Formed carbonate ions react with calcium forming calcium carbonate deposits between the resin beads and on the surface of anion-exchange membranes.

$$OH^- + HCO_3^- \rightarrow CO_3^{2-} + H_2O \tag{3.8}$$

$$CO_3^{2-} + Ca^{2+} \rightarrow CaCO_3 \downarrow \tag{3.9}$$

When feed water is free of calcium, then hydrogen ions freely pass through the cation-exchange membrane and hydroxyl ions pass through the anion-exchange membrane into concentrate compartments where they recombine into water molecules. Therefore, to avoid the formation of hardness deposits on the membrane surface processes and minimize the energy consumption, EDI units should be operated avoiding excessive applied voltages and currents.

Currently there is no method allowing resins washing. Moreover, repairing of EDI cells is very difficult, because the cells are filled with resin during assembly. There is also no effective method for unloading and loading the resin. EDI cell must first be disassembled, and the resin layer is removed and again reassembled. This process inevitably leads to the damage of the membranes. Therefore, feed water supplied to the EDI unit must have a very low level of suspended solids and pretreated for example by ultrafiltration, nanofiltration, or reverse osmosis. Moreover, EDI units should be operated avoiding excessive applied currents and voltages.

Another variation of EDI method, allowing removal of weakly ionized pollutants from a mixture of strong and weak electrolytes, is fractional electrodeionization (FEDI, Fig. 3.5). The method is patented by the QUA Group LLC [5]. FEDI is a two-stage process applying different voltages/currents to remove alternately strongly and weakly ionized pollutants. At first, electrolyte solution is fed to the part of FEDI unit where lower currents/voltages are applied to drive the process. As a result, strongly ionized pollutants such as Ca^{2+}, Mg^{2+}, SO_4^{2-}, Na^+, and Cl^- are removed in this stage. The product water after the first stage passes to the second stage of FEDI unit where elevated voltage/currents that are able to remove weakly ionized impurities such as boron and silica are applied. Highly alkaline conditions of feed water entering the second stage of FEDI unit additionally ensure removal of silica and boron. The removal of weakly ionized compounds can be described by the following reactions [3].

$$SiO_2 + OH^- \rightarrow HSiO_3^- \tag{3.10}$$

$$H_3BO_3 + OH^- \rightarrow B(OH)_4^- \tag{3.11}$$

$$NH_3 + H^+ \rightarrow NH_4^+ \tag{3.12}$$

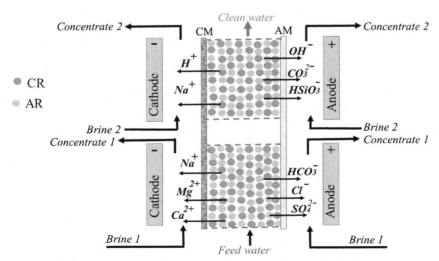

FIGURE 3.5 Working principle of fractional electrodeionization unit. *AM*, anion-exchange membrane; *AR*, anion resin; *CM*, cation-exchange membrane; *CR*, cation resin.

Concentrate solutions from stage 1 and stage 2 are removed separately to avoid precipitation of strong electrolytes in the zone of elevated voltage/current.

EDI units can be used to obtain ultrapure water with resistivity of up to 15−18 MΩ cm when changing the position of ion-exchange membranes inside EDI unit and resin type in different compartments. Higher values of water resistivity is difficult to obtain due to the fact that the efficiency of the cation- and anion-exchange membranes never reaches 100%. Therefore, insignificant amount of the ions manages to migrate from the concentrate flow through the ion selective membrane in the desalinating compartment. Typically, this occurs at the outlet of the water flow from EDI cell, at the time when the ions leave the system with the purified water before they are able to rediffuse into the resin layer and undergo the exchange reaction. Therefore, to obtain water of higher quality, permeate after EDI units should be filtrated through a mixed bed filter containing filtering layer of mixed cation- and anion-exchange resins of nuclear grade.

The only condition, which is always satisfied in EDI units, is that cathode is faced to the cation-exchange membrane and anode is faced to the anion-exchange membrane. Let us consider one of the possible setups suitable for ultrapure water production (Fig. 3.6). EDI unit consists of five compartments, i.e., two electrode compartments, two concentrate compartments, and one diluate compartment. Cathodic compartment is filled with a cation-exchange resin and separated from a concentrate chamber of cathode side by a cation-exchange membrane. Anodic compartment is filled with an anion-exchange resin and separated from the concentrate chamber on anode side by an

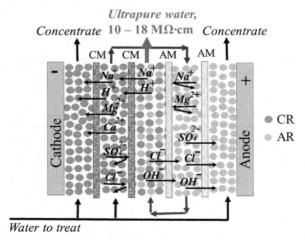

FIGURE 3.6 Electrodeionization setup used for production of ultrapure water. *AM*, anion-exchange membrane; *AR*, anion resin; *CM*, cation-exchange membrane; *CR*, cation resin.

anion-exchange membrane. Concentrate compartments on cathodic and anodic sides are packed with cation and anion resins, respectively. The diluate compartment is filled with a mixture of cation and anion resin and detached from concentrate compartments by cation-exchange membrane on cathode side and anion-exchange membrane on anode side. Water is supplied to the electrode compartments as well as to the concentrate chamber on cathodic side. When potential difference is applied between electrodes, cations from cathodic concentrate chamber move to the cathodic compartment through the cation-exchange membrane. At the same time anions cannot pass to the diluate compartment because they are retained by cation-exchange membrane and washed out with the solution, which is further redirected to the concentrate compartment on the side of anode. Anions pass through the anion-exchange membrane to the anodic chamber. The effluent containing traces of anions and cations, which were not removed in the former two compartments, is fed to the diluate compartment. Traces of ions left in the solution are completely purified in the compartment giving ultrapure water as the final product of EDI.

The mechanism of ions removal in cation-exchange and then anion-exchange resins can be described by the following reactions when using an example of sodium chloride salt removal:

CR:

$$R-H + NaCl \rightarrow R-Na + HCl \qquad (3.13)$$

AR:

$$R-OH + HCl \rightarrow R-Cl + H_2O \qquad (3.14)$$

The mechanism of simultaneous regeneration of resins by H^+ and OH^- ions coming from the water splitting occurring on the surface of resin beads can be represented as follows:

CR:

$$R-Na + H^+ \rightarrow R-H + Na^+ \qquad (3.15)$$

AR:

$$R-Cl + OH^- \rightarrow R-OH + Cl^- \qquad (3.16)$$

Because EDI is usually used as a tertiary treatment method for product water obtained in RO applications, it is often combined with the RO in the same unit. In such cases EDI is placed behind the RO and concentrate solution is usually recovered and fed to the RO part of the unit. This is explained by the fact that quality of concentrate solution is often better than the quality of feeding RO water and enables a significant reduction of wastewater from EDI part of the unit to less than 1%.

Practical Applications of EDI in Reduction of Silica, Carbon Dioxide, TOC, and Other Parameters

The first study on EDI treatment date back to 1955, when EDI reactor was proposed to treat and concentrate radioactive wastes [6] followed by the EDI theory development in the late 1950s and production of high-purity water from tap and brackish water at the beginning of 1970s [7]. Over the past 50 years, the process has been used successfully for water treatment from metal ions, boron, silica, carbon dioxide, total organic carbon (TOC), etc.

Removal of silica can reach up to 95% and carbon dioxide up to 99% in EDI process. A reason for a lower efficiency of silica removal is that the pH of water layer surrounding the grains of the anion resin is low at the beginning of EDI process. Therefore, silica present in the water is only partially ionized to the anion state and the ion-exchange process almost does not occur. Moreover, when electric field is applied, the process of salt ions transfer is initiated and water splitting consumes insignificant amount of electric power at the beginning of EDI. As process is progressed, the majority of salts is removed in 30%—40% of the total time spent by water in the EDI cell. Priority of electric power consumption is shifted toward water splitting. As a result, the pH value increases and silica is ionized to anions of polysilicon acids and enters into an exchange reaction with the anion resin. A similar process occurs for the carbon dioxide that is transformed into the form of carbonate anions [8]. However, this process occurs at much lower pH values, resulting into ionization of carbon dioxide at the very beginning of the flow in EDI cell.

The removal of TOC is usually significantly lower than those of silica, boron, and other ions and is in the range between 30 and 60%; however, in rare cases it can be reduced by up to 90%. The lower removal rates of TOC are explained by the nature of organic molecules, which are either nonionized or weakly ionized.

EDI can be successfully used for removal of toxic and radioactive metals such as cobalt, nickel, strontium, cadmium, and copper from mono- and multicomponent solutions. For example, Co, Cs, and Sr ions at an initial concentration of 10 g/L can be removed with efficiency of 99.98%, 99.99%, and 99.95%, respectively, while filling CDEI unit with 50% strongly acidic cation-exchange resin and 30% of strongly basic anion-exchange resin [9]. In general, if hardness ions such as calcium and magnesium are present in solution along with other metals, Ca^{2+} and Mg^{2+} are removed preferably. The separation sequence can be written as follows: Ca^{2+}, Mg^{2+}, $K^+ > NO_3^- > Cl^- > Na^+$. Among toxic metal ions of the same valence in a mixture the removal efficiency was found more prominent in the following sequence: $Pb^{2+} >> Cd^{2+} > Cu^{2+} \geq Zn^{2+}$ [4]. The latest is partly explained by the selectivity of ion-exchange resins toward ions of different hydrated radii. The smaller the hydrated radius of ions, i.e., greater ionic radius, the higher mobility of ions is observed, which leads to the enhanced exchange of these ions with the counterions of resin and higher efficiency of their removal. In general, hydrated radii of ions decrease from the top downward in the same group of the periodic table. Exchange of metal ions with hydrogen ions of a resin can be described by the following reaction:

$$nH^+_{resin} + Me^{n+}_{solution} \rightarrow Me^{n+}_{resin} + nH^+_{solution} \qquad (3.17)$$

Table 3.1 summarizes efficiencies of EDI for the removal of different pollutants.

TABLE 3.1 Efficiency of Electrodeionization (EDI) Process in Different Wastewater Treatment Applications

Application	Feed Water Quality	Working Parameters of EDI Unit	Product Water Quality/ Removal Efficiency	References
Power plant	SiO_2: 110 µg/L λ: 4.56 µS/cm Na^+: 849 µg/L Cl^-: 135 µg/L BO_3^{3-}: 14 µg/L	Q: 400 L/h	17 MΩ·cm / 99.1% 99.9% 99.9% 99.3% 97.1%	[3]
Microchip manufacturing plant	SiO_2: 42.5 µg/L λ: 3.4 µS/cm Na^+: 553 µg/L Cl^-: 62 µg/L TOC: 61.5 µg/L	Q: 800 L/h	16–17 MΩ cm / 97.6% 99.9% 99.8% 98.4% 36.6%	
Model water for nuclear power plant after RO	λ: 2.3 µS/cm SiO_2: 1030 µg/L TOC: 108 µg/L Ca^{2+}: 250 µg/L Mg^{2+}: 63 µg/L Na^+: 35 µg/L Cl^-: 12 µg/L NO_3^-: 13 µg/L SO_4^{2-}: 7 µg/L	$T = 25°C$ pH 6.4 Q: 0.6 L/h $j = 20$ A/m^2	16.7 MΩ cm / 100% 18.5% 98.4% 96.8% 100% 66.7% 100% 100%	[10]
Groundwater treatment	NO_3^- – N: 200 mg/L Ca^{2+}: 50 mg/L Mg^{2+}: 50 mg/L	A: Ti/Ta$_2$O$_5$–IrO$_2$ C: Stainless steel Q: 5.4 L/h $j = 3.2$ A/m^2	96.2% 96.7% 96.1%	[11]

Continued

TABLE 3.1 Efficiency of Electrodeionization (EDI) Process in Different Wastewater Treatment Applications—cont'd

Application	Feed Water Quality	Working Parameters of EDI Unit	Product Water Quality/ Removal Efficiency	References
Low radioactive wastewater	Cs^+: 50 mg/L	pH 7 Q: 6 L/h $j = 36.4$ A/m^2	99.9%	[12]
Multicomponent synthetic solution	Ni^{2+}: 0.3 mol/m^3 Ca^{2+}: 0.9 mol/m^3 Mg^{2+}: 0.3 mol/m^3	pH 3 Q: 0.54 L/h $E_{cell} = 5$ V	92.8% 99.3% 99.4%	[13]
Rinse solution from galvanic plant	Ni^{2+}: 5 mg/L	$T = 25°C$ pH 2 Q: 14–60 L/h $E_{cell} = 5$ V	96%	[14]
Cr(VI) synthetic wastewater	Cr^{6+}: 40– 100 mg/L	A: Ti/ IrO_2–SnO_2–Sb_2O_5 C: Ti $T = 25°C$ Q: 0.8 L/h $j = 5$ mA/cm^2	99.5–99.8% EC: 4.1– 7.3 kWh/mol	[15]
Simulated dilute industrial wastewater	Cu^{2+}: 100 mg/L	pH 3.2 $j = 20$ mA/cm^2 t: 10 h Q: 0.6 L/h $T = 22°C$	41%	[16]
Primary coolant of a nuclear power plant	Co^{2+}: 0.34 mM	pH 5.7 Q: 0.3 L/h $j = 1.7$ mA/cm^2 t: 1 h	99%	[17]

A, anode; C, cathode; EC, energy consumption; j, current density; NO_3^- – N, nitrate nitrogen; Q, flow rate; RO, reverse osmosis; t, treatment time; TOC, total organic carbon; λ, conductivity (56 μS/cm).

Main Calculation Parameters of Electrodeionization Process

At the first approximation the mechanism of EDI process is determined by the ion transfer in solid and liquid phases. Because the concentration and mobility of ions in solution is much lower than in a solid phase, the efficiency of EDI process will be mostly controlled by the rate of ion transfer from solution to solid surface.

The mean interstitial velocity (U_e) of the fluid in EDI unit is calculated as follows [10].

$$U_e = \frac{Q}{\varepsilon_{V,avg}A} = \frac{U}{\varepsilon_{V,avg}} \tag{3.18}$$

where Q is the volume flow rate, A is the cross-sectional area of the diluate compartment, $\varepsilon_{V,avg}$ is the average void fraction of the packed ion-exchange resin, and U is the superficial speed of the fluid.

The void fraction (ε) of resin in the diluate compartment can be determined as follows:

$$\varepsilon_V = \frac{V_t - V_s}{V_t} = \frac{V_t - \dfrac{W_m}{\rho}}{V_t} \tag{3.19}$$

where V_t is the total volume of the diluate chamber, V_s is the total volume taken by the ion-exchange resin in the diluate compartment, W_m is the weight of the ion-exchange resin in the diluate compartment, and ρ is the resin density.

The specific surface area (A_{sp}, cm^2/g) of ion-exchange resin is estimated as follows:

$$A_{sp} = 3\frac{\sum r^2 x(r)}{\rho \sum r^3 x(r)} \tag{3.20}$$

where $x(r)$ is the fraction of resin beads with radius r (cm) and density ρ (g/cm^3) [18].

The ionic conductivity of i substance (λ_i, 1/(Ohm·cm)) is determined using the Nernst−Einstein equation

$$\lambda_i = \frac{z_i^2 F^2}{RT} D_i C_{i,v} \tag{3.21}$$

where D_i (cm^2/s) is the diffusion coefficient and $C_{i,v}$ (mmol/mL) is the volumetric concentration of current carriers.

The EDI cell potential is determined using the following equation:

$$E_{cell} = I_{stack}(R_{solution} + R_{resins} + R_{membranes}) + \phi_{water\ splitting} \tag{3.22}$$

where I_{stack} is the EDI module ("stack") current, $R_{solution}$ is the resistance of interstitial solution, R_{resins} resistance of ion-exchange resins, and $\phi_{water\ splitting}$ is the potential of water splitting reactions.

While neglecting the membrane resistance due to their thickness, the change of EDI cell potential with time can be calculated as follows [10]:

$$\Delta E_{cell} = I_{stack}(\Delta R_{solution} + \Delta R_{resins}) + \Delta\phi_{water\ splitting} \tag{3.23}$$

where $\Delta R_{solution}$, ΔR_{resins}, and $\Delta \phi_{water\ splitting}$ are changes in resistance of interstitial solution, resistance of resins, and potential causing water splitting.

When changes in water splitting potential and solution resistance are constant, i.e., they do not vary, then

$$\Delta E_{cell} = I_{stack} \Delta R_{resins} \tag{3.24}$$

The CE (%) of the process can be determined using the following equation [19]:

$$CE = \frac{z_i \cdot (C_0 - C_t) \cdot Q \cdot F}{1000 \cdot 60 \cdot M \cdot I} \cdot 100\% \tag{3.25}$$

where z is the charge number/valence of i-species; C_t and C_0 are concentrations of i-species in the outlet and inlet water, respectively (mg/L); F is the Faraday constant; Q is the deionized stream flow rate (L/min); M is the molar mass of i-species (g/mol); and I is the applied current, A.

The energy consumption (EC, kWh/mol) of EDI process can be found using the following calculation [15]:

$$EC = \frac{M \cdot I \cdot \overline{E_{cell}}}{(C_0 - C_t) \cdot Q} \tag{3.26}$$

where $\overline{E_{cell}}$ is the mean cell potential (V).

The total flux of ions through the EDI system is determined by the Nernst–Plank equation

$$J_i = -\frac{D_i \cdot C_i}{RT} \frac{d\mu_i}{dx} - \frac{z_i \cdot F \cdot D_i \cdot C_i}{RT} \frac{d\phi}{dx} + C_i \cdot u \tag{3.27}$$

where D_i is the diffusion coefficient of i-ion (m^2/s); z_i is the valence of i-species; F is the Faraday constant (A·s/mol); R is the gas constant (8.314 J/mol K) and T is the temperature (K); ϕ is the electric potential (V); dx is the distance (m); C_i is the concentration of ions (mol/m^3); and μ is the chemical potential (J/mol).

However, the flux of ions flowing through the ion exchanger in EDI unit is mostly determined by migration processes; therefore, the total flux of ions (J) at time $t = 0$ can be determined as follows:

$$\overline{J_\iota} = -\overline{u_\iota} \overline{C_\iota} \Delta \overline{E} \tag{3.28}$$

where $\overline{u_\iota}$ is the ion mobility in the bed of ion exchanger (m^2/V s) and $\Delta \overline{E} = \frac{d\phi}{dx}$ is the change of potential through ion exchanger bed (V) [1].

$$\overline{u_\iota} = \frac{\overline{D_\iota} z_\iota F}{RT} \tag{3.29}$$

3.2 CAPACITIVE DEIONIZATION

Capacitive deionization (CDI) is another emerged method for water desalination providing the highest cost and energy efficiency at a salt content in water below 10 g/L. CDI is successfully used for sea and brackish water desalination (TDS is below 4000 mg/L), water softening, wastewater treatment, etc. The first works of CDI date back to the late 1960s—early 1970s, when American researchers first studied electrochemical water desalination using porous carbon electrodes [20,21]. However, the intensive adaptation of CDI technology began from the middle of 1990s when carbon aerogel electrodes were developed by Farmer and others [8].

In CDI cell (Fig. 3.7) electrolyte solution passes through an electrochemical cell between two porous carbon electrodes of high surface area and thickness of about 100—150 μm, when potential difference of about 0.8—1.5 V is applied between those electrodes. Anode and cathode are usually made of the same material and have the same size. Because of the established electrical field, positive ions are adsorbed in intraparticle pores of negative electrode and negative ions are adsorbed in intraparticle pores of positive electrode, thus desalinating the solution. When adsorbing, ions move through the interparticle pores to the intraparticle pores where they deposit in the electrical double layers (EDLs), thus charging them. Adsorption capacity of electrodes gradually decreases during the operation cycle, and conductivity of product water gradually increases [22]. When pores become saturated with the cations and anions, the capacitive storage of electrodes is reached and they undergo regeneration. Regeneration is conducted by means of lowering down the applied potential difference or completely reversing the polarity of electrodes. This forces ions to diffuse back to the stream, resulting in the formation of

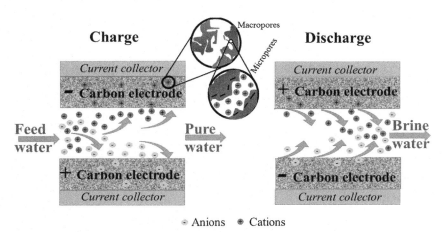

FIGURE 3.7 Working principle of capacitive deionization process.

concentrated solution. In general, the process can be divided into two steps, the step of electrodeposition or charge where water is desalinated and concentration step or discharge step when electrodes are regenerated. Energy of discharged ions can be recuperated and used for example to maintain the work of the next CDI unit.

The main advantages of CDI are as follows:

- Low operation voltage and pressures. Because applied potentials are low, the separation process does not include electrode reaction mechanism and are fully based on the effect of electrostatic adsorption of ions.
- Possibility to operate at different flow rates and low pressures.
- Chemically stable electrode materials not containing precious or toxic metals.
- Recuperation of electric energy is possible during the electrode regeneration step.
- Chemical-free process and no membranes required to desalinate brackish water.

However, the following disadvantages can be identified:

- Formation of concentrate solution during electrode regeneration, which requires treatment and utilization
- Necessity to pretreat the water from organic compounds, which can be intensively adsorbed by porous carbon electrodes and cause electrode fouling by microorganisms

CDI rarely undergoes fouling by inorganic compounds, such as hydroxides depositions or others; however, it is sensitive to biological built-up and fouling by organic molecules.

- Relatively high price of mesoporous carbon electrodes

CDI cells can be operated either in a single-pass mode, when water flows through the system at certain flow rate and is discarded after the passing the cell, or in a batch mode, when water is recirculated through the unit until the electrodes are saturated with the adsorbed ions. Under ideal conditions and at the established dynamic equilibrium the amount of adsorbed salt during the single-pass operation mode should be equal to the amount of desorbed salt and the amount of accumulated charge should be equal to the charge discharged. In practice, there can be insignificant current leakage, which will lead to the decrease of discharged current compared to the charge current.

Parameters Influencing the Efficiency of Capacitive Deionization

There are many parameters, such as pH and composition of the feed water including oxygen content, material of electrodes, and the value of applied voltage/current, which influence the efficiency of CDI unit work. In a first approximation, all parameters affecting CDI process can be divided into two groups. They are Faradic, connected with the transfer of electrons due to electrochemical reactions, and non-Faradic effects. The first group includes capacitive ion storage, ion kinetics, and chemical surface charge effects, where the first two processes contribute the most to the performance of CDI. The effect of capacitive ion storage is described by the phenomena of EDL formation and is a crucial point of CDI process. Counterions adsorbed electrostatically in EDL of electrode micropores are stored capacitively in the diffuse layer of the EDL. Ion kinetics takes into account different ion mobility. For example, high mobility of H^+ and OH^- ions present in the feed solution can significantly reduce the adsorption capacity of electrode materials toward other ions. Moreover, fluctuations of H^+ and OH^- content in the purified water can significantly change the acidity of the product water. Chemical surface charge effect refers to the charge of surface groups such as carboxylic and amine groups, which are naturally present at the carbon interface [8]. The dominance of certain groups and respectively the surface charge of carbon electrode strongly depend on the pH of solution at the surface of electrodes and in electrode micropores.

The second group of Faradic effects includes carbon redox reactions, water chemical reactions, and carbon oxidation reactions. Generally, consumption of electrons and protons by side reactions during CDI leads to the electrical current leakage. Carbon redox reactions refer to the reaction occurring with functional groups at the surface of carbon electrode. In such reactions, electrons and protons are consumed by redox reactions of surface groups preventing adsorption of pollutant ions. Water chemical reactions are attributed to all kinds of electrode redox reactions occurring between the compounds of water phase and leading to the formation of reactions products staying in the water phase. The most common water electrochemical reactions are water electrolysis reactions, chlorine gas evolution, and formation of hydrochloric and hydrochlorous acids. Theoretically, CDI should be conducted at applied voltage of below than 1.23 V to eliminate the loss of current to the oxygen and hydrogen evolution reactions. Carbon oxidation reactions include oxidation of electrode carbon material to CO_2 gas, which leads to the loss of carbon electrodes material and deterioration of CDI process.

Electrodes Used in Capacitive Deionization

The most available carbon-based electrodes are activated carbon, carbon nanotubes and aerogels, carbon black, Ti-modified carbon, etc. The majority of these electrodes are commonly used in conventional FBCDI process and have pore diameters below 50 nm, thus providing a high hydraulic resistance to water flow. Consequently, conventional carbon electrode materials would have low efficiency in flow-through electrode capacitive deionization (FTECDI) applications. The higher the porosity (total pore volume, pores connectivity, and pore size) of electrode the better performance CDI process can provide. To reduce hydraulic resistance of carbon electrodes, new monolithic materials were developed. Hierarchical carbon aerogel monoliths (HCAMs) are one type of the new materials having dual pore structure with pore diameters laying in both micrometer and nanometer regions. Nanopores are usually built in micropores by means of thermal activation etching, thus creating high specific capacitance of electrodes (>100 F/g).

Quality of carbon electrode materials plays an important role in the efficiency of CDI process. The main conditions, which CDI electrodes should satisfy, are as follows:

- High specific surface area, electronic conductivity, and sensitivity to the adsorbed ions
- Chemical and electrochemical stability and resistance to fouling
- Low costs and ease to shape
- High hydrophilic properties
- Good processability
- Low contact resistance between electrode pores and current collector
- High wettability
- High mobility within the electrode pores

Conditionally all pores can be divided into two categories of macro-, meso- and micropores. Electroneutrality of charge in macropores (size > 50 nm) is achieved by compensation with ion concentration ($C_{cations} = C_{anions}$), while electrons additionally balance charge in micropores (size < 2 nm). The concentration of cations and anions in micropores can be different [23]. Interparticle macropores creating macroporosity (p_{mA}) serve only as transport canals for ions, and micropores are involved in formation microporosity (p_{mi}) and EDL and electrodes charging (Fig. 3.7). Table 3.2 shows some examples of different electrode materials used in CDI processes for desalination and water treatment purposes.

Capacitive Deionization Cell Types Depending on a Distribution of Feed Water Flow

Depending on the feed water supply mode, CDI cells can be divided into four categories [8]. The first one is a conventional flow between or flow by (FB) process (Fig. 3.8A), where water flows between two electrodes through either an open channel (thickness starting from 1 mm) or a porous separator layer (flow spacer with thickness of about 100–300 μm). Separator should allow free flow of feed water through its structure as well as prevent a contact between electrodes to avoid short circuits. Efficiency of conventional CDI process is limited by the diffusion of electroactive species. In this regard, it often becomes difficult to find an optimal size of separator maintaining both

TABLE 3.2 Electrodes Used in CDI Applications for Water Treatment and Desalination Applications

Process/ Electrode Type	Pollutant/ Concentration in Feed Water	Working Parameters	Efficiency of Salt Removal/ Electrosorptive Capacity	References
MCDI with ACC electrode (1117 m²/g specific surface area)	NaCl 1000 mg/L	$E = 1.2$ V $Q = 2.4$ L/h	25%	[22]
	Daily wastewater from thermal power plant: Ca^{2+}: 441 mg/L Mg^{2+}: 108 mg/L Na^+: 2299 mg/L K^+: 30 mg/L Cl^-: 2797 mg/L SO_4^{2-}: 150 mg/L NO_3^-: 144 mg/L	$E = 1.2$ V pH 7.1 Q: 2.4 L/h	EC: 1.96 Wh/L 92% 92% 95% 87% 95% 95% 85%	
CDI 10-cell CDI MCDI CNTs-CNFs electrode (0.1415 cm³/ g total pore volume)	NaCl solution 50 µS/cm	$E = 1.2$ V $t = 120$ min $T = 25°C$ Q: 2.4 L/h	41.6% 89.8% 90.8%	[24]
CDI and MCD with AC (1260 m²/g specific surface area)	NaCl: 200 mg/L 420 µS/cm	$E_{sorption} = 1.2$ V $E_{desorption} = 0$ V $t_{sorption} = 3$ min $t_{desorption} = 2$ min Q: 1.2 L/h	67%$_{salt\ removal}$ 74%$_{conductivity}$ 85%$_{salt\ removal}$ 82.5%$_{conductivity}$	[25]
BM CDI GNFs	NaCl 25 mg/L	$E = 2$ V $t = 35$ min	12 µmol/g	[26]
CDI 0-MCDI r-MCDI Carbon electrodes with 37% micropores porosity	NaCl 1169 mg/L	$E_{sorption} = 1.2$ V $t_{sorption} = 10$ min	0.17 mmol/g 0.21 mmol/g 0.22 mmol/g	[23]

Continued

TABLE 3.2 Electrodes Used in CDI Applications for Water Treatment and Desalination Applications—cont'd

Process/ Electrode Type	Pollutant/ Concentration in Feed Water	Working Parameters	Efficiency of Salt Removal/ Electrosorptive Capacity	References
CDI RGO-RF electrodes	NaCl 65 mg/L	$E = 2$ V $Q = 1.2$ L/h $t = 30$ min	38%$_{conductivity}$ 3.2 mg/g	[27]
BM MCDI AC electrodes (70.7 nm pore size)	TDS 2000 mg/L $[Na^+] = [Ca^{2+}]$	$E_{sorption} = 1$ V $E_{desorption} = -1$ V $t_{sorption} = 2$ min $t_{desorption} = 2$ min $Q = 1.8$ L/h	10% Ca^{2+} 25% Na^+	[28]
CDI ce-MoS$_2$ electrode	NaCl 400 mg/L	$E = 1.2$ V $t = 60$ min	35%$_{conductivity}$ 9 mg/g	[29]
CDI TiO$_2$/CNTs electrodes	NaCl (50 µS/cm)	$E = 2$ V $t = 15$ min $Q = 4.8$ L/h pH 6.5	40%$_{conductivity}$	[30]

0-MCDI, zero-voltage desorption MCDI; *AC*, activated carbon; *ACC*, activated carbon cloth; *BM*, batch mode; *CDI*, capacitive deionization; *ce-MoS$_2$*, chemically exfoliated MoS$_2$ electrode, *CNTs-CNFs*, carbon nanotubes and carbon nanofibers composite; *EC*, energy consumption; *GNFs*, graphene-like nanoflakes; *MCDI*, membrane capacitive deionization; *RGO-RF*, reduced graphite oxidate-resol like material; *r-MCDI*, reverse-voltage desorption MCDI; *TDS*, total dissolved substances.

sufficient flow of water and minimizing cell electrical resistance [31]. Moreover, desalination rate and amount of removed salt per charge are relatively slow.

The second type of CDI process is the so-called FTECDI process (Fig. 3.8B) where water flows directly through porous carbon electrodes along with the direction of applied electrical field [31]. In addition to the water flow mode difference, FTECDI process differs from a conventional FBCDI process by a separator thickness, which is about 1% of the total cell size. Significantly small achieved thickness of separator in FTECDI units is explained by its secondary role in flow transport. Therefore, FTECDI process is not limited by the diffusion process, thus providing faster desalination rates (4 to 10 times higher) and greater amount of removed ions per charge (about 3.5 times) comparing to conventional CDI process. Regardless the feed supply mode and analogous to EDI units, both FB and FTECDI cells are combined in series to build FB and FTECDI unit stacks, respectively.

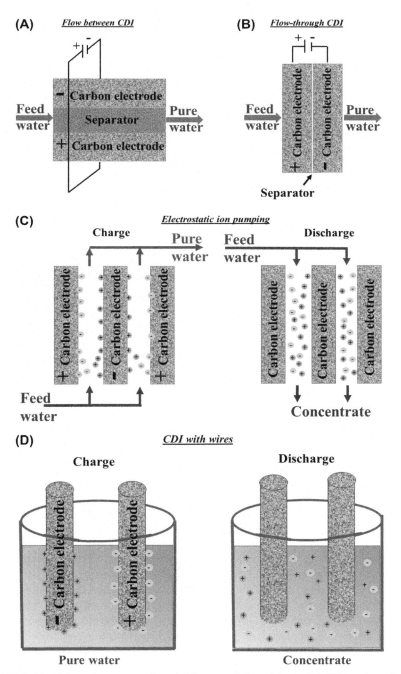

FIGURE 3.8 Schematic representation of different capacitive deionization (CDI) unit configurations depending on the distribution of feed water flow. (A) The flow between CDI, (B) the flow-through CDI, (C) the electrostatic ion pumping CDI, and (D) the CDI with wires.

Another type of CDI cells are called electrostatic ion pumping (EIP) and are shown in Fig. 3.8C [32]. In this configuration, there are two channels for feed and concentrated water input, which are fluidly joined by a series of interelectrode canals. Electrodes play the major role of a link transferring ions from the feed to concentrate solution. Feed water can be alternately supplied to one of the channels at different speeds depending on the semicontinuously applied electric field. Feed and pure water as well as feed and concentrate water are withdrawn from different ends of CDI unit. Electric field is applied in such manner that electrodes are alternately charged either positively or negatively. In a charge phase, feed water passes through interelectrode canal, counterions become adsorbed on the surface of electrodes, and water is purified. Before reaching the maximum sorption capacity, the electric field is turned off and ions are desorbed into the concentrate solution allowing regeneration of electrodes (Fig. 3.8C). In general, interelectrode spacing width is kept in the range of 0.1−2 mm and interelectrode canal length is in the range of 0.1 mm−10 cm. EIP CDI allows faster desalination rates compared to the conventional FBCDI process.

Another type of CDI cells of different water flow distribution is based on desalination process using wires/rods made of carbon material and serving as electrodes (Fig. 3.8D). Wires can be either plane or coated with an ion-exchange membrane similarly to the membrane capacitive deionization (MCDI) process described below. Electrodes are movable. When applying electric field, electrodes become saturated with ions. When electric field is turned off, electrodes are lifted from a purified solution and shifted to a water stream, which should become a brine/concentrate solution. The regeneration step is conducted similar to other CDI processes by lowering down or shifting the polarity of electrodes.

Membrane Capacitive Deionization

MCDI (Fig. 3.9) combines CDI process and ion-exchange membranes, thus enhancing the desalination efficiency of the whole process. Ion-exchange membranes can be placed in front of either one or both electrodes; however, the most prominent enhancement of water treatment is obtained while placing ion-exchange membranes in front of both electrodes. Cation-exchange membrane is placed on the side of cathode, thus allowing only cations to pass through it and be adsorbed inside of cathode micropores. In turn, anion-exchange membrane is placed on the side of anode letting anions to migrate inside the pores of the electrode. The main difference between conventional CDI and MCDI processes occurs during the electrosorption and regeneration phases. Because pores of carbon electrodes are filled with

Charge

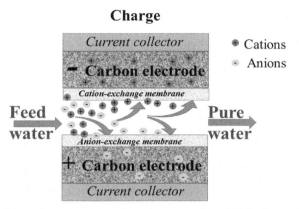

FIGURE 3.9   Schematic representation of membrane capacitive deionization working principle during charging phase.

electrolyte solution in a potential/current free mode, imposition of potential difference between electrodes forces co-ions to migrate from the electrodes, thus contaminating purification water stream. Moreover, contamination of electrodes with salt solution reduces adsorption capacitance of electrodes during potential/current application. The use of membranes on the side of electrodes prevents expulsion of co-ions in purified water stream by trapping them in intraparticle macropores of electrode [33]. While changing the polarity of electrode in conventional CDI process, ions start to desorb from electrodes forming the concentrate solution. Simultaneously, the desorbed ions can be again adsorbed on the surface of oppositely charged electrodes. In opposite to conventional CDI, ion desorbed from electrodes in MCDI process cannot be readsorbed on the surface of oppositely charged electrode because near-electrode membrane prevents ions from passing to the electrode surface. Therefore, enhanced regeneration of electrodes, adsorption and desorption of ions, and consequently enhanced demineralization are achieved in MCDI compared to the CDI process. It was found that desalination efficiency of one MCDI cell can be as similar as up to 10 pairs of conventional CDI cells [24]. Moreover, regeneration cycles of MCDI process are shorter than that of MCDI. In general, MCDI was found more efficient for the removal of monovalent ions than for higher valence ions, thus making the technology compatible for desalination of seawater and wastewater, containing up to 85 and 90% of sodium and chloride ions, respectively [22].

Constant Current Versus Constant Voltage Operation Mode of CDI and MCDI Units

The majority of CDI/MCDI studies use the constant voltage (CV) operation mode, when any applied to the electrodes potential above zero is attributed to the water desalination step and applied potential of 0 V is connected with the desorption or regeneration of electrodes. However, there are a number of studies where different voltages are applied to conduct electrosorption and desorption steps. In any case, more positive potential is always attributed to the sorption stage and more negative potential is connected with desorption (Fig. 3.10). Let us consider what happens with the ion concentration and current when operating CDI cell at a CV. At the first moment when voltage is applied across the cell, solution, containing ions, starts to move through the intraparticle macropores of electrode material to the micropores (mi). Because the macropores (mA) serve only as transport channels, the concentration of ions in the spacer channel (c_{sp}) where electrolyte solution was fed is equal to the ion concentration in the electrode macropores (c_{mA}). When ions of electrolyte solution reach the micropore volume, they become adsorbed fast in the EDL of micropores, thus reducing drastically the concentration of salt in the effluent electrolyte solution. EDL of micropores becomes saturated with ions and adsorption capacity of electrodes decreases, which leads to the gradual increase of ions in the effluent until the value of initial ion concentration, when maximum saturation of EDL is achieved. The next step involves the reduction of applied voltage to the zero value, which facilitates the fast desorption of ions first from the near surface micropores followed by gradual decrease of desorption and effluent concentration stabilization to the feed water values. Let us now consider the current fluctuation during a working cycle of CDI cell. When CV is applied across the CDI cell, flow of ions gradually decreases from a maximum value to a minimum, thus explaining gradual decrease of current during the charge step. When voltage is reduced to zero or reversed, ions start to be desorbed and current flows in the opposite direction gradually attenuating to the zero value at the end of desorption step.

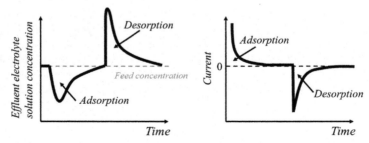

FIGURE 3.10 Change of effluent electrolyte concentration and current during the adsorption and desorption steps of capacitive deionization process at applied constant voltage.

Reverse-voltage operation mode is also implied to conduct discharge of electrodes. When comparing regeneration of electrodes at a zero-voltage desorption (ZVD) compared to the reverse-voltage desorption (RVD), the latter can provide the complete removal of counterions from the EDL in electrode micropores since coions start to attract to the EDL as countercharged ions. This allows higher efficiency of regenerated electrodes in consequent sorption stages. In general, the efficiency of different CDI processes at CV can be placed in the following series: MCDI-CV-RVD > MCDI-CV-ZVD > CDI-CV-RVD > CDI-CV-ZVD [8].

Constant current (CC) operation mode is used in MCDI applications where product water of a constant quality is required. CC-CDI units contribute to a nonconstant composition of effluent water due to the fact that ions electrosorption changes as a function of time.

The Mechanism of Capacitive Deionization

CDI process is analogous to the process occurring in supercapacitors, which use the same type of electrodes. The only difference between supercapacitors and CDI is that supercapacitors are used as energy storage devices or backup power sources and CDI is used for water desalination purposes. In turn, the mechanism of supercapacitors work is based on the working principles of parallel plate capacitors. When potential difference ($\Delta\phi$) is applied between electrodes, electrons flow from anode to cathode, thus causing to a positive charge at the anode and negative charge at the cathode [34]. Potential difference between two points in electric field is often called voltage (U_V). From another side, if two insulated conductors are charged, there is a potential difference between them, which can be gathered. The case when the charges of conductors are equal in magnitude and opposite in sign: $q_1 = -q_2 = q$ is of the greatest practical interest. In this case, it is possible to introduce the concept of capacitance.

Capacitance of the system consisting of two conductors is defined as the ratio of the charge q of one of the conductors to the potential difference $\Delta\phi$ between them and is measured in Farads (F).

$$C = \frac{q}{\Delta\phi} = \frac{q}{U_V} \tag{3.30}$$

Capacitors can be connected together either in parallel or in series to form capacitor stacks (Fig. 3.11). When capacitors are combined in parallel, voltages across each capacitor are equal ($U_1 = U_2 = U$) and charges can be found as follows: $q_1 = C_1 \cdot U$ and $q_2 = C_2 \cdot U$. Such a system can be considered as a single capacitor with a capacitance $C = C_1 + C_2$ and charge $q = q_1 + q_2$ at a voltage between the electrodes equal to U. Therefore, the

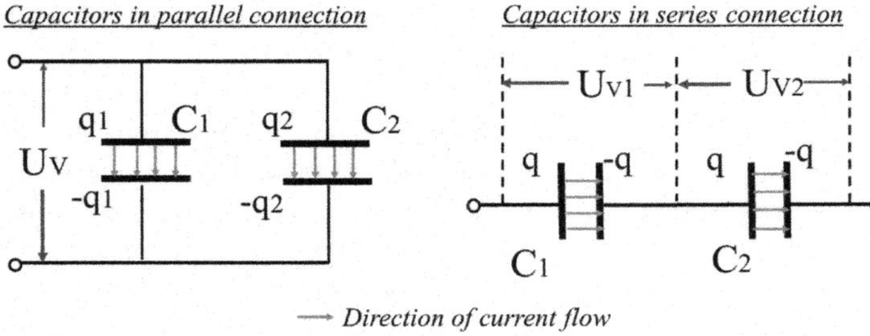

Capacitors in parallel connection *Capacitors in series connection*

⟶ *Direction of current flow*

FIGURE 3.11 Parallel connection of two plate capacitors and connection in series of a capacitor couple.

capacitance of the stack formed of capacitors connected in parallel can be found as follows.

$$C = \frac{q_1 + q_2}{U_V} \tag{3.31}$$

When capacitors are connected in series, charges of capacitors are equal $q_1 = q_2 = q$ and voltages can be found as follows: $U_{V1} = q/C_1$ and $U_{V2} = q/C_2$. Such a system can be considered as a single capacitor with a charge Q and the voltage between the electrodes equal to $U_V = U_{V1} + U_{V2}$. Therefore, to obtain a required voltage, capacitors are connected in series. Series connection is also applicable for CDI units, when electric energy is meant to be recuperated. Moreover, connection of CDI unit in series prevents excess voltages across separate CDI units, which can cause decomposition of electrolyte solution and excessive current. Capacitance of the stack formed of capacitors connected in series can be found as follows:

$$C = \frac{q}{U_{V1} + U_{V2}} \tag{3.32}$$

and

$$\frac{1}{C} = \frac{1}{C_1} + \frac{1}{C_2} \tag{3.33}$$

The maximum energy, which can be stored in capacitors, is found using the following equation:

$$W_{max} = \frac{C \cdot U_V^2}{2} \tag{3.34}$$

EDL is a characteristic of any interface between solid surface and solution in contact with the surface. Because of a similarity of EDL with a capacitor, it can be described using the capacitance electrostatic theory. The mechanism of EDL formation in parallel plate capacitors were first described by Helmholtz in 1883 and Perrin in 1904 [35]. According to their theory, electrode attracts oppositely charged ions, thus forming EDLs. The layers of oppositely charged ions are separated by a hydration sheath. The model is linear and assumes that capacitance is stable during reaction, i.e., it does not undergo changes while changing the applied potential. Capacitance of parallel plate capacitors depends on the shape and dimensions of the conductors and properties of the dielectric, which separates conductors and can be calculated using the following equation:

$$C = \frac{\varepsilon \cdot \varepsilon_0 \cdot A}{d} \tag{3.35}$$

where ε is the relative dielectric permittivity of the medium between conductors, ε_0 is the electrical permittivity of vacuum, A is the plate surface area of conductor (electrode), and d is the EDL thickness. However, the Helmholtz theory is not applicable for the porous electrodes materials. Interfacial surface tension of solution (γ) has parabolic dependence of the applied potential in parallel plate capacitors.

$$\gamma = \gamma_{max} - \frac{\varepsilon \varepsilon_0 A E^2}{2d} \tag{3.36}$$

As it is seen from Eq. (3.35), to increase capacitance of a device or CDI unit, it is necessary to either further increase the surface area of electrodes and permittivity of dielectric between electrodes or decrease the distance between electrodes. These conditions are applied in supercapacitors, where plate conductors are replaced with porous electrodes, dielectric separator is replaced with electrolyte solution of high permittivity, and distance between electrodes is adjusted to the minimum value.

Gouy—Chapman theory determines the capacity change on applied potential differences; however, it neglects ion-to-ion interactions and assumes point-charge ions. The system as a whole always obeys the rule of electroneutrality; however, within the EDLs there are regions of excess charge (q). In a case of ideally polarized electrode and absence of specific adsorption on the surface of electrode, the excess charge at the outer Helmholtz plane (OHP, the border between the Stern and diffuse layers) can be calculated using the following equation [36].

$$q = \left(\frac{2RT\varepsilon C_b}{\pi}\right)^{1/2} \cdot \sin h\left(\frac{zF\phi_\delta}{2RT}\right) \tag{3.37}$$

where C_b is the bulk concentration (M) and ϕ_δ is the Stern potential or potential at OHP (V).

The modern theory combines both Helmholtz and Gouy—Chapman theories into one Helmholtz—Gouy—Chapman—Stern theory and determines a relation between amount of stored ions or stored charge and potential difference across EDLs [8]. The total electrical capacity of the interface (C_T) is composed of electrical capacity of the compact double layer (C_{E-H}, μF/cm^2, the layer located between the electrode surface E and OHP, H) and diffuse layer (C_{H-S}, μF/cm^2, the layer starting from the OHP into the solution S) and described as follows [36].

$$\frac{1}{C_T} = \frac{1}{C_{E-H}} + \frac{1}{C_{H-S}} \tag{3.38}$$

where

$$C_{H-S} = zF \left(\frac{\varepsilon C_b}{2RT\pi} \right)^{1/2} \cdot \cos h \left(\frac{zF\phi_\delta}{2RT} \right) \tag{3.39}$$

The Stern theory considers ions to be in close proximity to each other; however, adsorption remains negligible for Na$^+$ and F$^-$ ions, which is explained by a large hydration layer attached to these ions.

Desalination ratio (DR) in CDI process can be determined as follows:

$$DR = 1 - \frac{C_t}{C_0} = 1 - \frac{C_p}{C_f} \tag{3.40}$$

where C_t and C_0 are final outlet and initial inlet concentrations of electrolyte solution. For the batch process C_t and C_0 can be represented through the product and feed concentration, C_p and C_f, respectively.

CE of both CDI and MCDI can be found using the following equation [25,37].

$$\text{CE} = \frac{(C_0 - C_t) \cdot V \cdot F}{\int I \cdot dt} \cdot 100\% \tag{3.41}$$

Capacitance of CDI system per unit of electrode mass is called specific capacitance (C_s) and can be found as follows:

$$C_s = \frac{I \cdot \Delta t}{m \cdot \Delta \phi} \tag{3.42}$$

where m is the mass of electrode material and $\Delta \phi$ is the potential window. Moreover, specific capacitance of an electrode can also be estimated from experimental data of electrochemical impedance spectrum and cyclic voltammetry respectively as follows [38,39].

$$C_s = \left| \frac{1}{\omega \cdot Z''} \right| \tag{3.43}$$

$$C_s = \frac{\int j d\phi}{2 \cdot v \cdot \Delta \phi \cdot m} \tag{3.44}$$

where ω is the angular frequency of the applied AC signal, Z'' is the imaginary part of impedance, j is the current density, and v is the scan rate.

Electrosorptive capacity (mg/g) of electrodes is determined similarly to the conventional adsorption process [27].

$$\text{Electrosorptive capacity} = \frac{(C_0 - C_t) \cdot V}{m} \tag{3.45}$$

where C_0 and C_t are the initial and final concentrations of salt.

In general, CDI and MCDI processes are very complex and described by a number of models. At the first approximation, the overall mechanism of CDI/MCDI process can be described by two main models. The first one is the improved modified Donnan (i-mD) dealing with the mechanism EDL formation and the second one is the transport theory describing ion transfer in the pores of the electrode, in ion-exchange membrane characterized by the Nernst–Plank equation, at the membrane/solution interface characterized by voltage drop and in the separator channel between two electrodes [23].

Improved Modified Donnan Model

Donnan theory considers the formation of EDLs inside the micropores. Because the Debye length (the length where electric potential will decrease by 1/e magnitude) is significantly larger than the micropore size, Donnan theory assumes the complete overlapping of two diffuse EDLs of opposite side walls in micropores. This means that diffuse layer potentials do not dissipate but jump from a value of Stern potential on one side of micropore to the value of Stern potential on the opposite side of micropore (Fig. 3.12). So it is assumed that diffuse layer potential is constant inside the micropores [40].

The i-mD model describes the formation of EDLs, accumulation of charge, and salt adsorption in the EDL of electrode pores for CDI and MCDI

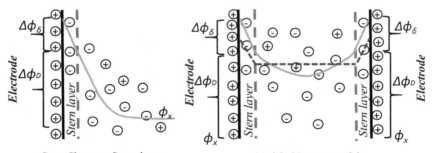

Gouy-Chapman-Stern theory   *Modified Donnan model*

FIGURE 3.12 Classical Gouy–Chapman–Stern theory versus modified Donnan model. *Solid yellow line* (Light gray in print visions) shows potential profile and *dashed red line* (Dark gray in print visions) is the potential profile according to the Donnan model. $\Delta\phi_\delta$ is the Stern potential, $\Delta\phi_D$ is the Donnan potential, and ϕ_x is the potential at distance x from the surface.

processes. Because formation of EDLs occurs in the micropores, it is necessary to determine the concentration of ions in these pores ($C_{mi,i}$)

$$C_{mi,i} = C_{mA} \cdot \exp\left(-z_i \cdot \Delta\phi_D + \mu_{att}\right) \tag{3.46}$$

where μ_{att} is the excess attractive chemical potential and $\Delta\phi_D$ is the Donnan potential (dimensionless). Donnan potentials are set when the two phases containing a number of different ions are in contact with each other such as the membrane and electrolyte solution. At the same time, one of the ions is able to pass through the phase boundary, and transition of another ion is restricted.

Concentration of ions in micropores can be also determined through the micropore charge density $\sigma_{mi}\left(\sigma_{mi} = \sum_i z_i \cdot C_{mi}\right)$ according to the following equation.

$$C_{mi,i}^2 = \sigma_{mi}^2 + \left(2 \cdot C_{mA} \cdot \exp\left(\mu_{att}\right)\right)^2 \tag{3.47}$$

The Stern potential (ϕ_δ) determines the ionic charge in micropores, which can be found also as follows [41].

$$\sigma_{mi} = -C_{E-H,\text{vol}} \cdot \Delta\phi_\delta \cdot U_T \tag{3.48}$$

where $C_{E-H,\text{vol}} = C_{E-H,\text{vol},0} + \beta\sigma_{mi}^2$ is the volumetric capacitance of the Stern layer (F/m^3); $C_{E-H,\text{vol},0}$ is the Stern capacity in the zero-charge limit; $U_T = RT/F = 25.7$ mV is the thermal voltage; and β is a parameter correlating the Stern capacitance to the micropore charge density.

At the equilibrium in the CDI cell, the charge density and concentration of ions in the micropores are dependent on the salt concentration outside the electrode (c_∞).

$$\sigma_{mi} = C_{\text{cation},mi} - C_{\text{anion},mi} = -2 \cdot c_\infty \cdot \exp\left(\mu_{att}\right) \cdot \sin h(\Delta\phi_D) \tag{3.49}$$

$$C_{\text{ions},mi} = C_{\text{cation},mi} + C_{\text{anion},mi} = 2 \cdot C_\infty \cdot \exp\left(\mu_{att}\right) \cdot \cos h(\Delta\phi_D) \tag{3.50}$$

The cell voltage can be defined through the Donnan and Stern potentials.

$$U_{\text{cell}} = 2 \cdot U_T \cdot |\Delta\phi_D + \Delta\phi_\delta| \tag{3.51}$$

The charge density ($\sum F, C/g$) and salt removal capacity (adsorption, Γ_{salt}, mg/g or mol/g) in micropores are determined as follows:

$$\sum F = \frac{F \cdot V_{mi} \cdot \left|\sigma_{mi}^{\text{ch}} - \sigma_{mi}^{\text{disch}}\right|}{2} \tag{3.52}$$

$$\Gamma_{\text{salt}} = \frac{V_{mi} \cdot \left|C_{\text{ions},mi}^{\text{ch}} - C_{\text{ions},mi}^{\text{disch}}\right|}{2} \tag{3.53}$$

where V_{mi} is the volume of micropores (mL/g); $C_{\text{ions},mi}^{\text{ch}}$ and $C_{\text{ions},mi}^{\text{disch}}$ are ion concentrations in micropores at the end of charge and discharge steps, respectively; and σ_{mi}^{ch} and σ_{mi}^{disch} are micropore charge densities at the end of charge and discharge steps, respectively.

Charge efficiency is one of the most important characteristics of CDI process, which shows a dependence of the amount of adsorbed ions in the micropores on the charge applied. In an ideal case, each unit of charge transferred to an electrode should adsorb an oppositely charged ion. However, because electrode pores contain both counter- and co-ions in the absence of electric field when being in contact with solution, part of the charge is consumed to remove co-ions from the pores when electric field is established. Therefore, the charge efficiency is never 100%. Generally, higher applied voltages are associated with the higher charge efficiencies. The theoretical charge efficiency (Λ) of the system can be calculated as follows [41].

$$\Lambda = \frac{C_{ions,mi}^{ads} - C_{ions,mi}^{des}}{\left|\sigma_{mi}^{ads} - \sigma_{mi}^{des}\right|} = \frac{2 \cdot z \cdot \Gamma_{salt}}{Q} = \frac{\Gamma_{salt}}{\sum F} \tag{3.54}$$

where $C_{ions,mi}^{ads}$, $C_{ions,mi}^{des}$ are the concentrations of ions in micropores at the end of adsorption and desorption steps, respectively, and σ_{mi}^{ads}, σ_{mi}^{des} are micropore charge densities at the end of adsorption and desorption steps, respectively.

Ion Transport Model

The total flux (J_i, mol·s/m²) of ions in macropores is characterized by the migration and diffusion processes and can be expressed using the Nernst–Planck equation [41].

$$J_i = -D_i \cdot \left(\frac{dC_{mA,i}}{dx} + z_i \cdot c_{mA,i} \cdot \frac{d\phi_{mA,i}}{dx}\right) \tag{3.55}$$

where D_i is the diffusion coefficient (m²/s); $C_{mA,i}$ is the concentration of i-type ions in the macropores (mol/m³); ϕ_{mA} is the dimensionless potential in the macropores and dx is the coordinate in the plane between membrane/electrode and electrode/current collector interfaces; and z_i is the valence of ions.

The total mass balance of ions across the electrode pores can be found as follows:

$$\frac{d}{dt}\left((p_{mA} + p_{meso}) \cdot c_{mA,i} + p_{mi} \cdot c_{mi}\right) = -p_{mA} \cdot \frac{dJ_i}{dx} \tag{3.56}$$

where p_{mA}, p_{meso}, and p_{mi} are the macro-, meso-, and microporosities of the electrode. In turn, the micropore salt balance equation can be defined as follows:

$$p_{mi} \cdot \frac{d\sigma_{mi}}{dt} = 2 \cdot p_{mA} \cdot D \cdot \frac{d}{dx} \cdot \left(C_{mA} \cdot \frac{d\phi_{mA,i}}{dx}\right) \tag{3.57}$$

where C_{mi} and C_{mA} are ion concentrations in micropores and macropores, respectively.

The overall mass balance of salt across the electrode can be determined using the following equation.

$$\frac{d}{dt}\left(2\cdot(p_{mA}+p_{meso})\cdot c_{mA}+p_{mi}\cdot c_{mi}\right)=2\cdot p_{mA}\cdot D\cdot\frac{d^2 c_{mA}}{dx^2} \qquad (3.58)$$

Taking into account the external electronic resistance (R_{EE}, $\Omega\cdot m^2$), which is located in the cables, current collectors, etc., the CDI cell voltage can be determined as follows [41].

$$U_{\text{cell}} = 2\cdot\phi_1\cdot U_T + I\cdot R_{EE} \qquad (3.59)$$

where ϕ_1 is the potential in the electrode matrix and $U_T = R\cdot T/F$ is the thermal voltage, R is the gas constant, F is the Faraday constant, and T is the temperature.

The mass balance of ions in separator channel (sp) is described by the following equation:

$$p_{sp}\cdot\frac{dC_{sp}}{dt} = p_{sp}\cdot D\cdot\frac{d^2 c_{sp}}{dx^2} + \Gamma_{sp} \qquad (3.60)$$

where C_{sp} and p_{sp} are concentrations of ions and porosity of separator channel, respectively, and Γ_{sp} is the characteristics of diffusive transport of salt in the separator.

$$\Gamma_{sp} = \nu_{sp}\cdot(C_{\text{out}} - C_{sp}) \qquad (3.61)$$

for models without electrolyte flow through the spacer channel and

$$\Gamma_{sp} = \frac{\Phi_v\cdot N_m}{A\cdot L_{sp}}\cdot(C_{\text{inflow}} - C_{sp}) \qquad (3.62)$$

for models where electrolyte flows through the spacer channel, where ν_{sp} is the transport coefficient of salt from outer volume into the spacer channel (mol/s); C_{out} and C_{inflow} are the output and inflow salt concentrations in the spacer channel; Φ_v is the feeding water flow rate through the cell (m^3/h); N_m is the number of mathematical subcells in series in stack (conditional spacer channel division to a number of parts with different C_{sp} and J_i); and L_{sp} is the thickness of the spacer channel (m).

The resistance in the spacer channel (R_{sp}, $\Omega\cdot m^2$) is equal to

$$R_{sp} = \frac{2\cdot\Delta\phi_{sp}\cdot U_T}{I} = \frac{L_{sp}\cdot U_T}{2\cdot p_{sp}\cdot D\cdot F\cdot\overline{C_{sp}}} \qquad (3.63)$$

where $\Delta\phi_{sp}$ is the potential difference across a half of a spacer channel and $\overline{C_{sp}}$ is the average concentration of ions in the spacer channel.

The salt mass balance across the membranes considering the total concentration of counter- and co-ions $(C_{T,m})$ in the membrane and expressed as follows [41].

$$p_m \cdot \frac{dC_{T,m}}{dt} = -p_m \sum_i \frac{dJ_{i,m}}{dx} \quad (3.64)$$

where p_m is the membrane porosity and dx is the direction of ion flow starting from the membrane/separator (S/M) interface toward membrane/electrode (M/E) interface.

The ionic resistance of membranes $(R_m, \Omega \cdot m^2)$ in MCDI cell unit can be found using the following dependence:

$$R_m = \frac{2 \cdot \Delta\phi_m \cdot U_T}{I} \quad (3.65)$$

where $\Delta\phi_m = \phi_{m,M/E} - \phi_{m,S/M}$ is the potential difference across the membrane.

The selectivity of CDI process for multicomponent systems is mostly dependent on the hydration radius and charge affinity of ions. The smaller the hydration radius and the higher the charge affinity of the ion, the better the selectivity of CDI electrodes toward the removal of these ions. In this regard, the selectivity of CDI can be placed in the following sequences: $Fe^{3+} > Ca^{2+} > Mg^{2+} > Na^+$ for cations and for anions $SO_4^{2-} > I^- > Br^- > Cl^- > F^- > NO_3^-$ [35]. Consequently, the general trend of sorption capacity can be placed as follows: $Na^+ \gg Ca^{2+} > Mg^{2+}$ for cations and $Cl^- > Br^- > I^-$ for anions. The efficiency of CDI process is usually enhanced at higher applied potentials.

3.3 ELECTRO-FENTON METHODS

Electro-Fenton (EF) process is an advanced oxidation process (AOP) that is able to mineralize organic pollutants such as pharmaceuticals, pesticides, dyes, phenols, and phenolic compounds through radical reactions. It is considered that hydroxyl radicals contribute the most to the degradation of compounds because of its high standard reduction potential $\left(E^0._{OH/H_2O} = 2.8 \text{ V vs. SHE}\right)$. EF process is a modification of conventional Fenton reaction (a synergetic action of H_2O_2 and iron catalysts) by means on in situ electrogeneration of Fenton's reagent. Hydrogen peroxide can be generated at the cathode when bubbling oxygen gas through the acidic solution followed by its reduction to H_2O_2 (3.67) [42]. Electrochemical generation of Fe(II) ion occurs through the reaction of anodic material dissolution (3.69) or cathodic reduction of Fe(III) ions. In general, there are four combinations of how the Fenton reaction can be carried out electrochemically [43]. They are as follows:

1. **Conventional EF process:** cathodic in situ generation of H_2O_2 and external addition of Fe(II) ions. This option is considered as a conventional EF process, in which anodic and cathodic processes can be described by the following reactions:
   Anode:

   $$2H_2O \rightarrow 4H^+ + O_2 + 2e^- \qquad (3.66)$$

   Cathode:

   $$O_2 + 2H^+ + 2e^- \rightarrow \mathbf{H_2O_2} \qquad (3.67)$$

2. Cathodic in situ generation of Fe^{2+} and external addition of H_2O_2. The process is described as follows:
   Anode:

   $$2H_2O \rightarrow 4H^+ + O_2 + 2e^-$$

   Cathode:

   $$Fe^{3+} + e^- \rightarrow \mathbf{Fe^{2+}} \qquad (3.68)$$

3. **Electrochemical peroxidation:** anodic in situ generation of Fe^{2+} and external addition of H_2O_2.
   Anode:

   $$Fe^0 \rightarrow \mathbf{Fe^{2+}} + 2e^- \qquad (3.69)$$

   Cathode:

   $$2H_2O + 2e^- \rightarrow H_2 + 2OH^- \qquad (3.70)$$

4. **Peroxicoagulation process:** anodic electrogeneration of Fe^{2+} and cathodic generation of H_2O_2.
   Anode:

   $$Fe^0 \rightarrow \mathbf{Fe^{2+}} + 2e^- \qquad (3.71)$$

   Cathode:

   $$O_2 + 2H^+ + 2e^- \rightarrow \mathbf{H_2O_2} \qquad (3.72)$$

Reaction of cathodic H_2O_2 formation in acidic conditions can compete with the reaction of oxygen reduction to water by the following half-reaction:

$$O_2 + 2H^+ + 4e^- \rightarrow 2H_2O \qquad (3.72a)$$

Moreover, electrogenerated hydrogen peroxide can undergo further reduction at the cathode surface or disproportion to oxygen gas and water in accordance with the following reactions:

$$H_2O_2 + 2H^+ + 2e^- \rightarrow 2H_2O \qquad (3.73)$$

$$2H_2O_2 \rightarrow 2H_2O + O_2 \tag{3.74}$$

In undivided cells, H_2O_2 can be also consumed by parasitic anodic reactions such as

$$H_2O_2 \rightarrow HO_2^{\bullet} + H^+ + e^- \tag{3.75}$$

$$2H_2O_2 \rightarrow O_2 + H^+ + e^- \tag{3.76}$$

Moreover, Fe(II) can be oxidized to Fe(III) in undivided cells

$$Fe^{2+} \rightarrow Fe^{3+} + e^- \tag{3.77}$$

CE of H_2O_2 electrogeneration can be calculated as follows:

$$CE = \frac{z \cdot F \cdot C_{H_2O_2} \cdot V}{M_{H_2O_2} \cdot q} \tag{3.78}$$

where $C_{H_2O_2}$ is the concentration of accumulated H_2O_2; V is the volume of treated solution; M is the molar mass of H_2O_2; and q is the consumed charge.

EF process has a number of advantages over convention Fenton reaction. They are as follows:

- In situ generation of reaction reagents eliminates necessity to have storage and dosage facilities for those chemicals.
- Anodic dissolution of iron electrode is possible to conduct at neutral pH; however, the problem of excessive ferric hydroxide sludge generation will still be present.
- Ease of implementation and automation.
- Continuous cathodic regeneration of Fe^{3+} to Fe^{2+}.
- Higher mineralization rates of organic compounds compared with conventional Fenton reaction process due to additional $^{\bullet}OH$ radical electrocatalytic generation at the anode.

Among disadvantages of EF process, the following can be the highlighted ones:

- Corrosive acidic environment of the process, which requires corrosion resistant electrodes. Acidic media is a preferable condition for electrogeneration of H_2O_2 because in alkaline media the reduction of O_2 occurs through the following reactions:

$$O_2 + H_2O + 2e^- \rightarrow HO_2^- + OH^- \tag{3.79}$$

$$O_2 + 2H_2O + 4e^- \rightarrow 4OH^- \tag{3.80}$$

- Formation of iron(III) oxide-hydroxide sludge. In this regard, the option of in situ reduction of Fe(III) to Fe(II) would be the most preferable to avoid excessive sludge production. However, the reaction rates are only reasonable at pH below 2.5. Another way to reduce sludge generation is to use solid iron catalyst in the form of sieves, particles, and iron oxides, which can be easily removed from treated solution.

- Relatively low rates of electrochemical generation of H_2O_2, which however can be improved by some type of electrodes, as for example porous gas dispersion electrodes.
- High acidity of the treated water, which should be neutralized and requires additional chemicals use.

The first studies on EF-like reactions date from the late 1970s to early 1980s when reaction of hydroxylation of organic compounds such as toluene, cyclohexane, benzene, phenol, and others was conducted in acidic media with addition of different catalyst including Fe(II) [44,45]. However, Japanese scientists Sudoh et al. were the first, who applied EF process for water treatment, in 1986. They studied decomposition of phenol at platinum (Pt) and graphite electrodes while applying cathodic potential of -0.6 V and adding Fe^{2+} catalyst [46].

Efficiency of EF process depends on many parameters such as pH and temperature of electrolyte solution, electrochemical cell configuration, type of electrode materials, value of applied potential or current, concentration of iron catalyst, amount of electrogenerated hydrogen peroxide, type of iron catalyst, stirring rate, organic pollutant concentration, and degree of electrolyte solution saturation with O_2 gas. To keep high saturation degree of solution by oxygen, pure O_2 gas or air is continuously bubbled through the solution. The higher the gas flow rate as well as stirring of the solution, the better degree of homogenization, i.e., greater H_2O_2 evolution, organic compounds decomposition and efficiency of the EF process. The temperature of electrolyte solution in the range of $35-40°C$ is considered an optimal temperature for EF process.

Fenton's Reaction

The Fenton's reaction was named after Henry John Horstman Fenton who discovered the specific electron transfer in some metals. The first study conducted by Fenton was devoted to the degradation of tartaric acid by H_2O_2 in the presence of Fe(II) in 1876 [47,48]. Conventional Fenton process is used for the oxidation of pollutants in water and based on the use of Fenton's reagent (a mixture of hydrogen peroxide acting as an oxidant and iron(II) salt acting as a catalyst). While interacting, Fe(II) is oxidized to Fe(III) forming hydroxyl radicals, which further reacts with either organic compounds decomposing them or Fe(III) and H_2O_2 forming intermediate products [43].

$$Fe^{2+} + H_2O_2 \rightarrow Fe^{3+} + {}^{\cdot}OH + OH^- \tag{3.81}$$

$$Fe^{2+} + OH^{\cdot} \rightarrow Fe^{3+} + OH^- \tag{3.82}$$

$$^{\bullet}OH + H_2O_2 \rightarrow HO_2^{\bullet} + H_2O \tag{3.83}$$

$$Fe^{3+} + HO_2^{\bullet} \rightarrow Fe^{2+} + H^+ + O_2 \tag{3.84}$$

$$Fe^{3+} + O_2 \rightarrow Fe^{2+} + O_2 \tag{3.85}$$

$$Fe^{3+} + O_2^{\bullet-} \rightarrow Fe^{2+} + O_2 \tag{3.86}$$

$$Fe^{3+} + R^{\bullet} \rightarrow Fe^{2+} + R^+ \tag{3.87}$$

$$Fe^{2+} + HO_2^{\bullet} \rightarrow Fe^{3+} + HO_2^- \tag{3.88}$$

$$Fe^{3+} + H_2O_2 \rightarrow Fe^{2+} + HO_2^{\bullet} + H^+ \tag{3.89}$$

$$^{\bullet}OH + {}^{\bullet}OH \rightarrow H_2O_2 \tag{3.90}$$

$$HO_2^{\bullet} + HO_2^{\bullet} \rightarrow H_2O_2 + O_2 \tag{3.91}$$

$$HO_2^{\bullet} \leftrightarrow O_2^{\bullet-} + H^+ \tag{3.92}$$

The optimum conditions for the above transformations shown are acidic media with pH from 2.8 to 4.0 and $[H_2O_2]/[Fe^{2+}] \sim 0.5/25$. When $[Fe^{2+}] > [H_2O_2]$, Fenton mechanism includes only oxidation of Fe(II) to Fe(III) according to reactions (3.81 and 2.82). When $[H_2O_2]/[Fe^{2+}]$ is in the range between 200 and 10^4, then decomposition of H_2O_2 takes place mainly through the reaction (3.81 and 3.83−3.85) with the release of molecular oxygen and simultaneously rapid accumulation of Fe^{3+} in solution. When $[H_2O_2]/[Fe^{2+}] > 10^5$, then in addition to reaction (3.81−3.85) there is an allocation of molecular oxygen according to the following reaction:

$$2H_2O_2 \rightarrow H_2O + O_2 \tag{3.93}$$

$$HO_2^{\bullet} + H_2O_2 \rightarrow H_2O + O_2 + {}^{\bullet}OH \tag{3.94}$$

$$O_2^{\bullet-} + H_2O_2 \rightarrow H_2O + O_2 + {}^{\bullet}OH \tag{3.95}$$

Even though the optimum acidity conditions to carry out the Fenton process is in the range between pH 2.8 and 4, it is considered that pH 2.8 provides the highest rates of Fenton reaction. So precise value of pH is explained by the fact that Fe^{3+} ions are present in water in the highest amount at pH 0. Subsequent increase of pH values leads to the decrease of Fe^{3+} species in water and increase of iron hydroxo complexes, which can only interact with other species through the mechanism of electron transfer, thus causing inhibition of Fe^{2+} regeneration. Such, at pH 2.8, iron(III) in water consists of mostly Fe^{3+} and $[Fe(OH)]^{2+}$ ions. Further decrease of solution acidity till pH 4 leads to complete substitution of Fe^{3+} ions with $[Fe(OH)]^{2+}$, $[Fe(OH)_2]^+$, and $[Fe_2(OH)_2]^{4+}$ ions. When solution pH $> 4-5$, Fe(II) is oxidized to Fe(III) and precipitates in the form of $Fe(OH)_3$, while hydrogen peroxide is mainly decomposed to molecular oxygen.

Hydrogen peroxide dosage influences the most removal of COD and TOC values, whereas iron catalyst concentration has a greater influence on the kinetics of pollutant oxidation. However, high levels of H_2O_2 concentration also inhibit the reaction kinetics. Increasing the dose of Fe^{2+} and H_2O_2 increases in the degradation rate constant. Dosage of Fe(II) ions required to conduct Fenton reaction is usually much less (below 1 mM) than the amount of H_2O_2. This is explained by the fact of continuous regeneration of Fe^{2+} ions from Fe^{3+} according to reactions (3.84−3.87). Low amounts of Fe(II) used in Fenton process is also explained by wasting reactions (3.82, 3.86, 3.88, 3.89 and 3.96−3.98), which significantly decrease the efficiency of the process.

$$Fe^{2+} + O_2^{\cdot -} + 2H^+ \rightarrow Fe^{3+} + H_2O_2 \tag{3.96}$$

$$O_2^{\cdot -} + HO_2^{\cdot} + H^+ \rightarrow H_2O_2 + O_2 \tag{3.97}$$

$$O_2^{\cdot -} + {}^{\cdot}OH \rightarrow OH^- + O_2 \tag{3.98}$$

The presence of anions in the solution can have a significant role in the Fenton process. When using $FeCl_2$ as a catalytic source of Fe^{2+} ions, decomposition rates of organic compounds are significantly lower if $FeSO_4$ and $Fe(NH_4)_2(SO_4)_2$ would be used. If solution contains non−complex-forming ions such as ClO_4^- и NO_3^- anions, then ferric ion exists predominantly in the form of Fe^{3+} and $FeOH^{2+}$. However, when electrolyte solution contains SO_4^{2-} and Cl^- ions, then ferric iron forms complexes $FeA^{(3-z)+}$ and $FeA_2^{(3-2z)+}$, where A is Cl^- or SO_4^{2-} anions. Inhibition of Fenton reaction in the presence of sulfate ions is explained by the decrease of activity of iron(III) ions due to complexation, whereas the presence of chloride ions in electrolyte solution reduces efficiency of Fenton process due to competing interaction of Cl^- ions with the hydroxyl radicals.

$$Cl^- + {}^{\cdot}OH \rightarrow HOCl^- \tag{3.99}$$

Hydroxyl radicals produced during the Fenton's reaction can interact with pollutants through the following mechanisms:

- Hydroxilation (addition of −OH group followed by hydrogen atom elimination) [49]. The reaction of hydroxylation is the main reaction allowing gradual conversion of aromatic compounds into carbon dioxide.

$$C_6H_6 + {}^{\cdot}OH \rightarrow C_6H_6(OH) \tag{3.100}$$

$$C_6H_6(OH) \rightarrow C_6H_5(OH) + H^{\cdot} \tag{3.101}$$

$$RH + {}^{\cdot}OH \rightarrow R^{\cdot} + H_2O \tag{3.102}$$

$$R^+ + H_2O \rightarrow ROH + H^+ \tag{3.103}$$

- Hydrogen abstraction (dehydrogenation)

$$CH_3OH + \,^{\bullet}OH \rightarrow CH_2OH + H_2O \tag{3.104}$$

- Electron transfer (redox reaction)

$$[Fe(CN)_6]^{4-} + \,^{\bullet}OH \rightarrow [Fe(CN)_6]^{3-} + OH^- \tag{3.105}$$

Partial reduction of Fenton reaction efficiency can also be observed in the presence of radical scavengers such as chloride, nitrates, and others.

Electrode Materials Used in Electro-Fenton Process

As mentioned earlier, electrodes should have high chemical stability toward corrosive acidic environment. Moreover, electrodes should have high overpotential toward hydrogen and oxygen evolution reaction, high surface area, good mechanical stability, high sorption capacity, and conductivity. The higher the surface are of the electrodes, the faster the H_2O_2 and Fe^{2+} generation and consequently the greater generation of hydroxyl radicals.

Carbon electrodes such as graphite, activated carbon, carbon nanotubes and sponges, vitreous carbon, and gas diffusion electrode (GDE) are among the most commonly used materials for cathodes. This is explained by the chemical inertness, nontoxicity, high conductivity properties, high overpotential toward HER, low activity toward H_2O_2 decomposition, and high porosity of these electrodes. GDE is a thin porous carbon plate, which allows gas bubbles pass through its porous structure, thus significantly intensifying the generation of hydrogen peroxide.

Boron-doped diamond electrode (BDD), mixed-metal oxide, Pt, and graphite electrodes are among the most commonly used anodes.

Types of Electrochemical Cells Configurations in Electro-Fenton Process

Electrogeneration of Fenton reagents can be conducted in three- or two-electrode systems in either undivided (Fig. 3.13A,C and E) or divided (Fig. 3.13B,D and F) cell types under potentiostatic or galvanostatic conditions [42]. Potentiostatic electrolysis is conducted under constant potential, which is set up by a potentiostat or power supply, whereas galvanostatic electrolysis is conducted under a CC, which is applied to the system by means of a power supply. Advantage of potentiostatic mode is the generation of high amounts of H_2O_2 (3.67). Furthermore, the optimum value of applied potential depends on the electrode material and is determined experimentally. For example, the optimum potential for carbon cathode is -0.5 V versus SHE, graphite electrode is -0.9 V versus SHE, and reticulated vitreous carbon electrode is -1.6 V versus SHE [50–52]. Nevertheless, applied potential is only

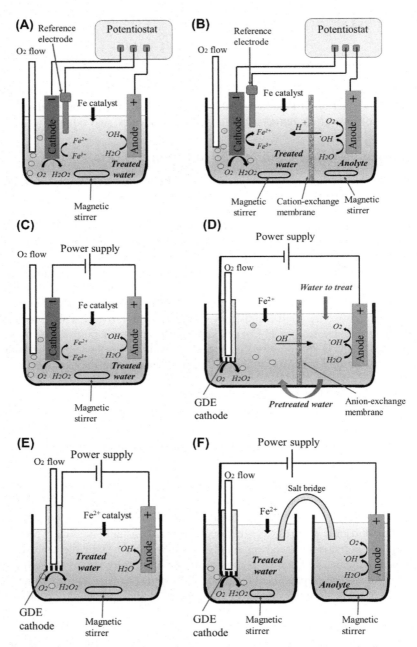

FIGURE 3.13 Schematic representation of divided and undivided electrochemical cells set-up to conduct Electro-Fenton process. *GDE*, gas diffusion electrode. (A) Undivided three-electrode system, (B) divided by a cation-exchange membrane three-electrode system, (C) undivided two-electrode system, (D) divided by an anion-exchange membrane two-electrode system with GDE cathode, (E) undivided two-electrode system with a GDE cathode, and (F) divided by a salt bridge two-electrode system with a GDE cathode.

possible to optimize in three-electrode systems. Therefore, galvanostatic mode is mostly used in two-electrode cells and provides higher degradation rates of organic compounds at higher current density applied. However, higher currents lead to smaller current efficiencies. There is an optimum value of current density, above which degradation rates start to decrease. An ion-exchange membrane or a salt bridge usually plays a role of separator in divided electrochemical cells. Oxygen gas is bubbled near to or through the surface of cathode depending on its material and configuration.

When using undivided cells or recirculating electrolyte solution from the anodic compartment of divided cells to the cathodic one, enhanced EF degradation rates of organic pollutant can be achieved. This condition is fulfilled when anode materials (M) with high overpotential toward oxygen evolution reaction (BDD, some mixed-metal oxide electrodes) are used. The enhanced mineralization of organic pollutants is explained then by the additional electrocatalytic generation of hydroxyl radicals at the surface of such electrodes through the following reaction:

$$H_2O + M \rightarrow M(\cdot OH) + H^+ + e^- \qquad (3.106)$$

The more detailed information on mechanism of electrocatalytic $\cdot OH$ radical formation can be found in the second chapter of this book under the Electrochemical Oxidation section.

As it is seen in Fig. 3.13, cathode materials can be made in the form of plates, and then water passes by such electrodes, or in the form of GDEs and water flowing through such electrodes. In the former case, there are two simultaneous reactions, which take place at the cathode. They are oxygen reduction to hydrogen peroxide (3.67, $E^0 = 0.695$ V vs. SHE) and iron(III) ion reduction to iron(II) ion (3.68, $E^0 = 0.77$ V vs. SHE). Occurrence of two reduction reactions at the same time is possible due to the close values of standard redox potentials of those reactions and nonselectivity of such electrodes to a particular reaction. This behavior is especially prominent for three-dimensional carbon materials, thus making them universal electrodes able to run EF reaction using both Fe^{2+} and Fe^{3+} as catalysts. In the latter case, mostly cathodic reduction of O_2 to H_2O_2 takes place. This is explained by the fact that O_2 reduction reaction is slightly thermodynamically more favorable than iron reduction and low activity of GDEs toward electrochemical Fe^{3+} reduction. Nevertheless, the type of reaction is not limited by the redox potential and depends on many factors such as electrodes surface area, applied potential, pH, temperature of solution, and electrocatalytic activity of electrodes.

Fenton-Type Processes

When conventional EF process includes electrogeneration of H_2O_2 in situ and addition of iron catalyst in the form or Fe^{2+} salt or iron particles to the system, there are other Fenton-type electrochemical processes, which can provide efficient degradation of organic compounds. As mentioned earlier, it is possible to eliminate the step of iron catalyst addition to the system from external sources and electrogenerate Fe^{2+} in situ by using sacrificial iron anodes (3.69). To maintain high removal rates of pollutants, acidity of solution in such process should be controlled by addition of diluted acid, for example H_2SO_4. This is explained by the fact of continuous increase of solution pH (3.81) that is not compensated by anodically generated H^+ ions, which is the case with insoluble anodes (3.66). Increase of solution alkalinity leads to precipitation of $Fe(OH)_3$ and simultaneous coagulation of pollutants trapped into hydroxide flakes. Because degradation of pollutants occurs due to both the oxidation by hydroxyl radicals and coagulation by iron hydroxides, the process where H_2O_2 and Fe^{2+} are electrogenerated in situ is called peroxicoagulation process. When optimized, *peroxicoagulation process* allows achieving high removal efficiencies of organic compounds.

The process where iron(II) ions are generated as a result of electrochemical anode dissolution (3.69) and hydrogen peroxide is added to the system from external sources is called electrochemical peroxidation (ECP). Hydrogen peroxide is dosed into reactor either continuously or at equal intervals. The optimum ratio of $[Fe^{2+}]/[H_2O_2]$ for the removal of persistent organic compounds is 1:5 to 1:10. Another kind of ECP process is anodic Fenton process where anodic and cathodic compartments are separated by either an ion-exchange membrane (both anion- and cation-exchange membranes can be used) or salt bridge. Hydrogen peroxide is added to anodic compartment, containing treated solution. Separation of cathodic and anodic compartments allows reducing the negative effect of cathodic OH^- ions generation (3.70), which significantly increases pH of solution in undivided cells, thus shifting process conditions from optimal values.

Another type of Fenton-alike process is *the use of any alternative redox couple* $M^{(z+1)+}/M^{z+}$ such as Co^{3+}/Co^{2+} $\left(E^0_{Co^{3+}/Co^{2+}} = 1.92 \ V \right)$, Cu^{2+}/Cu^+ $\left(E^0_{Cu^{2+}/Cu^+} = 0.16 \ V \right)$, or Mn^{3+}/Mn^{2+} $\left(E^0_{Mn^{3+}/Mn^{2+}} = 1.5 \ V \right)$ ions as a replacement of Fe^{3+}/Fe^{2+} $\left(E^0_{Fe^{3+}/Fe^{2+}} = 0.77 \ V \right)$ catalyst. The overall reaction representing the mechanism of Fenton-alike process remains the same as in the case of Fe^{2+} (3.81) catalysts and described as follows [53].

$$M^{z+} + H_2O_2 \rightarrow M^{(z+1)+} + {}^{\bullet}OH + OH^- \qquad (3.107)$$

Among mentioned catalytic couples Co^{3+}/Co^{2+} and Fe^{3+}/Fe^{2+} pairs provide greater mineralization of organic compounds at concentrations of Fe^{2+}

and Co^{2+} ions than Cu^{2+}/Cu^+ and Mn^{3+}/Mn^{2+} couples. This is explained by the fact that copper and manganese catalysts partly deposit at the cathode and anode, respectively. Therefore, the amount of Cu^{2+} and Mn^{2+} catalysts required is significantly greater (dozens of times) than the amount of Co^{2+} and Fe^{2+} catalysts. Moreover, high concentrations of the later ions cause scavenging reaction of the ions with $^{\cdot}OH$ radicals and as a result decrease of the treatment process efficiency. Deposition reactions of Cu^{2+} and Mn^{2+} and scavenging reactions of Co^{2+} and Fe^{2+} are represented in the following reactions [53].

$$Cu^{2+} + 2e^- \rightarrow Cu_{(s)} \tag{3.108}$$

$$Mn^{2+} + 2H_2O \rightarrow MnO_{2(s)} + 4H^+ + 2e^- \tag{3.109}$$

$$Fe^{2+} + {}^{\cdot}OH \rightarrow FeOH^{2+} \tag{3.110}$$

$$Co^{2+} + {}^{\cdot}OH \rightarrow CoOH^{2+} \tag{3.111}$$

Another type of Fenton-alike process is the *Fered—Fenton process*, when both iron (II or III) salt and hydrogen peroxides are added continuously to electrochemical cell containing acidic electrolyte solution. Concentration of iron catalyst and process parameters should be optimized for each particular process and usually kept at high amounts. H_2O_2 is added to the system in an amount, which exceeds theoretical value 1.2 times. Fered—Fenton process mechanism includes cathodic reduction of Fe^{3+} to Fe^{2+} (3.68), water reduction to hydrogen (3.70), anodic generation of $^{\cdot}OH$ radicals at the surface of electrode (3.106), and partial oxidation of Fe^{2+} ions to Fe^{3+} (3.77). Reduced at the surface of cathode when Fe^{2+} reacts with H_2O_2 producing $^{\cdot}OH$ radicals according to reaction (3.81).

Efficiency and Energy Consumption Determination

Calculations of organic compounds' degradation efficiency, EC, and CE (see Chapter 1, Eq. 1.2) are similar to those in the electrochemical oxidation (EO) process and based on COD and TOC values decrease as well as applied current or potential values. Here there are few specific calculations to determine some of these parameters. So, for instance, mineralization current efficiency (MCE, %) is calculated by substitution COD value in a conventional CE calculation for TOC value as follows [42].

$$MCE = \frac{\Delta TOC \cdot z \cdot F \cdot V_s}{4.23 \cdot 10^7 \cdot n \cdot I \cdot t} \tag{3.112}$$

where ΔTOC is the TOC decrease (mg carbon/L) during degradation of pollutants at time t (s); n is the number of carbon atoms in the studied molecule; and $4.23 \cdot 10^{-7}$ is the conversion factor to homogenize the units (3600 s/h $\cdot$ $1.2 \cdot 10^4$ mg carbon/mol).

Specific EC per unit of removed COD or TOC (kWh/g$_{COD}$ or kWh/g$_{TOC}$, respectively) can be calculated as follows:

$$EC_{COD} = \frac{E_{cell} \cdot I \cdot t}{\Delta COD \cdot V_s} \tag{3.113}$$

$$EC_{TOC} = \frac{E_{cell} \cdot I \cdot t}{\Delta TOC \cdot V_s} \tag{3.114}$$

where E_{cell} is the cell voltage/potential (V); t is the electrolysis time (h); and ΔCOD (mg O$_2$/L) and ΔTOC (mg carbon/L) are COD and TOC values decrease, respectively.

3.4 MICROBIAL FUEL CELLS

Microbial fuel cell (MFC) is a part of microbial electrochemical technologies where cathode and anode combined with microorganisms serve for different purposes (Fig. 3.14). MFC is an emerging biotechnological device that converts the chemical energy of organic matter into electricity by means of microorganisms. Similar to fuel cells MFC is theoretically highly efficient device able to produce electrical energy. However, in contrast to the fuel cell running on hydrogen or methanol, MFC can use wastewater simultaneously treating it and producing electric power. The first study on the electrochemical effect occurring during microorganism activity was published in 1911 [54]. However, the intensive studies on the topic began only 10—15 years ago. MFC is based on the metabolism of the bacteria and its ability to reduce redox active compounds. MFC is the only technology allowing generation of electricity directly from the solid and liquid organic wastes. Bacteria play the role of biocatalysts.

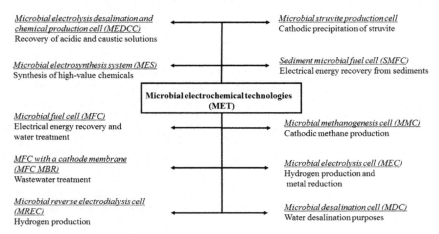

FIGURE 3.14 Examples of different METs.

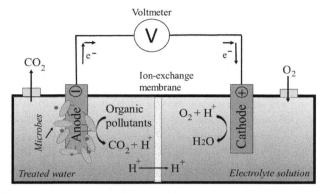

FIGURE 3.15 The working principle of conventional microbial fuel cell.

Conventional MFC consists of three basic parts such as an anodic and cathodic compartments and an ion-exchange membrane separating them, allowing hydrogen protons to pass only in one direction from the anodic to the cathodic compartment (Fig. 3.15). Anode is a negative electrode and cathode is the positive electrode in MFC. Carbon electrodes are commonly used as the anode, and catalytic materials are used for the cathode to provide better reduction of oxygen to water. Cathodic material and membrane are crucial elements of MFC significantly contributing to the cost of the treatment. Despite the fact that electrical energy can be recovered, high cost of materials result in high cost of electrical energy, which can be obtained by MFC and about 10 times higher per 1 kWh than 1 kWh of energy obtained by conventional methods. Therefore, search for a cheaper cathodic material electrocatalytically active for oxygen reduction reaction and development of new proton-exchange membranes is of great interest and ongoing.

Microorganisms that produce electricity are placed into the anodic compartment, wherein anaerobic conditions are maintained. The cathode is kept under aerobic conditions, which is provided by oxygen bubbling through the cathodic compartment or keeping the compartment and cathode exposed to the air. Microorganisms attached to the surface of anodic compartment (biofilm) receive a carbon source of energy (nutrients) in the form of wastewater, for example, that is necessary for them to grow and sustain life. Because bacteria are isolated in electrode compartment, the only way for them to survive is to process the organic substrate, which is fed for them and perform anaerobic respiration through the electrode. The principle of MFE operation is the detachment of electrons from the nutrient by microorganisms and electron transfer to the anode. Anode is connected with the cathode by a wire/electric circuit. Because of the difference of redox potentials, electrons start to move toward the cathode, where oxygen reduction occurs to form water. Electrons moving from the negative electrode to the positive generate the electric current produced by MFC. When electrons are detached from the nutrient, hydrogen protons are formed in the anodic compartment. The generated hydrogen ions pass from the anodic compartment through the ion-exchange membrane into the cathodic one where they are combined with oxygen to form water.

Double-chamber MFCs have widely been tested in laboratory scale for simultaneous production of electrical power and wastewater treatment. However, in real wastewater treatment the use of such configuration of MFCs is inconvenient due to the necessity of continuous aeration of cathodic compartment. Aeration of catholyte is an energy intensive process, which cannot be compensated by the power recovered from the MFC. For example, aeration of wastewater with oxygen in aerobic digestion systems contributes to nearly 50% of total electrical energy required for the maintenance of aeration tanks. As an alternative to double-chamber MFCs, single-chamber MFCs with air cathode can be an efficient replacement (Fig. 3.16). Air cathode in single-chamber MFCs can be separated from electrolyte solution by a proton-exchange membrane (Fig. 3.16B) or can be membrane free (Fig. 3.16A). Single-chamber MFCs provide higher removal efficiencies of organic compounds from wastewater and solid wastes. However, power recuperation in single-chamber MFCs is lower than that in double-chamber cells. Moreover, among single-compartment cells, greater powers can be recovered from cells

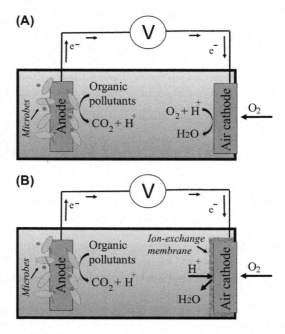

FIGURE 3.16 Single-compartment microbial fuel cells.

without proton-exchange membranes, which is explained by the reduced internal resistance of the system. It is worth to mention that resistance of ion-exchange membranes is about 10 times higher than the resistance of air cathode. However, the CE is lower in the latter case due to oxygen diffusion to the anode [55]. Increase of power recuperation and wastewater treatment efficiency can be also achieved by using biocathodes, i.e., cathodes inoculated with microorganisms. In addition to high performance, biocathodes allow cost reduction, because microorganisms play the role of biocatalysts replacing expensive catalysts required for oxygen reduction.

Air cathode is made of porous material, usually carbon or nickel, and can be conditionally divided into sintered and bonded electrodes. Sintered electrodes consist of two–three layers of different pore sizes (Fig. 3.17). Air electrode should simultaneously be in contact with liquid and gas phases and should prevent gas transfer into the near-electrode space. This condition is achieved in two-layer porous electrode due to different capillary pressure in pores of different size. The capillary pressure can be calculated using the Young–Laplace equation below

$$p_c = \frac{2 \cdot \gamma \cdot \cos \theta}{r} \tag{3.115}$$

where γ is the interfacial tension, θ is the wetting angle, and r is the pore radius.

When porous electrode is supplied with air, the pore distribution between gas and electrolyte is determined by gas pressure (p_g) and the capillary pressure (p_c) of the liquid. All pores, where $p_c \geq p_g$, will be filled with electrolyte solution. In contrast, pores where $p_c < p_g$ will be filled with gas. Electrodes made of two layers with different pore size and wetting angle creates an interface between electrolyte and gas inside the electrode. Electrode layer facing the electrolyte solution has smaller pores than the electrode layer facing the gas. Capillary pressure in fine pores is higher than the pressure drop

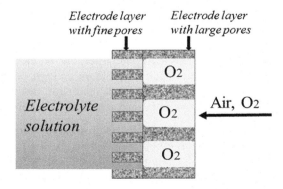

FIGURE 3.17 Working principle of air cathode.

between the gas and electrolyte solution; therefore, the layer with fine pores is filled with electrolyte. Capillary pressure in wide pores is lower than the pressure drop between the gas and electrolyte solution; therefore, wide pores are filled with gas. Electrode layer with wide pores often contains catalysts; therefore, it is called active or work layer. Electrochemical reactions take place at the interface of pore crossing layers. Two-layer electrodes are made of hydrophilic materials.

Bonded electrodes can be made through the hydrophobization of the electrode layer facing the gas by compounds, which are not wetted by the electrolyte solution, for example paraffin, polystyrene, and polytetrafluoroethylene (Teflon).

The main advantages of MFC process are as follows:

- Simultaneous treatment of wastewater and recovery of electric energy. However, there are some issues with renewable power losses at higher scales.
- MFC can work in a wide range of temperatures and external conditions such as humidity and intensity of light including its complete absence. Almost any kind of organic can serve as a fuel for MFC.
- High efficiency of the process.
- No need for aeration.
- Low operational costs.
- Ease of automation.
- Low carbon footprint compared to the CO_2 formation from the fossil fuel applications.

Among the disadvantages of MFC are

- The formation of excessive biological sludge needed to be utilized.
- Limited available commercial applications.
- Low efficiency of anode materials in regard of a balanced biochemical—electrochemical reaction ratio.
- Problem of electrode contamination with biofilms inhibiting processes of effective mass transport.
- Formation of toxic electrode reaction products inhibiting activity of biological species.
- Absence of efficient mediators for the transport of electrons between biological cells and anode. Usually mediators of high performance are toxic for microorganisms, whereas nontoxic mediators have low efficiency.
- Problems of cathode's overpotential and formation of hydrogen peroxide.

To facilitate oxygen reduction to water, electrocatalytic cathodic materials should be used. Pt and platinum-group metals have the lowest overpotential and the highest catalytic activity toward oxygen reduction. When platinum and platinum-group metals are used as the cathode, they can contribute up to 50% of the total cost of MFC reactor. It was estimated that to make MFC process economically feasible, the cost of electrode materials should be below 110 USD/m^2 [56].

- Relatively high price of membranes, which should be periodically replaced, limits their use in large applications.
- Low power capacity of existing MFC.

As a substitution of microorganism, enzymes can also be used as biological catalysts for redox reactions and electron transfer to the electrode. Such fuel cells are called enzymatic fuel cells (EFC). Oxidoreductase is the most commonly studied enzyme for EFC. EFC can provide more complete oxidation of organic compounds to carbon dioxide, which means that more chemical energy will be released and can be gathered as an electrical energy. However, enzymes have higher selectivity to a type of substrate that makes it even more difficult to find proper enzymes for multicomponent wastewater. Moreover, enzymes are more expensive to obtain and more difficult to attach to the surface of electrode.

So far, MFC process has limited applications in real systems and was mainly tested only in small pilot-scale applications with low flow rates of wastewater. The first practical application of MFC appeared in 2008 when a battery of MFC was used as an energy source for powering a meteorological buoy. The average annual consumption of power by the buoy was about 36 mW. Anode of the device is placed on the bottom of a water body, which is under anaerobic conditions, and cathode is placed in the upper aerated water layer. Such construction provides a potential difference between electrodes. Buoy uses electrogenic microorganisms and organic matter from marine sediments and can measure water pH, temperature, wind speed, etc., and send the data to the centre [57]. Another commercial application of MFC is the Electrogenic Bio Reactors (EBR) from Israeli company Emefcy. EBR are designed for wastewater with BOD load starting from 100 kg/day, maximum sulfate content of about 5 g/L, salinity of about 5%, and working at a pH range of 6.5−9.5. The process is promised to contribute to the decrease of excess sludge amounts, energy savings, and production, and have a payback period of 5 years. MFC has a potential to be used in toilets, detached houses, and farms for processing of household and farm wastes, respectively.

Parameters Influencing Microbial Fuel Cell Efficiency

Efficiency of MFC depends on many parameters such as type of electrode materials, composition of electrolyte solution, and distance between electrodes.

The *choice of microorganisms* for MFC is one of the crucial points in the process efficiency. Many biological species have limitations in terms of consumption of different organic substrate and sensitivity to oxygen and other compounds. For example, *Geobacteraceae* can utilize only acetate, ethanol, and fatty acids, thus limiting the application area of MFC using this biological species. Partially, the problem can be solved by inhabiting anode with different species at the same time. Mixed culture of microorganisms provides higher power recovery, higher resistance against process disturbances, higher rates of substrate consumption, and smaller substrate specificity. Thermophilic bacteria having optimum living conditions at 45−50°C is more preferable for the use in MFC than usual microorganisms growing at 20−30°C [58]. Common composition of mixed cultures in wastewater and sludge from wastewater plants includes species such as *Geobacteraceae, Proteobacteria, Clostridia, Bacteroides, Desulfuromonas, Alcaligenes faecalis, Enterococcus faecium, Pseudomonas aeruginosa*, and *Aeromonas*. The increased temperature of MFC operation can enhance the mass transport in the cell. Moreover, elevated temperatures reduce the solubility of oxygen, thus keeping anaerobic conditions in anodic chamber, and prevent the contamination of cell with mesophilic microorganisms.

The *distance between electrodes* is another parameter affecting the efficiency of water treatment and amount of recovered power. Unlike the traditional electrochemical applications where the distance between electrodes should be as small as possible to reduce solution resistance and prevent voltage dissipation, too close placement of electrodes in undivided MFCs (membrane-free MFCs) leads to power reduction due to interruption of bacteria activity by oxygen generated at the cathode. In such cells, a cloth separator for example could reduce the effect of oxygen crossover [56]. However, it should be noticed that electrodes should not be placed at a distance where oxygen interaction with anode is minimum, however not far beyond that point, because resistance of solution will rise too high and decrease power production efficiency.

It was earlier mentioned that *material of electrode* plays an important role in efficiency of the process. For example anodes with higher specific surface area, i.e., surface area per volume of the electrode, such as graphite brush, carbon felt, and foam (Fig. 3.18) provides more stable power generation comparing to flat electrodes. Moreover, electrodes with high specific surface area are less prone to fouling. This can be explained by the fact that oxygen is

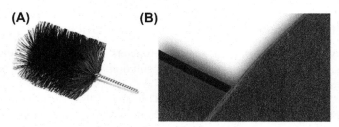

(A)　　　　**(B)**

FIGURE 3.18 Different shapes of carbon-based electrodes. (A) Graphite brush [59] and (B) graphite felt [60].

consumed by bacteria on the surface of such electrodes and bacteria, which are located inside thick electrodes and stay in anaerobic conditions, thus processing more organic matter and transferring electrons.

Cathode materials should have low overpotential toward oxygen reduction to water and have surface area significantly greater than surface area of anodes. This allows achieving greater power recovery from MFC. This condition can partly be met in single- or double chamber tubular reactors (Fig. 3.19). However, it is often difficult to implement in large-scale applications because of high cost of catalytic cathodes and difficulties in production of large cathodes.

The performance of MFC depends also on the *age and type of microorganisms* used in the process. Age of microorganisms is directly related to the thickness of biofilm formed at the anode surface, the properties of which and surrounding environment determine the degree of adhesion of bacteria to that surface. Once attached to the surface, bacteria start to grow covering the electrode with a thin film of microbial colony. While growing, biofilm attracts other microbial species providing them a shelter from the antimicrobials [61]. When anode is completely covered with biofilm, faster respiration of the bacteria, higher current, and electron transfer occur at the surface of the anode. However, while maturing and ageing, the thickness of biofilm can exceed the optimum value and prevent the penetration of organic substrate inside the film, thus reducing the produced current and efficiency of the process. Moreover, too thick and old biofilm can deteriorate due to big hollow areas inside the biofilm colony. Biofilm growth cycle is shown in Fig. 3.20.

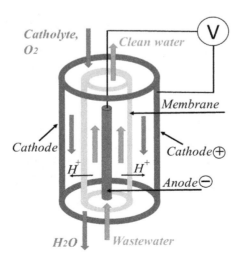

FIGURE 3.19 Double-chamber tubular microbial fuel cell.

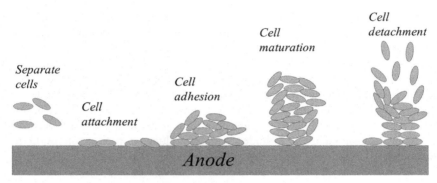

FIGURE 3.20 Biofilm growth cycle.

Acidity of solution has a direct impact on the performance of microorganisms. The best conditions for microbial growth are in the range of neutral pH. However, slightly alkaline conditions are also acceptable and even can improve pollutant degradation and power generation due to the presence of free OH^- ions, which improve the conductivity of solution. High salinity of wastewater also contributes to the greater removal efficiencies for the same reason.

Principles or Organic Compounds Decomposition by Microorganisms

Substance conversion processes occurring within the microbial cells play a major role in wastewater treatment. These processes end up with compound oxidation, energy release, and synthesis of new compounds. A large amount of chemical transformations inside microbial cells occurs at high rates. The rate of these transformations depends on the type and content of enzymes inside microbial cells. Enzymes are biocatalysts, which only accelerate reactions occurring spontaneously but at very slow rates.

Enzymes are complex protein-type compounds having molecular mass up to hundreds of thousands and even millions. Based on chemical composition, enzymes can be divided into two categories. They are simple proteins (homoproteins), for example, collagen, albumin, or keratin, consisting of only amino acids, and conjugated proteins (heteroproteins) made up of protein part (apoenzyme) and nonprotein part or coenzyme. Coenzymes possess catalytic activity, which is facilitated by apoenzyme.

Catalytic reactions take place on the surface of enzyme molecules at the places where active sites are formed. When comparing enzymes with conventional catalysts, enzymes work under mild conditions, i.e., close to standard temperatures and pressures, and neutral pH. Moreover, enzymes react only with a particular chemical compound and catalyze only one type of reaction with that particular compound. It means that in the case if substrate (wastewater composition) changes, different types of enzymes are required. Product of one reaction can be a substrate for another reaction.

The rate of enzymatic reactions increases with the temperature increase, however only until an optimal value, above which reaction rate decreases. To decompose a complex mixture of organic compounds, 80–100 different enzymes are required in average. There are compounds, which can facilitate (enzyme activators) or inhibit (enzyme inhibitors) enzymatic reactions. The most common enzymatic activators are some vitamins and cations of Ca^{2+}, Mg^{2+}, and Mn^{2+}. At the same time, toxic metal ions, antibiotics, hydrocyanic acid, etc., block the active site or enzymes, thus significantly reducing enzyme activity. Inhibitors can be reversible and irreversible. Irreversible inhibitors usually undergo chemical reaction forming a strong bond with enzymes, whereas reversible inhibition can be suppressed by dilution or dialysis. The simplest case of reversible enzyme inhibition is competitive inhibition shown in Fig. 3.21. As it can be seen in the figure, in competitive inhibition an inhibitor binds at the active site of the enzyme and compete for the active site with substrate.

Kinetics of enzymatic reaction is described by the Michaelis–Menten reaction (3.116) and determines the speed of the reaction inside the cells of microorganisms.

$$E + S \leftrightarrows ES \rightarrow E + P \tag{3.116}$$

where E is the enzyme, S is the substrate, ES is the formed enzyme-substrate complex, and P is the generated product.

The rate of product formation (ϑ, mol/min) is determined by the eponymous equation of Michaelis–Menten (3.117)

$$\vartheta = \frac{\vartheta_{max} \cdot [S]}{K_M + [S]} \tag{3.117}$$

where ϑ_{max} is the maximum rate of product formation, which is attained when all active sites are saturated with substrate (mol/min); $[S]$ is the substrate concentration (mol/L); and K_M is the Michaelis–Menten constant (mol/L). If

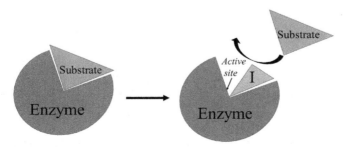

FIGURE 3.21 An example of competitive inhibition of the enzyme. I, inhibitor.

$\vartheta = 0.5 \cdot \vartheta_{max}$, then $K_M = [S]$, which means that K_M is numerically equal to the substrate concentration at which the rate of enzymatic reaction is half of the maximum possible. In other words, K_M shows the dependence of enzymatic process rates on the concentration of substrate. When concentration of substrate is very low, then $K_M >> S$ and the enzymatic reaction rate is directly proportional to the concentration of substrate $\left(\vartheta = \frac{\vartheta_{max} \cdot [S]}{K_M} \right)$. When substrate concentration is high, i.e., $S >> K_M$, then $\vartheta = \vartheta_{max}$, which means that the rate of enzymatic reaction is not dependent on the substrate concentration.

To determine the K_M value in practice, it is necessary to plot the dependence of reaction rates on substrate concentrations in a wide range. The ordinate axis represents the value of $1/\vartheta$ and the abscissa axis represents the value $1/[S]$. If the Michaelis–Menten equation is applicable to the studied reaction, the dependence of $1/\vartheta = f(1/[S])$ will be represented by a straight line (Fig. 3.22). Moreover, straight line cuts segments equal to $1/\vartheta_{max}$ on the ordinate axis and $1/K_M$ at the abscissa axis.

The specific growth rate of microorganisms (μ, h^{-1}) is determined by the Monod equation (3.118) and specific f or different microorganisms

$$\mu = \frac{\mu_{max} \cdot [S]}{K_M + [S]} \tag{3.118}$$

where μ_{max} is the maximum specific growth rate of the microorganisms. K_M in the Monod equation is always positive. When $[S] > K_M$, μ is close to μ_{max}.

Working Principles of Microbial Fuel Cell

MFC is attributed to galvanic cell; therefore, standard potential of the MFC is determined as a difference between anodic and cathodic standard reaction potentials (3.119).

$$E^0_{cell} = E^0_{red,cathode} - E^0_{red,anode} \tag{3.119}$$

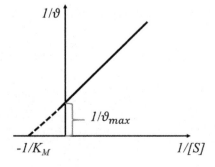

FIGURE 3.22 Graphical determination of the Michaelis–Menten constant.

Theoretical MFC potential can be thermodynamically predicted using the Nernst equation (Chapter 1, Eq. 1.7)

$$E_{cell,t} = E^0_{cell} - \frac{RT}{zF} ln\ Q_r \qquad (3.120)$$

Chemical reactions in galvanic cells are spontaneous, i.e., occur with a release of energy. Release of energy in MFC occurs in the form of electrons; therefore, the Gibbs free energy (ΔG, J/mol) determining the maximum useful work, which is possible to obtain from the system, is always negative in MFC and MFC cell potential is always positive (Chapter 1, Eq. 1.11).

$$\Delta G = -z \cdot F \cdot E_{cell,t} \qquad (3.121)$$

or

$$\Delta G^0 = -z \cdot F \cdot E^0_{cell} \text{ under standard conditions} \qquad (3.122)$$

Because in real systems there are losses of energy due to undesired reaction, heat, resistance of electrolyte solution, etc., the real potential, which can be obtained from the cell, is determined taking into account those losses as follows [62].

$$E_{cell} = E_{cell,t} - \sum \eta_i = E_{cell,t} - (\eta_c + \eta_a + \eta_{ohm}) \qquad (3.123)$$

where η_c, η_a, and η_{ohm} are concentrations due to mass transport, activation due to kinetic limitations, and ohmic due to electronic ionic resistance losses.

The power (P, W) production in MFC can be calculated as follows [63].

$$P = I \cdot E_{cell} \qquad (3.124)$$

where I is the applied current (A) and E_{cell} is the cell voltages established at applied current (V). The power obtained form MFC is usually expressed in units of power density (W/cm^2), i.e., power, which is normalized per area of the anode.

The working volume (V_{MFC}, m^3) of MFC reactor can be calculated based on the wastewater flow rate (Q, m^3/h) and hydraulic retention time (t_{HRT}, h), i.e., the time required to achieve desired removal of organic compounds.

$$V_{MFC} = Q \cdot t_{HRT} \qquad (3.125)$$

MFC can be mediated (electrochemically inactive) and nonmediated (electrochemically active). The electron transfer from microorganisms to the electrode in mediated MFC is facilitated by external or internal electron carriers (mediators) such as methylene blue, humic acid, thionine, ferricyanide, and methyl viologen, the majority of which are toxic and expensive [64]. Electron transfer in nonmediated MFC is directly conducted by biological species such as *Geobacter sulfurreducens*, *Shewanella putrefaciens*, and *Rhodoferax ferrireducens*, which have external cytochromes that can work as a nanowire, transporting electrons to any external surface. Some microorganisms use endogenous (synthesized by these microorganisms) electron carriers. However, the mechanism of electron transfer by microorganism is still not clear. It is only speculated that biofilm conductivity and electron transfer is conducted through either microbial nanowires or a network of extracellular cytochromes [54]. It is assumed that sulfate-reducing bacteria can transfer electrons from organic substrate to sulfate ions, thus reducing them to sulfides. Sulfide is oxidized spontaneously back to sulfate at anode and donate the electrons to the anode. Briefly, those transformations can be described by the following equations [65].

Biological reaction:

$$SO_4^{2-} + 2CH_2O \rightarrow S^{2-} + 2CO_2 + 2H_2O \tag{3.126}$$

Anode reaction:

$$S^{2-} + 4H_2O \rightarrow SO_4^{2-} + 8H^+ + 8e^- \tag{3.127}$$

Cathode reaction:

$$2O_2 + 8H^+ + 8e^- \rightarrow 4H_2O \tag{3.128}$$

Anodic oxidation of sulfide can also result in sulfur generation according to the following reaction:

$$HS^- \rightarrow S^0 + H^+ + 2e^- \qquad E^0 = -0.23 \ V \tag{3.129}$$

Let us consider the working principle of MFC having glucose as an organic fuel. Theoretically, all electrons can be deducted from glucose at the anodic by the following reaction:

Anode:

$$C_6H_{12}O_6 + 6H_2O \xrightarrow{\text{bacteria enzymes}} 6CO_2 + 24H^+ + 24e^- \tag{3.130}$$

The deducted electrons will move to the cathode through the external circuit making useful work. Generated hydrogen ions will pass through the ion-exchange membrane and meet at the cathode with electrons and oxygen, forming water molecules (products of bacterial glucose combustion).

Cathode:

$$6O_2 + 24H^+ + 24e^- \rightarrow 12H_2O \tag{3.131}$$

Overall combustion reaction:

$$C_6H_{12}O_6 + 6O_2 \rightarrow 6CO_2 + 6H_2O + \text{electricity} \tag{3.132}$$

Standard redox potential for the glucose/CO_2 reaction is -0.41 V; however, taking into account the energy required to sustain bacterial life, the real redox potential established in the anodic part of MFC is higher and is equal to about -0.2 V. Standard redox potential of O_2/H_2O reaction is $+0.82$ V (pH $= 7$). Therefore, the maximum theoretical potential difference, which can be obtained between electrode of MFC, is equal to $E_{cell} = 0.82 - (-0.2) = 1.02$ V [65]. However, modern MFCs have nominal voltage not exceeding 0.6 V, giving power density up to 2 W/m^2 of anode area from a wastewater and CE of 15%−30%. In ideal conditions when high concentrated "fuel" is used as acetate, for example, it is possible to recover power densities of up to $17-19$ W/m^2; however, such densities are not obtainable in real wastewater systems.

If in addition to organic compounds wastewater contains inorganic compounds, which can accept electrons (for example nitrates or higher valence metals), those electron acceptors undergo chemical transformations at the cathode in case when MFC is not separated by a membrane. This allows for example additional removal of metal cations through electrodeposition process on the cathode. Some of cathodic reactions with reaction standard potentials versus SHE are shown below:

$$O_2 + 4H^+ + 4e^- \rightarrow 2H_2O \qquad E^0 = 1.23 \text{ V} \qquad (3.133)$$

$$O_2 + 2H^+ + 2e^- \rightarrow H_2O_2 \qquad E^0 = 0.69 \text{ V} \qquad (3.134)$$

$$NO_3^- + 3H^+ + 2e^- \rightarrow HNO_2 + H_2O \quad E^0 = 0.94 \text{ V} \qquad (3.135)$$

$$Fe^{3+} + e^- \rightarrow Fe^{2+} \qquad E^0 = 0.77 \text{ V} \qquad (3.136)$$

$$MnO_2 + 4H^+ + 2e^- \rightarrow Mn^{2+} + 2H_2O \quad E^0 = 1.22 \text{ V} \qquad (3.137)$$

$$Cu^{2+} + 2e^- \rightarrow Cu \qquad E^0 = 0.34 \text{ V} \qquad (3.138)$$

$$2Cu^{2+} + H_2O + 2e^- \rightarrow Cu_2O + 2H^+ \quad E^0 = 0.21 \text{ V} \qquad (3.139)$$

$$Cu_2O + 2H^+ + 2e^- \rightarrow 2Cu + H_2O \quad E^0 = 0.06 \text{ V} \qquad (3.140)$$

When biocathodes are used in a single-chamber MFC, in addition to reactions (3.133 and 3.134), oxygen can be reduced to hydroxyl ions through the following reaction:

$$O_2 + 2H_2O + 4e^- \rightarrow 4OH^- \quad E^0 = 0.4 \text{ V} \qquad (3.141)$$

As a result metal ions present in the solution can react with OH$^-$ ions forming sediments of metal hydroxides, as in an example of cobalt ions shown below:

$$Co^{2+} + 2OH^- \rightarrow Co(OH)_2 \qquad (3.142)$$

3.5 PHOTOELECTROCATALYSIS

Photoelectrocatalysis (PEC) is the acceleration of a chemical reaction caused by the joint action of a catalyst activated by light irradiation and electric field. Photoelectrochemical reactions occur at the interface of two conducting phases with different conductivity character, such as electronic and ionic conductivities. PEC in water treatment is used for efficient organic compounds oxidation, inorganic ions reduction, and disinfection [66–68]. During PEC, a positive potential bias is applied to photocatalytic electrode, thus driving electron–hole pairs generated photocatalytically in the opposite directions. This prevents high electron–hole pair recombination rates and low efficiency of light utilization typical for singular photocatalytic process [69]. As a result, high rates of organic compounds mineralization can be obtained. Moreover, PEC eliminates the step of photocatalyst separation from the treated solution, which is used in photocatalysis. In PEC, photocatalyst particles are attached to a conductive solid substrate, so there is no necessity for catalyst separation from the solution. Photocatalytic electrodes can be simply withdrawn from the solution or solution can be pumped out of the reactor.

Principles of Photocatalysis

Light has a dual nature. While spreading, light shows the wave properties (phenomena of interference and diffraction) as well as light behaves as an elementary particle of matter (a quantum) while interacting with matter (emission and absorption).

Light quanta or photons exist only in motion with the speed of light, and they have a zero rest mass comparing to other particles. The photon energy is determined by the Plank–Einstein equation as follows:

$$E = h \cdot v_l = \frac{h \cdot c_l}{\lambda} \tag{3.143}$$

where h is the Plank constant ($h = 6.62 \cdot 10^{-34}$ J·s $= 4.13 \cdot 10^{-15}$ eV·s), v_l is the frequency of light (Hz); c_l is the speed of light ($c_l = 3.00 \times 10^8$ m/s); and λ is the light wavelength (m).

Photocatalysis is the change of rate or excitement of chemical reactions under the influence of light in the presence of substances (photocatalysts). Photocatalysts absorb light quanta and participate in the chemical reactions of the reactants, repeatedly engaging with them in the interim interactions, and regenerate their chemical composition after each cycle of such interactions. The absorption of photonic energy can occur in the film films, quantum dots, nanocomposite materials, macromolecules, etc. The use of metal electrodes in photocatalysis is limited because the energy of electron excitation is instantly dissipated and transformed into heat because of the strong interactions in the electron gas. In metals, valence band (VB) and conduction band (CB) are partly overlapped.

Photocatalytic transformations occurring in semiconductor material are explained by their electronic structure because the VB is separated from the CB by a band gap. Band gap (Fig. 3.23) is the energy difference (E_{gap}) between the bottom (lower level) of the CB and the ceiling (upper level) of the VB. Electrons inside the semiconductor can be either in free state, i.e., they can move in the crystal lattice (within the CB), or in bound state, i.e., they are chemically bonded with ions of the crystal lattice (in VB). The main feature of a semiconductor is the increase in the electrical conductivity with increasing temperature. Transition of electron from a bound state of VB into the free state of CB is associated with the EC of at least energy gap energy (E_{gap}), which can be provided by the light or heat energy. As a result of light excitation, electrons can move from the fully occupied VB to the CB, leaving unfilled levels in the VB, i.e., positive holes. Interactions between the electronic states in the VB and CB are weakened because of the presence of the band gap. Therefore, nonequilibrated electrons in the CB and holes in the VB have a relatively long lifetime. While moving in the crystal lattice, electrons and holes are either

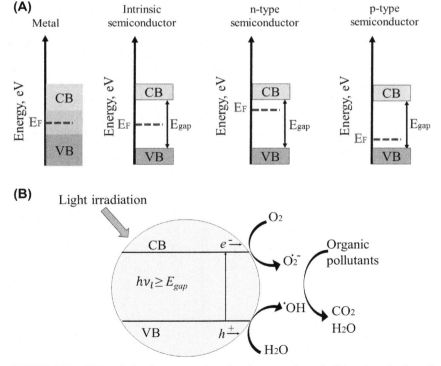

FIGURE 3.23 Electronic band structure of semiconductors and metals (A) and mechanism of photocatalytic pollutant degradation (B). *CB*, conduction band; E_F, the Fermi energy; E_{gap}, the band gap energy; *VB*, valence band.

partially recombined or come out to the surface. On the surface of semi-conductor, electrons and holes interact with oxygen, organic matter, and water, which leads to the formation of highly reactive oxygen-containing compounds and hydroxyl radicals. The latter can mineralize organic compounds adsorbed on the surface of catalysts. The presence of dissolved oxygen, which acts as an electron scavenger, is required to prevent their recombination. A graphic explanation of the mechanism of photocatalytic decomposition of organic compounds is shown in Fig. 3.23B. The band gap energy depends on the nature of the solid material. Typical values of band gap energy in semi-conductors comprise 0.1−4 eV (Fig. 3.24). The larger the band gap, the less likely the transfer of electron from the CB to the VB.

The Fermi level corresponds to the energy level (E_F), below which at 0 K all energy states are occupied by electrons, and above are free of electrons. The Fermi level lies within the band gap in semiconductors and plays a role of chemical potential of uncharged particles. The main meaning of the Fermi level is that at any temperature its occupation density by electrons is equal to 0.5. In other words, the Fermi energy (E_F) is the maximum value of energy, which an electron may have at absolute zero temperature. The Fermi level is located in the middle of the band gap in the semiconductor with intrinsic conductivity, i.e., undoped semiconductor, in which the concentration of excited electrons "n" is equal to the concentration of holes "p" (Si and Ge are examples of such semiconductors) (Fig. 3.23). Fermi level is shifted to the border of the major charge carriers in semiconductors with impurity conduction. In the case of n-type semiconductors, Fermi level is close to the bottom of CB and in p-type semiconductors, it is close to the ceiling of VB (Fig. 3.23).

As it was mentioned, irradiation of semiconductor (MO_x) by the light ($h \cdot \nu_l \geq E_{gap}$) facilitates generation of hole/electron pairs in semiconductor by

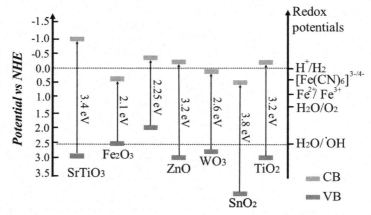

FIGURE 3.24 Typical band gap energies of some semiconductor materials. *CB*, conduction band; *NHE*, normal hydrogen electrode; *VB*, valence band.

excitation of electron from the VB to the CB. Schematically this process can be represented as follows:

$$MO_x \xrightarrow{h\nu} MO_x - e_{CB}^- + MO_x - h_{VB}^+ \tag{3.144}$$

Along with the useful reaction facilitated by the photoirradiation, free electron/hole can recombine in accordance with following reaction:

$$MO_x - e_{CB}^- + MO_x - h_{VB}^+ \rightarrow MO_x + \text{heat} \tag{3.145}$$

Semiconductors are divided into n-type and p-type depending on the type of conductivity. The semiconductor of n-type is impure in nature and conducts electricity like metals. The impurity elements, which are added to the structure of semiconductor to obtain n-type conductivity, are called donors. The term "n-type" comes from the word "negative," indicating a negative charge carried by the free electron. The n-type semiconductor can be obtained, for example, by doping pentavalent arsenic with tetravalent silicon. During the interaction, each arsenic atom forms a covalent bond with silicon atoms, but there is a fifth free atom of arsenic, which has no place in saturated valence bonds. The fifth arsenic atom shifts to the far electron orbital, where less energy is needed to separate electron from the atom. When electron is separated from the atom, it becomes capable of transferring charge. Thus, the charge transfer is carried out by electrons and not holes, which means that the n-type of semiconductors conduct electricity like metals. Oxygen vacancies act as donors in some oxides such as TiO_2 or Fe_2O_3. On the contrary, p-type semiconductors are characterized by conduction nature of holes ("p" means positive). Impurities, which are added to fabricate p-type semiconductor, are called acceptors in this case. For example, if tetravalent Si is doped with a small amount of trivalent indium, indium atoms will form a covalent bond with the three neighboring silicon atoms. Because silicon has one vacant bond, but indium atoms do not have a valence electron, it captures the valence electron from the covalent bond between the adjacent atoms of silicon and becomes a negatively charged ion, forming the so-called hole and thus a $p-n$ junction.

When placing semiconductor into electrolyte solution, the potential difference is established between the semiconductor and electrolyte due to the difference in their conductivities. Because electrolyte solution usually has higher conductivity than a semiconductor and because of the charge imbalance, band bending occurs mostly at the boundary layer of semiconductor and partly at the boundary layer of electrolyte solution.

Principles of Photoelectrocatalysis

PEC is one of the advanced oxidation methods, which allow efficient oxidation of organic compounds form wastewater, microorganism deactivation, CO_2 and ion reduction, etc. Reactions of hydroxyl radical formation are the main mechanism of organic compounds removal and water disinfection. Generation of hydroxyl radicals during PEC can be described by the following reactions [70].

Anode:

$$MO_x - h_{VB}^+ + H_2O_{ads} \rightarrow MO_x - HO_{ads}^\bullet + H^+ \tag{3.146}$$

$$MO_x - h_{VB}^+ + HO_{ads}^- \rightarrow MO_x - HO_{ads}^\bullet \tag{3.147}$$

$$MO_x - h_{VB}^+ + RX_{ads} \rightarrow MO_x + RX_{ads}^{\bullet+} \tag{3.148}$$

Cathode:

$$H_2O + e^- \rightarrow 1/2H_2 + OH^- \tag{3.149}$$

As it was mentioned earlier, PEC degradation of organic pollutants occurs because of simultaneous action of light irradiation and potential difference between the electrodes. PEC is based on the processes of photoexcitation of the electrode material, when valence electrons, adsorbed the light quantum, move to higher energy levels. It was also noticed that semiconductors are optimal materials to carry out the PEC process. The main benefits of the use of semiconductors as electrode materials are the following:

- Low concentration of free charge carriers contributes to the deep penetration of the external electric field in the electrode phase, thus forming a space charge layer in the junction region. SPL do not contain free charge carriers because they have been forced away by an electric field and it will be discussed later.
- Two types of free charge carriers (electrons and holes), where both can take part in electrode reactions.

Interactions at the Semiconductor/Electrolyte Solution Interface

Electrolyte solution containing some redox system ($Ox + ze^- \rightleftharpoons Red$) is characterized by a set of electron energy levels, which can be filled, i.e., correspond to reduced (Red) form, and free, corresponding to the oxidized (Ox) form. Reduction potential (E_{redox}) of redox reaction is described by the following equation:

$$E_{redox} = E^0 - \frac{kT}{z} ln \frac{C_{ox}}{C_{red}} \tag{3.150}$$

where C_{ox} and C_{red} are concentrations of oxidized and reduced forms; E^0 is the standard redox reaction potential (V); k is the Boltzmann constant ($k = 1.38 \cdot 10^{-23}$ J/K); T is the temperature (K); and z is the number of electrons, participating in the reaction. Redox potential is often called as "Fermi level of solutions." Similar to the Fermi level of electron conductors, E_{redox} value is called the "Fermi level of the solution." Actually, there is no Fermi distribution in the solution and no quasi-free electrons, but there are only bounded electrons in Red particles and free electron levels in Ox particles. When a semiconductor is placed into contact with electrolyte solution, the bands of semiconductor start to bend due to the flow of charge, which equilibrates E_F with E_{redox}. In this way, a junction of semiconductor/electrolyte interface (Schottky junction) is formed. If redox reaction in electrolyte solution is of good reversibility, then in dark conditions E_F becomes equal to E_{redox}. The amount of the band bending and direction of the bending depend on the difference between E_F and E_{redox} and type of semiconductor, respectively. In the case of n-type semiconductors there is a downward band bending. The band bending area is called the space charge layer (SCL). SCL is also called the *depletion layer*, when it is depleted with electrons and has an excess accumulation of positive charge, and *accumulation layer*, when it has an excess of electrons. When applying the potential bias to the system, it is possible to control the Fermi layer of semiconductor and provide better charge separation in the SCL. There is always an exact potential value for every semiconductor, which allows keeping the bands unbent or flat. This potential is called the flat-band potential (E_{fb}). When applying a potential greater than E_{fb} to n-type semiconductor, band bending will increase, and electrons will be depleted and holes will be accumulated in the SCL (Fig. 3.25A). When irradiating n-type semiconductor electrode with UV light, electrons become excited and shift to the conductance band and oxidizing holes are left at the semiconductor/electrolyte interface. In other words, simultaneous application of light irradiation and electric field generates free electrons and force them to move toward the cathode through the external circuit (Fig. 3.26). Holes gathered at the anode can oxidize water producing hydroxyl radicals and electrochemical species having redox potential below the potential of VB of the semiconductor. In turn, the electrons, concentrated at the cathode, can further react with oxygen producing superoxide radicals ($O_2^{-\cdot}$), which partially

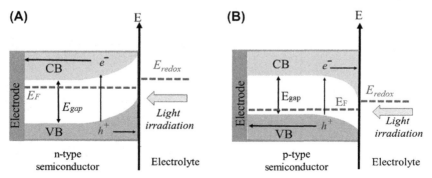

FIGURE 3.25 Band bending of n-type (A) and p-type (B) semiconductors under electric field and light irradiation. *CB*, conduction band; *E*, energy; E_F, Fermi level; E_{redox}, reduction potential; *VB*, valence band.

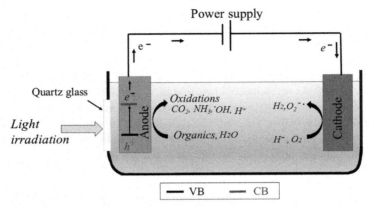

FIGURE 3.26 Schematic diagram of the photoelectrochemical cell setup. *CB*, conduction band; *VB*, valence band.

contributes to organic pollutant degradation. The general overall reaction of the oxidation of organic compounds at photocatalytic anode can be written as follows [71]:

$$C_yH_mO_jN_kX_q + (2y - j)H_2O$$
$$\rightarrow yCO_2 + qX^- + kNH_3 + (4y - 2j + m - 3k)H^+ + (4y - 2j + m - 3k - q)e^-$$

where X is a halogen atom.

The stronger the bending, the more efficient the electrochemical transformations, which is explained by better charge separation and lower electron–hole recombination. When external potential lower than E_{fb} is applied to p-type semiconductor (Fig. 3.25B), then the band bending will increase in the upward direction and holes are depleted in the SCL, while electrons are accumulated at the electrode/electrolyte interface. In this regard, n-type semiconductors such as TiO_2, SnO_2, or Fe_2O_3 are most suitable for the photoanodic oxidation reactions and p-type semiconductors, such as Cu_2O, NiO, or $CaFe_2O_4$, are suitable materials for photocathodic reduction. p-type semiconductors are less available comparing to the n-type materials.

TiO₂ and Parameters Influencing Its Efficiency in Photoelectrocatalysis

TiO_2 is the most commonly used material for photocatalysis as well as PEC. This is explained by its high chemical and thermal stability, low cost, and nontoxic composition. TiO_2 has a band gap of 3.2 eV, which means that the lowest wavelength able to activate the catalyst is equal to $\lambda = \frac{h \cdot c_l}{E} = \frac{4.13 \cdot 10^{-15} \cdot 3 \cdot 10^8}{4 \cdot 10^{-9}} = 3.87 \cdot 10^{-7} = 387$ nm or lower, i.e., the UV light. The share of UV light in the solar light is below 5%, which means that the efficiency of PEC is relatively low unless photocatalysts are exposed to the artificial source of UV irradiation (additional costs for the treatment). Therefore, it is necessary to decrease the band gap of semiconductors to make them active in the visible light. There are some attempts to modify TiO_2 catalyst by doping it with different elements (such as N, W, C, Zr, Si, Pt, Co) or using in composite photoelectrodes. Dopants replace oxygen in the TiO_2 structure. It was found that nitrogen introduction to TiO_2 allows reduction of band gap energy up to 2.7 eV. The band gap energy of carbon-doped TiO_2 could reach 2.3 eV and tungsten incorporated in the structure of TiO_2 led to the reduction of band gap energy up to 2.83 eV. When coating n-type TiO_2 with a p-type semiconductor of lower than TiO_2 band gap, it is possible to obtain a better separation of charges in composite photoanodes as well as to reduce the band gap of the composite. For example, Cu_2O/TiO_2 composite has the band gap energy of around 2.21 eV, when the band energy of TiO_2 is 3.2 eV and the band gap of Cu_2O is 2.2 eV. Composite materials consisting of TiO_2 and metal decrease the electron−hole recombination by trapping electrons in the metal part of the composite photoelectrode. Moreover, there are continuous studies on the search and development of TiO_2-free photoelectrodes.

The efficiency of organic compound degradation in PEC depends on many parameters such as pH and composition of electrolyte solution, type of photoelectrode and its morphology, applied potential, and the light wavelengths and intensity. The efficiency of the PEC is usually higher at greater values of the light intensity. Nanostructured materials have high surface area and therefore facilitate catalytic reactions and provide better efficiencies of PEC pollutant degradation. The pH of electrolyte solution and particularly the point of zero charge determine the charge of photocatalysts and hence their ability to adsorb different pollutant on the surface. For example, TiO_2 is charged positively in acidic solution and negatively in the basic ones, which can be represented by the following equations:

$$TiO_2 + H^+ \rightarrow TiOH_2^+ \qquad (3.151)$$

$$TiO_2 + OH^- \rightarrow TiO^- + H_2O \qquad (3.152)$$

The value of applied potential should exceed the flat band potential to provide separation of generated charges. The same is true for the applied current. The value of applied current should be adjusted in a way that ensures sufficient potential. Similar to other electrochemical processes, the efficiency of PEC treatment increases with the increase of applied current; however, there is an optimal value above which either there is no further improvement of the process or there is an inhibition of removal efficiency. This can be explained by either the maximum separation of carrier charges at an optimal applied current or deactivation of electrode due to corrosion or other mechanisms respectively.

The effect of different electrolytes on the efficiency of PEC process is similar to that in EO process (see Chapter 2). Moreover, photolysis can initiate the formation of different free radicals, such as chlorine radicals, which can contribute to the organic pollutant degradation reactions. Formation of chlorine radicals in PEC can be generated through reactions of chlorine gas formation from chloride ions at the anode with the following formation of hypochlorous acid, which in turn undergoes photolytic discharge to hydroxyl and chlorine radicals (3.153–3.157) [72].

$$2Cl^- - 2e^- \rightarrow Cl_2 \tag{3.153}$$

$$Cl_2 + H_2O \rightarrow HClO + HCl \tag{3.154}$$

$$HClO + h\nu_l \rightarrow OH^{\bullet} + Cl^{\bullet} \tag{3.155}$$

$$OCl^- + h\nu_l \rightarrow O^{\bullet-} + Cl^{\bullet} \tag{3.156}$$

$$O^{\bullet-} + H_2O \rightarrow OH^{\bullet} + OH^- \tag{3.157}$$

Presence of persulfate ions leads to the formation of sulfate free radicals, which can also contribute to the photolytic generation of chlorine radicals [73].

$$S_2O_8{}^{2-} + h\nu_l \rightarrow 2SO_4{}^{\bullet-} \tag{3.158}$$

$$SO_4{}^{\bullet-} + Cl^- \rightleftarrows SO_4{}^{2-} + Cl^{\bullet} \tag{3.159}$$

The Main Reactor Types Used in Photoelectrocatalysis

There are few reactor types used in PEC. In general, reactors can be rectangular (Fig. 3.26) or cylindrical and exposed to light irradiation from the outside of the reactor or from inside. Cylindrical reactors with the light irradiation source in the middle of the reactor provide better distribution of light over the cylindrical electrode (Fig. 3.27) and as result better excitation of the electrode and greater degradation of pollutants. Moreover, electrodes in PEC can be two dimensional with immobilized layer of photocatalyst on a

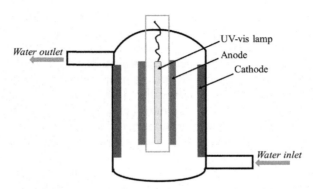

FIGURE 3.27 Cylindrical photoelectrocatalysis reactor.

conductive substrate and three-dimensional moving electrode. Three-dimensional electrode reactor provides higher degradation efficiencies of pollutants because of the enhanced mass transfer and surface area of photocatalyst. There are two types of three-dimensional electrode reactors. The first one is the three-dimensional electrode-slurry reactor, which is analogous to a fluidized bed reactor (Fig. 3.28A), and the second one is the three-dimensional electrode-packed bed reactor, which is analogous to a fixed bed reactor (Fig. 3.28B) [69]. Three-dimensional reactors consist of anode, cathode and particles of photocatalysts distributed between the electrodes. Photocatalytic particles in three-dimensional electrodes serve a role of microelectrodes owing to their polarization in electric field established between the anode and cathode. Compressed air is purged through the system to keep good mass transfer and maintain the fluidized bed in the reactor. Advantages of slurry-type reactor are the simplicity of construction, ease of catalyst replacement, and enhanced mass transfer. However, the main disadvantage of three-dimensional slurry-type reactors is the difficulty of catalyst separation from the treated water. On the opposite, three-dimensional electrode packed-bed reactor has a limited

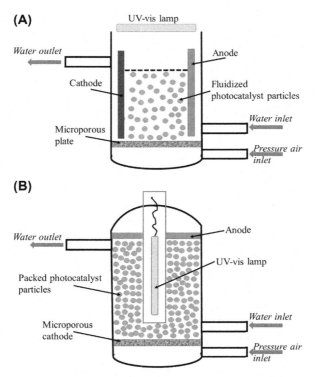

FIGURE 3.28 Three-dimensional photoelectrocatalysis reactors: (A) slurry-type reactor and (B) packed bed type reactor.

mass transfer and difficulties with catalyst replacement. However, degradation efficiencies in the packed-bed reactor are higher than in slurry reactor due to the better contact of treated solution with catalytic particles. Immobilized photoelectrodes in the form of plates allow overcoming the problem of catalyst separation from the treated solution and replacement typical for three-dimensional electrode systems; however, they have significantly lower surface area.

3.6 SONOELECTROCATALYSIS

One of the main disadvantages of electrochemical method is the polarization and passivation of electrodes due to poor mass transfer. Polarization can be also caused by gas accumulation near to the electrode surface and as a result depletion of pollutant in the electrode's boundary layer [74]. Passivation can be caused by the deposition of reaction products on the surface of electrodes, which results in diminishing of the process efficiency.

Ultrasound (US) technology is nonselective; therefore, it is suitable for the industrial wastewater treatment application. It does not require addition of chemicals to the treatment process and can be easy automated. However, ultrasonication alone often cannot achieve complete mineralization of organic pollutants and has low degradation rates. To provide synergistic effects and better degradation efficiency of organic pollutants, US is combined with other treatment methods such as Fenton, Fenton-like processes, ozonation, other oxidant addition, catalysis, photolysis, electrochemical remediation, and others. US, combined with electrochemical degradation process (sonoelectrochemical, or SEC), eliminates electrode contamination because of the continuous mechanical cleaning effect produced by the formation and collapse of acoustic cavitation bubbles near to the electrode surface. Moreover, US can reduce polarization by enhancing mass transfer of electroactive species near to the electrode surface and can activate catalytic surface of electrodes. The process is easy to automate. SEC decomposition of organic pollutant is a developing method. In addition to active radicals it can provide the complete oxidation of pollutants by formed oxidative species and thermal decomposition.

The intensive studies related to the use of SEC methods in pollutant degradation date back to the turn of the 21st century [75] where US assisted the electrochemical degradation of diuron herbicide, procion blue dye, and N,N-dimethyl-p-nitrosoaniline [76–78]. It was found that procion blue can be directly oxidized in US field on BDD anodes at potentials below 2.5 V versus SCE in acidic conditions since contamination of electrodes was enhanced in alkaline solutions [76]. US-assisted EC degradation was conducted to decompose diuron herbicide using glassy carbon plate anode and applying 35 and 20 kHz ultrasonic irradiation [78]. Glassy carbon anode was subjected to polarization at potentials higher than 1.4 V, therefore the electrolysis was conducted at lower potential of 1.16 V. After 8 h of SEC process, the degradation of diuron was equal to 72%. When conducting electrolysis in 35 kHz ultrasonic bath at 1.3 V applied voltage, the degradation efficiency did not exceed 43%. That was explained by the fact that the cleaning efficiency of the US horn is higher than in US bath thus preventing the passivation of the electrodes.

The Theory of Ultrasound

Sound is a mechanical longitudinal wave in which the particle oscillations are in the same plane as the direction of energy propagation (Fig. 3.29). The wave carries energy, but not the matter. In contrast to electromagnetic waves, such as light and radio waves, sound waves cannot propagate in a vacuum. Similar to all kinds of waves, sound can be described by the following parameters, such as frequency (f, Hz), wavelength (λ, m), propagation speed (c, m/s) in a medium, period (T, s), amplitude (A), and intensity (I, W/m^2). Frequency, period, amplitude, and intensity depend on the sound source; the propagation speed depends on the medium, and wavelength on both the sound source and the medium. Frequency is the number of complete cycles for a period of 1 s (Fig. 3.29). For example, the figure shows that two complete oscillations were made by sound wave for 1 s, which means that frequency of the sound wave depicted in Fig. 3.29 is equal to 2 Hz. The relationship between the frequency and the wave period is shown in the following equation:

$$f = \frac{1}{T_a} \tag{3.160}$$

The speed of sound in an ideal elastic material at a given temperature and pressure is constant. Relationship between the ultrasonic speed and wavelength is as follows:

$$\lambda = \frac{c}{f} \tag{3.161}$$

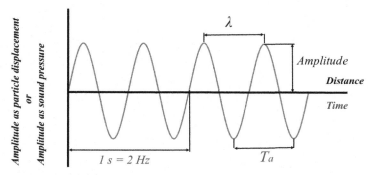

FIGURE 3.29 The main parameters of the ultrasonic wave. T_a, sound/acoustic period; λ, wavelength.

Sound intensity (acoustic intensity, I_a, W/m^2) is the average energy carried by a sound wave through a unit area (A) perpendicular to the direction of wave propagation. The intensity of the US is the value that expresses the power (P, W) of the acoustic field at the point.

$$I_a = \frac{P}{A} = \frac{p \cdot v}{2} = \frac{\rho \cdot c \cdot v^2}{2} \qquad (3.162)$$

where v is the sound particle velocity (physical speed of a parcel of fluid) (m/s) and ρ is the density of the medium (kg/m^3).

The intensity is proportional to the amplitude (sound pressure) squared ($I \sim p^2$) and pressure amplitude (sound pressure) is proportional to the square root of sound intensity ($p \sim \sqrt{I}$).

US waves are sound waves inaudible for humans having frequencies from 20 kHz and up to 10^6 kHz. US waves are transmitted in water via series of compression and rarefaction cycles, which lead to formation of high and low pressure zones in the medium respectively [79]. In rarefaction cycle a sudden drop in hydrostatic pressure in the liquid below a certain value can cause a rupture of liquid continuity, thus resulting into formation of cavitation bubbles filled with saturated liquid vapor [80]. Bubble undergo a continuous grow in rarefaction cycles and compression during subsequent compression cycles (Fig. 3.30).

While reaching an equilibrium size, stable cavitation bubbles continue to vibrate generating acoustic microflows. Transient cavitation bubbles further

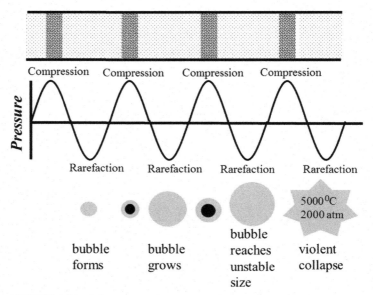

FIGURE 3.30 The mechanism of cavitation bubbles formation and collapse.

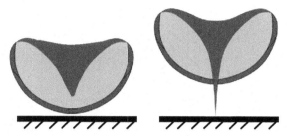

FIGURE 3.31 Formation of liquid microjet during asymmetric cavitation bubble collapse near a solid surface.

grow in size and then collapse violently. At the moment of collapse, gas pressure and temperature inside the bubbles reach up to 400 MPa and 2000−5000 K, respectively; and the bulk solution surrounding the bubble is heated up to 2000 K [81]. Moreover, attenuating spherical shock waves are transmitted in the liquid immediately after collapse. In addition to shock waves and acoustic microflows, the formation of liquid microjets (Fig. 3.31) having speed up to few hundred meters per second takes place during the asymmetric bubbles collapse on/or close to a solid surface. The abovementioned mechanical effects of US improve the mass transfer of electroactive species in the bulk solution, remove the passivation layer from the electrode surface, enhance electrolytic current mode and thus activate electrode surface, and assist electrochemical treatment process [82]. Moreover, US degassing of the generated gases from the electrodes surface significantly reduces polarization of electrodes. Static pressure equilibrium of bubbles in water can be determined as follows [83].

$$p_g + p_v = p_{hs} + \frac{2\gamma}{r} \qquad (3.163)$$

where p_g and p_v are pressure of gas and vapor inside the bubble, respectively; p_{hs} is the hydrostatic pressure; γ is the surface tension; r is the bubble radius. A sudden drop in hydrostatic pressure in the liquid below a certain value causes the formation of cavitation bubbles because of rupture of fluid continuity.

There are three theories explaining cavitation bubble collapse. These are (1) electrical, where electrical discharge is produced during asymmetric bubble implosion [84]; (2) plasma theory implying microplasma formation inside the cavitation bubble [85]; and (3) hot-spot theory, which is the most used due to its simplicity and the fact that other theoretical approaches could not describe various findings [86]. In the hot-spot theory there are three regions where chemical reactions take place (Fig. 3.32). A hot-spot hydrophobic gaseous nucleus (1) initiates high local temperatures inside the cavity and in interfacial region (2). Since temperature of hot spot is enormously high, the

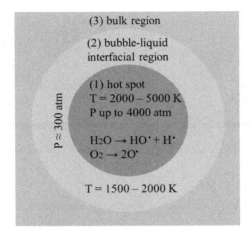

(3) bulk region

(2) bubble-liquid
interfacial region

(1) hot spot
T = 2000 – 5000 K
P up to 4000 atm

$H_2O \rightarrow HO^\bullet + H^\bullet$
$O_2 \rightarrow 2O^\bullet$

T = 1500 – 2000 K

$P \approx 300$ atm

FIGURE 3.32   The hot-spot model of cavitation.

heat disperses rapidly and warms up the bulk solution (3). It was suggested that degradation of compounds inside the cavity (hot spot) including solvent decomposition occurs through the thermal decomposition mechanism. Pyrolysis and hydroxylation are the main pathways of compounds decomposition in the interfacial region. Decomposition of compounds in the bulk solutions occurs by hydroxylation and reactions with hydroxyl radicals [87].

In addition to physical effect of US, chemical reactions described by the hot-spot theory and taking place at cavitation bubble collapse directly contribute to the degradation of pollutants. The extreme temperatures generated at cavitation bubbles collapse causing thermal decomposition reactions of pollutants and water vapor inside the cavitation bubble or hot spot. Water thermolysis leads to the formation of hydroxyl radicals, which can further react with organic pollutants oxidizing them or recombine forming hydrogen peroxide according to the following reactions:

$$H_2O \rightarrow {}^\bullet OH + H^\bullet \tag{3.164}$$

$$H^\bullet + O_2 \rightarrow HO_2{}^\bullet \tag{3.165}$$

$$2{}^\bullet OH \rightarrow H_2O_2 \tag{3.166}$$

$$2HO_2{}^\bullet \rightarrow H_2O_2 + O_2 \tag{3.167}$$

$$H^\bullet + O_2 \rightarrow {}^\bullet OH + O^\bullet \tag{3.168}$$

The presence of salts, gases, and other particles in the solution irradiated by US creates additional centers for the cavitation, leading to the enhanced production of ${}^\bullet OH$ and other secondary radicals, which in turn can increase the degradation efficiency of pollutants. Thus, for example, the presence of carbon tetrachloride, persulfate, and periodate ions initiates the formation of highly reactive radicals shown in the following reactions [88–91].

$$CCl_4 +))) \rightarrow {}^{\bullet}CCl_3 + {}^{\bullet}Cl \tag{3.169}$$

$${}^{\bullet}CCl_3 + H^{\bullet} \rightarrow HCCl_3 \tag{3.170}$$

$$HCCl_3 \rightarrow {}^{\bullet}HCCl_2 + {}^{\bullet}Cl \tag{3.171}$$

$${}^{\bullet}HCCl_2 + H^{\bullet} \rightarrow H_2CCl_2 \tag{3.172}$$

$$H_2CCl_2 +))) \rightarrow {}^{\bullet}H_2CCl + {}^{\bullet}Cl \tag{3.173}$$

$${}^{\bullet}Cl + {}^{\bullet}Cl \rightarrow Cl_2 \tag{3.174}$$

$$Cl_2 + H_2O \rightarrow HClO + HCl \tag{3.175}$$

$$S_2O_8^{2-} +))) \rightarrow 2SO_4^{-\bullet} \tag{3.176}$$

$$IO_4^- +))) \rightarrow IO_3^{\bullet} + O^{\bullet-} \tag{3.177}$$

$$O^{\bullet-} + H^+ \rightarrow {}^{\bullet}OH \tag{3.178}$$

$${}^{\bullet}OH + IO_4^- \rightarrow OH^- + IO_4^{\bullet} \tag{3.179}$$

There many parameters influencing the performance of US and as a result the efficiency of SEC process. They are ultrasonic frequency and intensity, temperature, composition of treated solution, etc. The main contributions of these parameters to cavitation effect are summarized in Table 3.3. In addition to the parameters listed in Table 3.3, transducer placement within the reactor and reactor geometry influence the distribution of ultrasonic field and, as a result, efficiency of sonochemical reactions. However, the modern ultrasonic systems are equipped with amplitude unit control, which allows automatic adjustment of amplitude to a constant value, and a unit for automatic over-lapping ultrasonic waves, which is achieved by sweeping the frequency in the range of ± 1 kHz, which eliminates the formation of dead zones and hot spots [92,93].

Principles of Sonoelectrocatalysis

SEC degradation of organic compounds is a relatively new developing tech-nique, which shows high mineralization rates in degradation of organic compounds. SEC degradation methods are based on the synergetic effect of separate mechanism of EO and US degradation, which were described in Chapter 2 and this section, respectively. The effects of sonication involving shock waves and shear forces promote almost all electrochemical processes. Acoustic streaming and liquid microjets activate an electrode surface by removing passivating impurities from the electrodes surface, enhancing elec-trolytic current mode, and facilitating mass transfer of electroactive species

TABLE 3.3 The Main Parameters That Influence the Cavitation

Parameter	Effect of Different Parameters of Cavitation Process
Frequency	Higher frequency leads to shorter acoustic cycles, smaller cavitation bubble size, shorter bubble collapse time, higher ˙OH formation. Lower frequencies provide more violent collapse.
Dissolved gases	Reduced tensile strength of the liquid initiates the cavitation process at lower pressures. The more gas nuclei in the liquid, the lower intensity of generated shock waves. Monatomic gases generate more energy on collapse than polyatomic gases, which is related to the heat conductivity of gases.
Intensity	There is a minimum level of intensity, which is required to initiate cavitation. Higher intensity generates more bubbles. However, too many bubbles may prevent the distribution of ultrasonic power into entire solution; therefore an optimal level of intensity should be found.
Temperature	Higher ambient temperature reduces viscosity of the liquid and surface tension, which results in cavitation threshold decrease and higher bubbles formation; however, bubble collapse is less violent. Too many bubbles may reduce the distribution of ultrasonic power into entire solution. Temperatures above the boiling point of the media significantly reduce the effect of sonochemical reactions.
Pressure	Higher ambient pressure reduces the number of bubble collapses at the same intensity; however, collapses are more violent.

[82]. Degassing effect of US can reduce polarization of electrodes by removing the generated gases from the electrodes surface. The extreme conditions generated at cavitation bubbles collapse provide additional mechanisms of pollutant degradation through generation of hydroxyl radicals and thermal decomposition. However, free radicals produced during SEC process can also be consumed by recombination reactions. When summarizing the advantages and disadvantages of EO and US treatments it can be concluded that combined SEC treatment is suitable for the degradation of highly toxic compounds regardless of their concentration and turbidity at ambient conditions.

Nevertheless despite the advantages of US observed in combination with electrocatalysis, there are a number of disadvantages, which should be taken into account. First of all, due to the strong cavitation and cleaning effect, US can initiate and enhance the corrosion of electrodes in the case when too high powers are used. The same could happen when corrosion of the electrodes has already started for some reasons, for example due to the enhanced oxygen and chlorine gas evolution or elevated applied current densities. In this regard, optimization of ultrasonical (frequency, power, reactor geometry, etc.) and electrochemical (applied current density and voltage) working parameters is required in SEC process. Usually decreased powers and voltages are required for SEC process compared to singular EO and US processes. It is worth to notice that lower ultrasonic frequencies promote higher cleaning and mechanical effect of US in enhancement of degradation processes with a domination of turbulent acoustic streaming [94]. Higher frequencies have major contribution to the degradation through free radicals formation [95]. A spatial setup of electrodes and transducer in the reaction vessel significantly influence sound distribution and the result has an effect on pollutant degradation efficiencies and electrode deterioration. The effect of US on decrease of decomposition voltage can be explained by the fact of reducing anodic reactions overpotential and increasing cathodic reactions overpotentials. This was shown on the example of $Ag(S_2O_3)_2^{3-}/Ag$ redox couple where shift of anodic potentials was observed at 20 and 500 kHz sonication of different intensity [96]. Cathodic polarization depression was obtained also in chromium deposition process under continuous acoustic irradiation [97]. The two-sided effect of US, which can both promote and suppress the corrosion of metals and metal alloys, was reported in a number of works [98−100].

The synergetic effect of combined EO and US processes can be calculated according to the following equation [101].

$$S = \frac{k_{US/EO}}{k_{EO} + k_{US}} \tag{3.180}$$

where S is the synergetic index; k_{EO}, k_{US}, and k_{SEC} are rate constants in EO, US, and SEC degradation processes, respectively.

Reactor Types Used in Sonoelectrochemical Degradation

All SEC reactors can be arranged in three major setups depending on the ultrasonic reactor type (Fig. 3.33). The first setup is an ultrasonic bath type with working electrodes immersed into the reactor. The distribution of acoustic field in bath type reactors is considered to be even [75]. The second group consists of working electrodes and ultrasonic horn placed in the reactor volume. The third setup uses a combination of working electrode and sonotrode in one device.

A special attention should be paid to the distance between US source in the case of horn-type sonotrodes and electrodes because it influences the intensity of mass transfer, etching of electrode surfaces, polarization intensity, etc. [102]. The approximate distance from the transducer to the vicinity of reactor

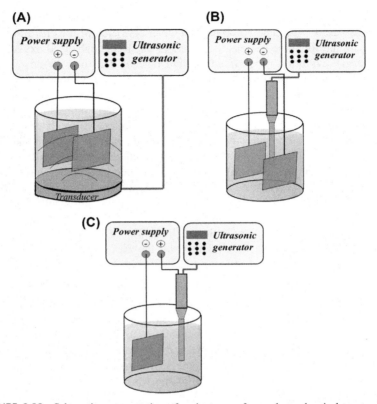

FIGURE 3.33 Schematic representation of main types of sonoelectrochemical reactors. (A) Ultrasonic bath type; (B) electrochemical reactor with separately placed ultrasonic horn; (C) electrochemical reactor with combined sonotrode and working electrode.

where the maximum peak of US pressure occurs is described by Eq. (3.181) [103]:

$$N = \frac{d^2 f}{4c} \tag{3.181}$$

where N is near field distance, d is transducer's diameter, f is frequency, and c is sound velocity in water.

Sonoelectrochemical Degradation of Organic Compounds

Over the last few years, ultrasonically assisted electrochemical methods were tested for the degradation of a range of different compounds such as pesticides, dyes, and pharmaceuticals. [104–108]. Table 3.4 contains a summary of extended works conducted in the field of SEC degradation of organic pollutants.

TABLE 3.4 Efficiency of SEC in Degradation of Different Organic Compounds in Water

Pollutant/Initial Concentration	Working Parameters		Removal Efficiency	Time/Conditions	References
	Electrochemical	Sonochemical			
Amaranth dye 100 mg/L	BDD anode $I = 35$ mA/cm^2	20 kHz 523 W/cm^2	95.1% TOC	1.5 h	[109]
Cl reactive black 8100 mg/L	Ti/RuO$_2$–IrO$_2$ anode $I = 31.7$ A/cm^2	20 kHz 100 W/L	32.4% COD	1.5 h pH 5.4	[110]
Pentachlorophenol 2 mg/L	Pt anode $E_{el} = 30$ V	22.5 W/L 35 kHz 170 kHz 300 kHz 500 kHz 700 kHz	75% 65% 50% 45% 40%	1 h	[111]
Perchloroethylene 60 mg/L	PbO$_2$ anode $I = 3.5$ mA/cm^2	20 kHz 0.27 W/mL	100% FC 56%	6 h	[112]
Phenol 0.5 mM	Stainless steel anode $E_{el} = 30$ V	850 kHz 170 W	98%	1 h	[113]
Sandolan yellow 50 mg/L	Pt anode $I = 60$ mA	40 kHz 0.011 W/mL	75% decolor	80 min	[104]
Reactive blue 19 dye 50 mg/L	PbO$_2$ anode $E_{el} = 10$ V	80 kHz 0.3 W/mL	90% decolor 56% TOC	2 h pH 8	[107]
Diuron 18–26 mg/L	BDD anode $I = 60$ mA/cm^2	20 kHz 750 W	100% 43% TOC	8 h 10°C pH 12	[114]
Rhodamine B 5 mg/L	Pt anode $E_{el} = 4$ V	22 kHz 400 W	91% decolor	6 min pH 6.5	[108]

Continued

TABLE 3.4 Efficiency of SEC in Degradation of Different Organic Compounds in Water—cont'd

Pollutant/Initial Concentration	Working Parameters		Removal Efficiency	Time/Conditions	References
	Electrochemical	Sonochemical			
Lissamine green B 20 mg/L	Graphite anode $E_{el} = 5$ V	20 kHz 1.05 W/mL	95% decolor 60% COD	2 h	[115]
Acid black 2475 mg/L			50.2% decolor 19% COD		
Methyl orange 20 mg/L			84% decolor 27% COD		
Reactive black 5 65 mg/L			73% decolor 22% COD		
Trupocor red 50 mg/L			78% decolor 24% COD		
Methyl paraben 100 mg/L	BDD 21.6 mA/cm²	20 kHz 523 W/cm²	50% TOC	2 h pH 5.7 T = 25°C	[116]
Methylene blue 200 mg/L	Ti/TiO₂–IrO₂–RuO₂ anode $I = 11$ mA/cm²	45 kHz 0.42 W/mL	92% TOC	1 h	[117]
Trichloroacetic acid 0.005 M	Pt/Ti anode $I = 4$ mA/cm²	24 kHz 0.032 W/mL	96%	6 h	[106]
Triclosan 10 mg/L	$E_{el} = 10$ V DCN anode Silicon carbide Stainless steel	850 kHz 0.425 W/mL	67% 61% 81%	15 min	[118]
2,4-dihydroxybenzoic acid 300 mg/L	Ti/Pt electrodes $I = 300$ A/m²	500 kHz 0.75 W/mL	47% TOC	1.5·A·h	[119]

BDD, boron-doped diamond electrode; *COD*, chemical oxygen demand; *SEC*, sonoelectrochemical; *TOC*, total organic carbon.

3.7 SUMMARY

The constant development of science and technology leads to the emergence of new electrochemical water treatment methods, such as EDI, CDI, PEC, SEC, treatment of water by EF processes, and MFCs. EDI is essentially and electrodialysis assisted by ion-exchange treatment by means of filling the intermembrane space with ion-exchange resins. Such an assembly increases the electrical conductivity by providing additional electrical path for ions, which allows to purify dilute solutions and, consequently, increases the efficiency of treatment. Moreover, the step of ion-exchange regeneration is eliminated in the process. While the diluate compartments should always be filled with a mixture of cation-exchange and anion-exchange resins, the concentrate compartments can either be filled with resin or free of it. However, the ion-exchange resins on both sides of the membranes provide higher demineralization efficiencies due to the enhanced ion transport. Because ion-exchange resins are selective toward different hydrated radii of ions initially separating ions with smaller hydrated radius, the efficiency of desalination will vary for different ions. The method of fractional EDI, conducted into in two stages with different magnitude of applied voltage or current intensity, partially solves this problem, leveling the degree of treatment for different ions.

CDI is another water desalination method used for water softening, brackish water desalination, wastewater treatment, etc. Anodes and cathodes are made of a porous carbon material, which sorb ions of different signs from electrolyte solution, thus purifying it, when potential difference is applied between the electrodes. When changing the polarity of electrodes, pollutants start to desorb forming the concentrate solution. To increase the efficiency of desalination, CDI is combined with ion-exchange membranes, which are tightly mounted to the electrode surface on the side of the electrolyte solution. Anion-exchange membrane on the anode side and cation-exchange membrane on the cathode side improve the efficiency of ion separation during electrosorption and regeneration steps of the treatment. The mechanism of CDI unit work is based on the mechanism of work of supercapacitors and therefore obeys similar laws, which are assisted by the theory of EDLs.

EF process is one of the methods of AOPs used for the degradation of organic pollutants by hydroxyl radicals formed from the electrogenerated Fenton's reagent (H_2O_2 and Fe(II) ions). EF provides higher mineralization efficiencies compared to conventional Fenton reaction. This is explained by additional •OH radical generation at the anode. Moreover, in situ generation of Fenton's reagent eliminates the step of chemicals dosage and reduces the amount of chemical required for the process. Electrodes for EF process should have high overpotentials toward hydrogen and oxygen generation as well as be inert to the corrosive environment of the process. Among the most used cathodes for EF treatment are various carbon and graphite-based materials.

BDD, mixed-metal oxide, Pt, and graphite electrodes are commonly used anodes for the process.

MFC technology is a combination of anaerobic biological water treatment conducted in the anodic compartment of the cell. Biofilm acting as a biocatalyst is attached to the anode surface. An ion-exchange membrane allowing maintaining different conditions in reactor compartments separates anodic and cathodic compartments. Anodic compartment is fed with polluted water and cathodic compartment is fed with electrolyte solution and purged with the air or oxygen gas. Microorganisms consume contaminants to maintain their vital activity and, as a result of this, activity electrons are released when substrates are degraded. Because of the difference of redox potentials, released electrons move to the cathode though the external electric circuit. Hence, MFC technology allows simultaneous water treatment and recovery of electrical energy. MFC process can also be performed in a single-compartment reactor when using air cathode. The efficiency of water treatment in MFCs strongly depends on the type of microorganisms and electrode materials used. To obtain higher degradation rates of pollutants, it is necessary to use mixed cultures of microorganism, high surface area anodes, and cathodes with low overpotential toward oxygen reduction.

PEC and SEC are another advanced oxidation methods, which can be used for the degradation of organic compounds, reduction of inorganic compounds. In PEC, semiconductor catalyst attached to a conductive substrate is activated by the joint action of light irradiation and electric field. The presence of a band gap in the electronic structure of semiconductors increases the lifetime of light-excited electrons in the CB and holes left in the VB. Additionally, positive potential bias applied to the electrode drives electron and holes generated due to the light excitation in the opposite directions. This makes possible the interaction of excited electrons and holes with water and pollutant molecules on the surface of the catalyst. As a result of such interactions, reactive radicals are formed and further consumed for the pollutant degradation reactions. SEC is a method of EO assisted by US. One of the main problems of EO process is the polarization and passivation of the electrodes. US allows solving this problem owing to the strong mechanical effect of cavitation. Shock waves and liquid microjets generating at the cavitation bubble collapse keep the surface of electrodes clean of contaminants and improve the mass transfer in the treated solution. Moreover, high temperatures and pressures formed at collapse contributes to water thermolysis and hence the hydroxyl radical formation, which also contribute to the enhanced efficiency of pollutant degradation.

SELF-CONTROL QUESTIONS

1. What is the working principle of MFC? What is the difference between mediated and nonmediated MFC?
2. What are the main advantages and disadvantages of MFC process?

3. What is the working principle of EDI? What are the typical reactor solutions used in EDI?
4. What are the pros and cons of EDI process? How electric conductivity can be improved in EDI units? What are the difference between EDI and fractional EDI processes?
5. How can ion-exchange resins be classified? Please specify in details.
6. What are the main principles of water desalination by CDI?
7. What are the main theories explaining the formation of electrical double layer inside the micropores of electrode in CDI treatment? Please specify essential provisions of these theories. What is the main difference between the classical Gouy–Chapman–Stern theory and modified Donnan model?
8. How does membrane capacitive deionization differ from the conventional CDI process?
9. What types of electrochemical cells are used in CDI depending on a distribution of feed water flow?
10. What are the main characteristics of sound waves? What is ultrasound? What are the main theories explaining cavitation effect?
11. How can ultrasonic irradiators be combined with electrochemical degradation in one reactor?
12. What is the mechanism of photocatalysis? What equation describes the photon energy?
13. What are the ways to decrease the band gap of semiconductors? Please provide some examples.
14. What kinds of reactor configurations are used in PEC?
15. What is the mechanism of electro-Fenton degradation of organic compounds?
16. What kinds of reactor configurations are used in electro-Fenton process?
17. What is capacitive deionization? What is the mechanism of water desalination by CDI?

REFERENCES

[1] Ö. Arar, Ü. Yüksel, N. Kabay, M. Yüksel, Various applications of electrodeionization (EDI) method for water, Desalination 342 (2014) 16–22.

[2] J. Marek, M. Simunkova, H. Parschova, Clogging of the electrodeionization chamber, Desalin. Water Treat. 56 (2015) 3259–3263.

[3] T. Prato, C. Gallagher, Using EDI to meet the needs of pure water production, GE Power Water. Water Process Technol. TP1078EN.doc (2010) 1–5.

[4] L. Alvarado, A. Chen, Electrodeionization: principles, strategies and applications, Electrochim. Acta 132 (2014) 583–597.

[5] Q.G. LLC, Fractional Electrodeionization (FEDI) Process, 2016. (Online). Available: http://quagroup.com/fedi/.

[6] W. Walters, D. Weiser, L. Marek, Concentration of radioactive aqueous wastes, Ind. Eng. Chem. 47 (1) (1955) 61−67.

[7] Z. Matějka, Continuous production of high-purity water by electro-deionisation electro-deionisation, J. Chem. Technol. Biotechnol. 21 (4) (1971) 117−120.

[8] S. Porada, R. Zhao, A. van der Wal, V. Presser, P. Biesheuvel, Review on the science and technology of water desalination by capacitive deionization, Prog. Mater. Sci. 58 (2013) 1388−1442.

[9] L. Fuzhi, X. Zhao, Method for Processing Low Radioactive Waste Liquid by Continuous Electrodeionization, March 10, 2010. China Patent CN101665277 A.

[10] J.-H. Song, K.-H. Yeon, J. Cho, S.-H. Moon, Effects of the operating parameters on the reverse osmosis-electrodeionization performance in the production of high purity water, Korean J. Chem. Eng. 22 (1) (2005) 108−114.

[11] Z. Zhang, A. Chen, Simultaneous removal of nitrate and hardness ions from groundwater using electrodeionization, Sep. Purif. Technol. 164 (2016) 107−113.

[12] Y. Zhang, L. Wang, S. Xuan, X. Lin, X. Luo, Variable effects on electrodeionization for removal of Cs^+ ions from simulated wastewater, Desalination 344 (2014) 212−218.

[13] Y.S. Dzyazko, L. Ponomaryova, L. Rozhdestvenskaya, S. Vasilyuk, V. Belyakov, Electrodeionization of low-concentrated multicomponent Ni^{2+}-containing solutions using organic−inorganic ion-exchanger, Desalination 342 (2014) 43−51.

[14] P. Spoor, L. Grabovska, L. Koene, L. Janssen, W.R. ter Veen, Pilot scale deionisation of a galvanic nickel solution using a hybrid ion-exchange/electrodialysis system, Chem. Eng. J. 89 (2002) 193−202.

[15] Y. Xing, X. Chen, P. Yao, D. Wang, Continuous electrodeionization for removal and recovery of Cr(VI) from wastewater, Sep. Purif. Technol. 67 (2009) 123−126.

[16] A. Mahmoud, A. Hoadley, An evaluation of a hybrid ion exchange electrodialysis process in the recovery of heavy metals from simulated dilute industrial wastewater, Water. Res. 46 (2012) 3364−3376.

[17] K.-H. Yeon, J.-H. Song, S.-H. Moon, A study on stack configuration of continuous electrodeionization for removal of heavy metal ions from the primary coolant of a nuclear power plant, Water Res. 38 (2004) 1911−1921.

[18] D. Golubenko, J. Křivčík, A. Yaroslavtsev, Evaluating the effectiveness of ion exchangers for the electrodeionization process, Pet. Chem. 55 (10) (2015) 769−775.

[19] Lopez-Rosa, Novel Separation Methods Using Electrodialysis/Electrodeionization for Product Recovery and Power Generation, University of Arkansas, 2015.

[20] D. Caudle, J. Tucker, J. Cooper, B. Arnold, A. Papastamataki, Electrochemical Demineralization of Water with Carbon Electrodes, Reserach Report, Oklahoma University Research Institute, Oklahoma, 1966.

[21] A. Johnson, J. Newman, Desalting by means of porous carbon electrodes, J. Electrochem. Soc. 118 (3) (1971) 510−517.

[22] J.-B. Lee, K.-K. Park, H.-M. Eum, C.-W. Lee, Desalination of a thermal power plant wastewater by membrane capacitive deionization, Desalination 196 (2006) 125−134.

[23] P. Biesheuvel, R. Zhao, S. Porada, A. van der Wal, Theory of membrane capacitive deionization including the effect of the electrode pore space, J. Colloid Interface Sci. 360 (2011) 239−248.

[24] H. Li, L. Gao, Y. Zhang, Y. Chen, Z. Sun, Electrosorptive desalination by carbon nanotubes and nanofibres electrodes and ion-exchange membranes, Water Res. 42 (20) (2008) 4923−4928.

[25] Y.-J. Kim, J.-H. Choi, Improvement of desalination efficiency in capacitive deionization using a carbon electrode coated with an ion-exchange polymer, Water Res. 44 (3) (2010) 99−996.

[26] H. Li, L. Zou, L. Pan, Z. Sun, Novel graphene-like electrodes for capacitive deionization, Environ. Sci. Technol. 44 (2010) 8692−8697.

[27] Z. Wang, L. Dou, G. Zheng, Z. Liu, Z. Hao, Effective desalination by capacitive deionization with functional graphene nanocomposite as novel electrode material, Desalination 299 (2012) 96−102.

[28] J. Choi, H. Lee, S. Hong, Capacitive deionization (CDI) integrated with monovalent cation selective membrane for producing divalent cation-rich solution, Desalination 400 (2016) 38−46.

[29] F. Xing, T. Li, J. Li, H. Zhu, N. Wang, X. Cao, Chemically exfoliated MoS_2 for capacitive deionization of saline water, Nano Energy 31 (2017) 590−595.

[30] H. Li, Y. Ma, R. Niu, Improved capacitive deionization performance by coupling TiO_2 nanoparticles with carbon nanotubes, Sep. Purif. Technol. 171 (2016) 93−100.

[31] M. Suss, T.F. Baumann, W. Bourcier, C.M. Spadaccini, J. Santiago, M. Stadermann, Capacitive desalination with flow-through electrodes, Energy Environ. Sci. (11) (2012).

[32] W. Bourcier, R. Aines, J. Haslam, C. Schaldach, K. O'Brien, E. Cussler, USA Patent US20070170060 A1, July 26, 2007.

[33] M. Andelman, G. Walker, Charge Barrier Flow-Through Capacitor, September 9, 2004. USA Patent US20040174657 A1.

[34] A. Heldenbrand, Development of a Predictive and Mechanistic Model for Capacitive Deionization, M.Sc. thesis, Rice University, Houston, TX, 2015.

[35] F. AlMarzooqi, A. Al Ghaferi, I. Saadat, N. Hilal, Application of capacitive deionisation in water desalination: a review, Desalination 342 (2014) 3−15.

[36] Y. Oren, Capacitive deionization (CDI) for desalination and water treatment—past, present and future (a review), Desalination 228 (2008) 10−29.

[37] J. Choi, Transport Phenomena in an Ion-Exchange Membrane at Under- and Over-Limiting Current Regions, Ph.D. thesis, GIST, Korea, 2002.

[38] H. Pröbstle, C. Schmitt, J. Fricke, Button cell supercapacitors with monolithic carbon aerogels, J. Power Sources 105 (2) (2002) 189−194.

[39] D. Zhang, T. Yan, L. Shi, Z. Peng, X. Wen, J. Zhang, Enhanced capacitive deionization performance of graphene/carbon nanotube, J. Mater. Chem. 22 (2012) 14696−14704.

[40] P. Biescheuvel, S. Porada, M. Levi, M. Bazant, Attractive forces in microporous carbon electrodes for capacitive deionization, J Solid State Electrochem. 18 (2014) 1365−1376.

[41] J. Dykstra, R. Zhao, P. Biesheuvel, A. van der Wal, Resistance identification and rational process design in Capacitive Deionization, Water Res. 88 (2016) 358−370.

[42] E. Brillas, I. Sires, M. Oturan, Electro-Fenton process and related electrochemical technologies based on Fenton's reaction chemistry, Chem. Rev. 109 (2009) 6570−6631.

[43] J. Pignatello, E. Oliveros, A. MacKay, Advanced oxidation processes for organic contaminant destruction based on the fenton reaction and related chemistry, Crit. Rev. Environ. Sci. Technol. 36 (2006) 1−84.

[44] R. Tomat, A. Rigo, Electrochemical production of hydroxyl radicals and their reaction with toluene, J. Appl. Electrochem. 6 (1976) 257−261.

[45] R. Tomat, A. Rigo, Oxidation of polymethylated benzenes promoted by hydroxyl radicals, J. Appl. Electrochem. 9 (1979) 301−305.

[46] M. Sudoh, T. Kodera, K. Sakai, J. Zhang, K. Koide, Oxidative degradation of aqueous phenol effluent with electrogenerated Fenton's reagent, J. Chem. Eng. Jpn. 6 (1986) 513−518.

[47] H. Fenton, On a new reaction of tartaric acid, Chem. News 33 (1876) 190.

[48] H. Fenton, Oxidation of tartaric acid in presence of iron, J. Chem. Soc. 65 (1894) 899−910.

[49] I. Tokmakov, M. Lin, Kinetics and mechanism of the OH + C_6H_6 reaction: a detailed analysis with first-principles calculations, J. Phys. Chem. A 106 (2002) 11309−11326.

[50] M. Oturan, J. Pinson, N. Oturan, D. Deprez, Hydroxylation of aromatic drugs by the electro-Fenton method. Formation and identification of the metabolites of Riluzole, New J. Chem. 23 (1999) 793−794.

[51] N. Daneshvar, S. Aber, V. Vatanpour, M. Rasoulirard, Electro-Fenton treatment of dye solution containing Orange II: influence of operational parameters, J.Electroanal. Chem. 615 (2) (2008) 165−174.

[52] C. Badellino, C. Rodrigues, R. Bertazzoli, Oxidation of herbicides by in situ synthesized hydrogen peroxide and fenton's reagent in an electrochemical flow reactor: study of the degradation of 2,4-dichlorophenoxyacetic acid, J. Appl. Electrochem. 37 (4) (2007) 451−459.

[53] M. Pimentel, N. Oturan, M. Dezotti, M. Oturan, Phenol degradation by advanced electrochemical oxidation process electro-Fenton using a carbon felt cathode, Appl. Cat. B Environ. 83 (1−2) (2008) 140−149.

[54] U. Schröder, Editorial: microbial fuel cells and microbial electrochemistry: into the next century!, ChemSusChem 5 (6) (2012) 959.

[55] A. Parkash, Microbial fuel cells: a source of bioenergy, J. Microb. Biochem. Technol. 8 (3) (2016) 247−255.

[56] B. Logan, M. Wallack, K.-Y. Kim, W. He, Y. Feng, P. Saikaly, Assessment of microbial fuel cell configurations and power densities, Environ. Sci. Technol. Lett. 2 (2015) 206−214.

[57] L. Tender, S. Gray, E. Groveman, J. Dobarro, The first demonstration of a microbial fuel cell as a viable power supply: powering a meteorological buoy, J. Power Sources 179 (2) (2008) 571−575.

[58] O. Bukach, L. Myakinkova, Microbal fuel cells: state of research and practical application, Innovatika i ekspertiza 2 (13) (2014) 51−59. (in Russian). Innovatics and Expert Examination.

[59] T. M. -R. Company, Carbon Fiber Brushes, The Mill-Rose Company, 2017.

[60] CeTech, Electrodes for VRB, CeTech Co., LTD., 2017.

[61] H. Saeed, G. Husseini, S. Yousef, J. Saif, S. Al-Asheh, A. Abu Fara, S. Azzam, R. Khawaga, A. aidan, Microbial desalination cell technology: a review and a case study, Desalination 359 (2015) 1−13.

[62] M. Lu, S. Yau Li, Cathode reactions and applications in microbial fuel cells: a review, Crit. Rev. Environ. Sci. Technol. 42 (2012) 2504−2525.

[63] R. Nastro, Microbial fuel cells in waste treatment: recent advances, IJPE 10 (4) (2014) 367−376.

[64] G. Delaney, H. Benetto, J. Mason, S. Roller, J. Stirling, C. Thurston, Electron-transfer coupling in microbial fuel cells. 2. Performance of fuel cells containing selected microorganism—mediator—substrate combinations, J. Chem. Technol. Biotechnol. 34 (1) (1984) 13−27.

[65] S. Kalyuzhnyi, V. Fedorovich, Microbial fuel cells, Himiya i zhizn 5 (2007) 36−39. (in Russian).

[66] R. Domínguez-Espíndola, J. Varia, A. Alvarez-Gallegos, M. Ortiz-Hernández, J. Peña-Camacho, S. Silva-Martínez, Photoelectrocatalytic inactivation of faecal coliform bacteria in urban wastewater using nanoparticulated films of TiO_2 and TiO_2/Ag, Environ. Technol. 38 (5) (2017) 606−614.

[67] F. Paschoal, G. Pepping, M. Boldin Zanoni, M. Anderson, Photoelectrocatalytic removal of bromate using Ti/TiO_2 coated as a photocathode, Environ. Sci. Technol. 43 (2009) 7496−7502.

[68] M. Brugnera, K. Rajeshwar, J. Cardoso, M. Boldin Zanoni, Bisphenol A removal from wastewater using self-organized TiO_2 nanotubular array electrodes, Chemosphere 78 (5) (2010) 569−575.

[69] X. Meng, Z. Zhang, X. Li, Synergetic photoelectrocatalytic reactors for environmental remediation: a review, J. Photochem. Photobiol. C: Photochem. Rev. 24 (2015) 83−101.

[70] G. Bessegato, T. Guaraldo, J. de Brito, M. Brugnera, M. Boldrin Zanoni, Achievements and trends in photoelectrocatalysis: from environmental to energy applications, Electrocatalysis 6 (2015) 415−441.

[71] J. Qiu, S. Zhang, H. Zhao, Nanostructured TiO_2 photocatalysts for the determination of organic pollutants, J. Hazard. Mater. 211−212 (2012) 381−388.

[72] A. El-Kalliny, Photocatalytic Oxidation in Drinking Water Treatment Using Hypochlorite and Titanium Dioxide, CPI Koninklijke Wöhrmann, The Netherlands, 2013.

[73] X. Yu, Z. Bao, J. Barker, Free radical reactions involving $Cl^{\cdot}$, $Cl_2^{-\cdot}$, and $SO_4^{-\cdot}$ in the 248 nm photolysis of aqueous solutions containing $S_2O_8{}^{2-}$ and Cl^-, J. Phys. Chem. A 108 (2004) 295−308.

[74] B.-S. Lee, H.-Y. Park, I. Choi, M. Cho, H.-J. Kim, S.J. Yoo, D. Henkensmeier, J. Kim, S. Nam, S. Park, K.-Y. Lee, J.H. Jang, Polarization characteristics of a low catalyst loading PEM water electrolyzer operating at elevated temperature, J. Power Sources 309 (2016) 127−134.

[75] B. Thokchom, A.B. Pandit, P. Qiu, B. Park, J. Choi, J. Khim, A review on sonoelectrochemical technology as an upcoming alternative for pollutant degradation, Ultrason. Sonochem. 27 (2015) 210−234.

[76] J. Foord, K. Holt, R. Compton, F. Marken, D. Kim, Mechanistics aspects of the sonoelectrochemical degradation of the reactive dye procion blue at boron-doped diamond electrodes. Diam. Relat. Mater. 10 (2001) 662−666.

[77] K. Holt, C. Forryan, R. Compton, J. Foord, F. Marken, Anodic activity of boron-doped diamond electrodes in bleaching processes: effects of ultrasound and surface states, New J. Chem. 27 (2003) 698−703.

[78] K. Macounova, J. Klima, C. Bernard, C. Degrand, Ultrasound-assisted anodic oxidation of diuron, J. Electroanal. Chem. 457 (1998) 141−147.

[79] T.J. Mason, E. Riera, A. Vercet, P. Lopez-Buesa, Application of ultrasound, in: Emerging Technologies for Food Processing, Academic Press, 2005, p. 325.

[80] N. Pokhrel, P. Vabbina, N. Pala, Sonochemistry: science and engineering, Ultrason. Sonochem. 29 (2016) 104−128.

[81] K. Suslick, The chemical effects of ultrasound, Sci. Am. 260 (1989) 80−86.

[82] T.J. Mason, J.P. Lorimer, D.J. Walton, Sonoelectrochemistry, Ultrasonics 28 (1990) 333−337.

[83] P. Eisenberg, Cavitation. Hydronautics Incorporated, Massachusetts Institute of Technology. Available: http://web.mit.edu/hml/ncfmf/16CAV.pdf.

[84] M.A. Margulis, I.A. Margulis, Mechanism of sonochemical reactions and sonoluminescence, High Energy Chem. 38 (5) (2004) 285–294.

[85] S.I. Nikitenko, Plasma formation during acoustic cavitation: toward a new paradigm for sonochemistry, Adv. Phys. Chem. 2014 (2014) 1–8.

[86] Y. Adewuyi, Reviews – sonochemistry: environmental science and engineering applications, Ind. Eng. Chem. Res. 40 (2001) 4681–4715.

[87] T. Sivasankar, V. Moholkar, Physical insights into the sonochemical degradation of recalcitrant organic pollutants with cavitation bubble dynamics. Ultrason. Sonochem. 16 (6) (2009) 769–781.

[88] S. Wang, N. Zhou, S. Wu, Q. Zhang, Z. Yang, Modeling the oxidation kinetics of sonoactivated persulfate's process on the degradation of humic acid, Ultrason. Sonochem. 23 (2015) 128–134.

[89] L. Weavers, I. Hua, M. Hoffmann, Degradation of triethanolamine and chemical oxygen demand reduction in wastewater by photoactivated periodate, Water Environ. Res. 69 (6) (1997) 1112–1119.

[90] X.-K. Wang, Y.-C. Wei, C. Wang, W.-L. Guo, J.-G. Wang, J.-X. Jiang, Ultrasonic degradation of reactive brilliant red K-2BP in water with CCl$_4$ enhancement: performance optimization and degradation mechanism, Sep. Purif. Technol. 81 (2011) 69–76.

[91] L. Wang, L. Zhu, W. Luo, Y. Wu, H. Tang, Drastically enhanced ultrasonic decolorization of methyl orange by adding CCl$_4$, Ultrason. Sonochem. 14 (2007) 253–258.

[92] F. Incorporated, Ultrasonic Cleaning Equipment for Ultimate Cleaning Power, Fusion Incorporated, (Online). Available: http://www.fusion-inc.com/wp-content/uploads/2013/08/Ultrasonic_Cleaning_Equipment.pdf.

[93] U. Meinhardt, Ultrasonic Transducer E/805/T02, (Online). Available: http://www.meinhardt-ultraschall.de/page03_e.html.

[94] F. Marken, R. Akkermans, R. Compton, Voltammetry in the presence of ultrasound: the limit of acoustic streaming induced diffusion layer thinning and the effect of solvent viscosity, J. Electroanal. Chem. 415 (1996) 55–63.

[95] Y. Adewuyi, Sonochemistry in environmental remediation. 1. Combinative and hybrid sonophotochemical oxidation processes for the treatment of pollutants in water, Environ. Sci. Technol. 39 (10) (2005) 3409–3420.

[96] B. Pollet, J. Lorimer, S. Phull, T. Mason, D. Walton, J. Hihn, V. Ligier, M. Wéry, The effect of ultrasonic frequency and intensity upon electrode kinetic parameters for the Ag(S$_2$O$_3$) 23–/Ag redox couple, J. Appl. Electrochem. 29 (12) (1999) 1359–1366.

[97] J. Dereska, Y. Ernest, F. Hovorka, Effects of acoustical waves on the electrodeposition of chromium, J. Acoust. Soc. Am. 29 (1957) 769.

[98] C. Kwok, F. Cheng, Man, Synergistic effect of cavitation erosion and corrosion of various engineering alloys in 3.5% NaCl solution, Mater. Sci. Eng. A 290 (1–2) (2000) 145–154.

[99] G. Whillock, B. Harvey, Ultrasonically enhanced corrosion of 304L stainless steel II: the effect of frequency, acoustic power and horn to specimen distance, Ultrason. Sonochem. 4 (1) (1997) 33–38.

[100] R. Wang, K. Nakasa, Effect of ultrasonic wave on the growth of corrosion pits on SUS304 stainless steel, Mater. Trans. 48 (5) (2007) 1017–1022.

[101] P. Finkbeiner, M. Franke, F. Anschuetz, A. Ignaszak, M. Stelter, P. Braeutigam, Sonoelectrochemical degradation of the anti-inflammatory drug diclofenac in water, Chem. Eng. J. 273 (2015) 214–222.

[102] S. Coleman, S. Roy, Effect of ultrasound on mass transfer during electrodeposition for electrodes separated by a narrow gap, Chem. Eng. J. 113 (2014) 35–44.

[103] M. Thein, A. Cheng, P. Khanna, C. Zhang, E.-J. Park, D. Ahmed, C.J. Goodrich, F. Asphahani, F. Wu, N. Smith, C. Dong, X. Jiang, M. Zhang, J. Xu, Site—specific sonoporation of human melanoma cells at the cellular level using high lateral—resolution ultrasonic micro—transducer arrays, Biosens. Bioelectron. 27 (1) (2011) 25—33.

[104] J. Lorimer, T. Mason, M. Plattes, S. Phull, Dye effluent decolourisation using ultrasonically assisted electro-oxidation. Ultrason. Sonochem. 7 (2000) 237—242.

[105] R. Alkire, S. Perusich, The effect of focused ultrasound on the electrochemical passivity of iron in sulfuric-acid. Corros. Sci. 1121—1132 (1983) 23.

[106] M. Esclapez, V. Saez, D. Milán-Yáñez, I. Tudela, O. Louisnard, J. González-García, Sonoelectrochemical treatment of water polluted with trichloroacetic acid: from sono-voltammetry to pre-pilot plant scale, Ultrason. Sonochem. 17 (2010) 1010—1020.

[107] M. Siddique, R. Farooq, Z. Khan, Z. Khan, S. Shaukat, Enhanced decomposition of reactive blue 19 dye in ultrasound assisted electrochemical reactor, Ultrason. Sonochem. 18 (2011) 190—196.

[108] Z. Ai, J. Li, L. Zhang, S. Lee, Rapid decolorization of azo dyes in aqueous solution by an ultrasound-assisted electrocatalytic oxidation process, Ultrason. Sonochem. 17 (2010) 370—375.

[109] J. Steter, W. Barros, M. Lanza, A. Motheo, Electrochemical and sonoelectrochemical processes applied to amaranth dye degradation, Chemosphere 117 (2014) 200—207.

[110] J. Wu, F. Liu, H. Zhang, J. Zhang, L. Li, Decolorization of CI Reactive Black 8 by electrochemical process with/without ultrasonic irradiation, Desalin. Water Treat. 44 (2012) 36—43.

[111] K. Kim, E. Cho, B. Thokchom, M. Cui, M. Jang, J. Khim, Synergistic sonoelectrochemical removal of substituted phenols: implications of ultrasonic parameters and physicochemical properties, Ultrason. Sonochem. 24 (2015) 172—177.

[112] V. Sáez, M. Esclapez, I. Tudela, P. Bonete, O. Louisnard, J. González-García, 20 kHz sonoelectrochemical degradation of perchloroethylene in sodium sulfate aqueous media: influence of the operational variables in batch mode, J. Hazard. Mater. 183 (2010) 648—654.

[113] Y.-Z. Ren, Z.-L. Wu, M. Franke, P. Braeutigam, B. Ondruschka, D. Comensky, P. King, Sonoelectrochemical degradation of phenol in aqueous solutions, Ultrason. Sonochem. 20 (2013) 715—721.

[114] E. Bringas, J. Saiz, I. Ortiz, Kinetics of ultrasound-enhanced electrochemical oxidation of diuron on boron-doped diamond electrodes, Chem. Eng. J. 172 (2—3) (2011) 1016—1022.

[115] M. Rivera, M. Pazos, M. Sanromán, Improvement of dye electrochemical treatment by combination with ultrasound technique, J. Chem. Technol. Biotechnol. 84 (8) (2009) 1118—1124.

[116] J. Steter, D. Dionisio, M. Lanza, A. Motheo, Electrochemical and sonoelectrochemical processes applied to the degradation of the endocrine disruptor methyl paraben, J. Appl. Electrochem. 44 (2014) 1317—1325.

[117] B. Yang, J. Zuo, X. Tang, F. Liu, X. Yu, X. Tang, H. Jiang, L. Gan, Effective ultrasound electrochemical degradation of methylene blue wastewater using a nanocoated electrode, Ultrason. Sonochem. 21 (2014) 1310—1317.

[118] Y.-Z. Ren, M. Franke, F. Anschuetz, B. Ondruschka, A. Ignaszak, Sonoelectrochemical degradation of triclosan in water, Ultrason. Sonochem. 21 (2014) 2020—2025.

[119] R. de Lima Leite, P. Cognet, A.-M. Wilhelm, H. Delmas, Anodic oxidation of 2,4-dihydroxybenzoic acid for wastewater treatment: study of ultrasound activation, Chem. Eng. Sci. 57 (2002) 767—778.

Chapter 4

Equipment for Electrochemical Water Treatment

Mika Sillanpää, Marina Shestakova

Lappeenranta University of Technology, Lappeenranta, Finland

NOMENCLATURE

Latin Alphabet

A	Area occupied by EC reactor	m^2
A_F	Area of flotation compartment	m^2
a	Side dimension of square cross section of the EC compartment	m
b	Width of electrode plate	m
B'	Gap between the electrode system and the wall of reactor	m
b_{st}	Thickness of the electrode stack	m
B	Reactor width	m
C	Initial concentration of pollutant in wastewater	g/m^3
d	Diameter	m
$d_{in.\ pipe}$	Diameter of the water inlet pipe	m
D	Dose of metal ions	g/m^3
G	Hourly metal ion consumption	g/h
h	Working height of electrode plate	m
H_0	Layer height of the treated liquid	m
H_r	Total reactor height	m
$h_1,\ h_2$	Height of the froth layer and the side of the apparatus above the froth layer	m
I	Applied current	A
j	Current density	A/m^2
k	Electrochemical equivalent	$g/A \cdot h$
l	Distance between electrodes	m
L	Reactor length	m
L_F	Length of flotation compartment	m
L_1	Length of the collector of treated water	m
m_a	Mass of anode	kg
N_{EC}	Total number of electrode plates in EC	
$N_{a,EF},\ N_{c,EF}$	Total number of anode and cathode plates in the EF process	

Electrochemical Water Treatment Methods. http://dx.doi.org/10.1016/B978-0-12-811462-9.00004-9
227

q_g	Deterioration of graphite during electrolysis	mg/A·h
Q	Water flow rate	m³/h
q_{Me}	Specific metal consumption	g/g
q_l	Specific electricity consumption	A·h/g
ρ_g	Gas density	kg/m³
ρ	Electrolytic conductivity (specific conductance)	1/Ω cm
ρ_l	Liquid density	kg/m³
S	Surface area	m²
S'	Surface area of one electrode	m²
t	Time of continuous operation hours of electrodes	h
τ_{EC}, τ_{EF}	Residence time of wastewater in EC and EF reactors, respectively	h
T	Solution temperature	K
V_a	Anode volume	m³
V_p	Volume of the premises	m³
V_r	Working volume of reactor	m³
$\vartheta_{in.\ pipe}$	Velocity of water movement in the inlet water pipe	m/s
W	Weight of the electrode system	kg
x	Thickness of dielectric spacer	m
γ	Density of electrode material	kg/m³
δ	Electrode thickness	m

Abbreviations

AO	Advanced oxidation
BOD	Biological oxygen demand
CE	Current efficiency
CF	Cash flow
COD	Chemical oxygen demand
EA	Electrochemical activation
ED	Electrochemical disinfection
EDI	Electrodeionization
ER	Electrochemical reactor
EC	Electrocoagulation
EF	Electroflotation
EO	Electrooxidation
EWS	Electro Water Separation
TSS	Total suspended solids
TDS	Total dissolved solids
WWTP	Wastewater treatment plant
3D	Three-dimensional

4.1 ELECTROCHEMICAL REACTORS

Electrochemical reactors (ERs) are devices in which electrochemical processes are carried out. Schematic diagram of electrochemical reaction operation was shown and discussed in Chapter 1 (Fig. 1.6). Various types of ER find applications in wastewater treatment, particularly widely in the treatment of wastewaters from electroplating industry. This is explained by the fact that wastewater from electroplating industry is a good electrolyte because it contains various metal ions.

To ensure high efficiency of metal recovery from diluted wastewaters, three-dimensional (3D) electrodes are used. There are three types of 3D electrodes. They are plate-type electrodes, flow-through voluminous porous electrodes, and fluidized electrodes. The operating principle of ER with the electrodes of the first two types is that the solution to be treated is passed through channels in the body 3D electrode. At the same time, electrode potential is maintained at a level that ensures the maximum removal rate of metal ions. An example of plate-type 3D cathode is shown in Fig. 4.1A. The electrode is a pack of plates separated by washers in such a way that there is a gap between adjacent plate allowing wastewater flow in a direction perpendicular to the flow of electric current. Such cathode is mounted perpendicular to the anode in reactors for metal recovery. The conditions of electrolysis are chosen in a way that allows obtaining powdered metals. Metals are removed then from the bottom of the reactor where particles detached from the cathode under the influence of gravity are precipitated.

Highly porous flow-through 3D electrodes are made of particulate conductive materials, metal pellets, particles, or graphite fibers, which are cold-pressed, hot-pressed, or sintered to form the electrode structure [1]. The electrode usually is a chamber with perforated walls between which a layer of graphite felt, graphite, carbon, or other particles is pressed or placed (Fig. 4.1B) [2]. Fiber carbon—graphite materials are obtained by calcination of textile materials based on rayon. Electrodes of such configuration are more suitable for microbial fuel as cathodes or diluted wastewaters because they are

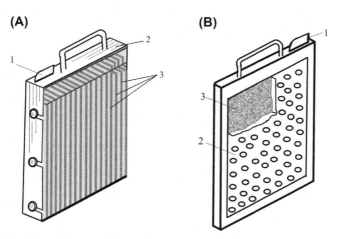

FIGURE 4.1 Three-dimensional electrodes of plate type (A) and flow-through type (B). 1, Electro-conductive bus; 2, frame; 3, electrodes.

prone to contamination. Wastewater passes through the voids between the pressed particles or fibers, which leads to metal deposition on such cathodes and organic compounds oxidation on such anodes.

Fluidized electrodes are conductive particulate materials (graphite, metal, etc.) moving between the current collectors and confining grid. The diameter of holes of current collectors and confining grid is less than the average particle size of the fluidized bed, thus preventing particle carryover. The operation principle of fluidized electrodes (Fig. 3.28) was discussed in Chapter 3 under photoelectrocatalysis.

Fig. 4.2 shows a diagram of electrochemical regeneration of spent iron—copper chloride etching solution, used in the process of printed circuit boards. As one of the simplest embodiments, the etching solution comprises ferric chloride salt and hydrochloric acid. During the etching process, copper is dissolved to copper(II) ions and iron(III) is reduced. The overall reaction of the etching process can be written as follows:

$$2FeCl_3 + Cu \rightarrow 2FeCl_2 + CuCl_2 \qquad (4.1)$$

As it is seen from the above reaction, spent etchant contains copper(II) ions, which are valuable components. To reduce the consumption of chemicals in the process and decrease the load onto environment, spent etching solution can be regenerated electrolytically (Fig. 4.2). The spent etching solution is delivered to heat exchanger where it is cooled. After heat exchangers, cooled water through the hydraulic locks and head tanks is delivered to electrolyzers. Under the influence of electric field, metallic copper is deposited at the

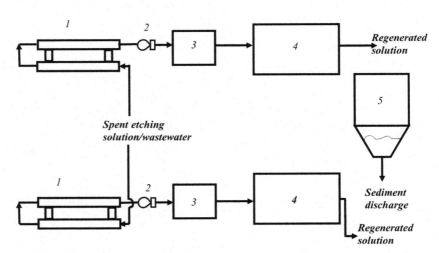

FIGURE 4.2 Diagram of electrochemical regeneration of spent iron—copper chloride etching solution. 1, Heat exchangers; 2, hydraulic locks; 3, head tank; 4, electrochemical reactors; 5, apparatus for copper precipitate removal.

cathode (4.2) and anodic reaction is accompanied by the oxidation of ferrous iron to ferric ions (4.3). As a result of anodic and cathodic reactions, solution can be recycled for reuse.

$$\text{Cathode:} \quad Cu^{2+} + 2e^- \rightarrow Cu \quad\quad (4.2)$$

$$\text{Anode:} \quad Fe^{2+} \rightarrow Fe^{3+} + e^- \quad\quad (4.3)$$

Copper metal is deposited at the cathode in the form of fine powder and periodically removed from the surface of cathodes. Copper deposition cycles last for 2 min, after which the stack of cathodes is removed and transferred into the unit provided with a mechanism for cleaning metal deposit from the electrode surface. Hydrochloric acid is often added to the apparatus of copper deposit removal to intensify the process. When cathodes are purified of deposits, they are returned back to ER. Duplication of treatment equipment in the scheme enables its continuous operation. When one line works for the deposition, another is purified from deposits. The operation of equipment is automated.

Electrochemical devices of this kind make it possible to solve the problems associated with the concentration of metal salts contained in the wastewater and return the concentrated solution for reuse. This is explained by the electrolyzer design. ER for conducting the electrolysis is divided into two compartments by an ion-exchange membrane or diaphragm (Fig. 4.3).

Both compartments contain stacks of plate electrodes. The polarity of electrodes in chambers is changed from time to time. Wastewater entering into the electrolyzer is divided into two streams. The major wastewater stream is delivered to the compartment 1 that is working in the cathodic mode (Fig. 4.3A). Discharge of metal ions takes place in this compartment leading to metal deposit generation and simultaneous water treatment. The second smaller part of water flow enters the compartment 2, where anodic dissolution of metal deposited in the previous cycle of the compartment operating as a cathode stack takes place (Fig. 4.3B). When polarity of electrodes is changed,

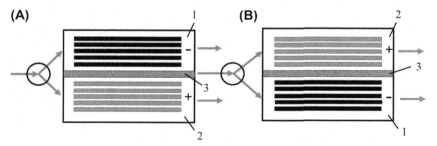

FIGURE 4.3 Working principle of electrochemical reactor for concentration of metal salts. 1, Cathodic compartment; 2, anodic compartment; 3, separator (ion-exchange membrane or diaphragm). (A) The mode of cathodic deposition of metals and water treatment and (B) the mode of anodic dissolution of deposited metals and concentration of metal salts.

metal previously dissolved at the cathode starts to dissolve. Thus, for example, it is possible to increase the content of metal salts from below 1 g/L to several tens and return the solution for reuse.

In addition to the removal and recovery of inorganic pollutants, electrochemical methods can be successfully used for wastewater treatment from organic compounds. Electrochemical decomposition of organic compounds can be carried out either in electromembrane reactors and diaphragm reactors or in reactors without ion-exchange membrane. The presence of the ion-exchange membranes and diaphragms between the electrodes leads to a significant voltage increase in the reactor and, consequently, to an increase in electricity consumption. It should also be noticed that electrolyzers with diaphragms have more complex construction. Usually materials used in industry for the separation of anodic and cathodic processes in wastewater treatment applications do not have 100% selectivity. This is explained by their high costs. Therefore, the process of electrochemical destruction of compounds does not occur in the desired mode. Moreover, during the electrolysis process, diaphragm and ion-exchange membranes change their structure and are often clogged with electrolysis products or mechanical impurities of electrolyte solutions, which cause the necessity of membrane replacement or regeneration. However, the use of diaphragms and ion-exchange membranes in the electrochemical decomposition of organic pollutants can be required in some cases. For example, the treatment of wastewater from nitro compounds is advantageously to carry out in diaphragm-type reactor in two stages. Reduction of nitro compounds to amino compounds takes place in the cathodic compartment at the first stage and the oxidation of generated amines to nontoxic products occurs in the anodic compartment at the second stage.

The structures of ERs free of diaphragms and ion-exchange membranes are very diverse. Fig. 4.4 shows an example membrane-free reactor for the oxidation of organic compounds. The reactor consists of a number of electrolysis cells formed by vertically arranged electrodes, which are separated by insulating gaskets. Water inlet pipe and purified water outlet pipe are welded to the bottom part of outermost electrodes that act as the cathodes. Additionally, the overflow pipe with a level gauge is connected to the outermost electrode on the side of the water inlet pipe. At the upper part of the reactor, electrodes are welded to the bar with holes for fastening the cover. Anodes and cathodes are made in the form of plates with apertures for the passage of wastewater. All electrodes are connected to the bus on the side, which supplies the stack with current. Electrodes and gaskets are assembled as a single stack with stud bolts. Reactor has legs for fixing it to the foundation. There are two gas pipes at the top of the reactor. One pipe is the air inlet pipe for dilution of electrolytic gases, and the other one is for discharge of diluted gases. The reactor operates as follows. Wastewater is supplied through the water inlet pipe to the first cell, where it moves from right to left. Next through the apertures on the left side of the second electrode, water enters the second cell, then the third, and so on.

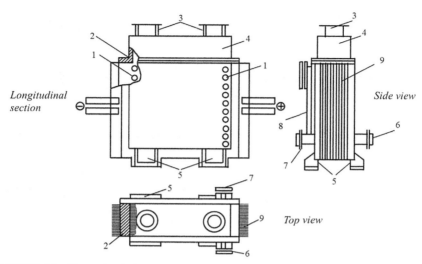

FIGURE 4.4 Electrochemical reactor for decomposition of organic compounds. 1, Apertures in electrodes; 2, insulating gaskets; 3, gas outlet and air inlet pipes; 4, cover; 5, legs; 6, purified water outlet pipe; 7, wastewater inlet pipe; 8, overflow pipe with level gauge; 9, electrodes.

Thus, the zigzag course of the treated liquid in a horizontal plane is obtained in the reactor. The treated water is discharged through the water outlet pipe from the electrolyzer.

Figs. 4.5 and 4.6 show examples of incorporation of electrochemical methods in the treatment schemes of surfactant-containing wastewaters and oily waters from petrochemical industry, respectively [3].

Synthetic surface-active substances (surfactants) are widely used for cleaning parts of machines and aggregates in technology. Spent cleaning solutions contain high concentrations of pollutants such as asphalt/tar oil components, corrosion products of equipment, surfactants, and dispersed minerals.

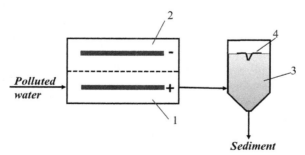

FIGURE 4.5 Membrane electrolyzer in the flow chart of wastewater treatment from surfactants. 1, Anode compartment; 2, cathode compartment; 3, settler; 4, oil trap.

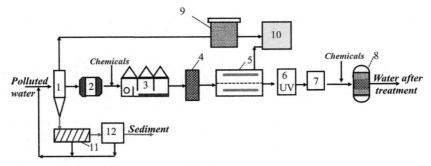

FIGURE 4.6 Schematic diagram of water treatment from industrial oil-containing wastewaters. 1, Hydrocyclone; 2, ferromagnetic filter; 3, three-compartment electroflotation reactor; 4, strainer; 5, electrochemical reactor for water treatment; 6, water treatment with UV irradiation; 7, water tank; 8, electrochemical filter; 9, hydrophobic filter; 10, accumulation tank; 11, precipitation tank; 12, filter press.

The treatment of such solution is complicated because of the high stability of surfactants. Destabilization of spent cleaning solutions take place in acidic media of pH 1−3. The spent cleaning solutions are passed through the anodic compartment of membrane ER (Fig. 4.5) followed by clarification in the settler, where oil products are separated from suspended solids. Oil products float up, accumulate in the upper part of the settler, and are removed from the system using an oil trap. The clarified water is neutralized by passing through the cathode compartment of ER.

The scheme in Fig. 4.6 combines chemical, mechanical, and advanced oxidation methods in one scheme. Wastewater sequentially passes through the hydrocyclone, ferromagnetic filter, combined pressure and electroflotation (EF) reactor, strainer, electrochemical reactor, UV-lamp treatment unit, clean water tank, and electrochemical filter. Oil and suspended solids are separated in the hydrocyclone. Electromagnetic forces acting in the ferromagnetic filter facilitate the retention of particles in the filtration layer. The main feature of the filter is its high solids take-up capacity and filtration speed. Iron and hydrogen sulfide react forming iron sulfide in the form of suspended substances, which are extracted by the filter. Pollutant flotation by gas bubbles formed by the electrode pair occurs in the second and third sections of the flotation reactor. Separation of mechanical impurities takes place in the strainer. Removal of organic compounds and metal ions occur in the ER. In addition to the assembled electrode, electrochemical reactor is filled with a mineral granular active material such as calcite. This allows increasing the efficiency of water treatment in the reactor due to the simultaneous action of redox reactions and filtration and sorption processes in the filled material. Disinfection of water from sulfate-reducing bacteria occurs by UV lamps. Oil separated by hydrocyclone and flotation reactors are concentrated in the hydrophobic filter and accumulated in the storage tank. The sediment from the hydrocyclone and the wash water of the ferromagnetic filter is further settled in the precipitation tank and dewatered by the filter press.

4.2 TECHNOLOGICAL SOLUTIONS AND EQUIPMENT USED IN ELECTROCOAGULATION PROCESS

Electrocoagulation (EC) reactors are intended for nonchemical coagulation and sedimentation of pollutants from wastewater. Dissolution of sacrificial electrodes in EC reactors leads to formation of polynuclear hydroxo complexes, which trap pollutants in their structure. In addition to coagulation process, processes such as electrophoresis, cathodic reduction of dissolved organic and inorganic substances, and chemical reactions between irons of wastewater and aluminum and iron ions can take place. Therefore, the efficiency of water treatment by EC in some cases is higher than efficiency of water treatment when metal salt coagulants are used. After EC reactor water usually passes to a clarifier where sedimentation of pollutant hydroxo complexes takes place, thus leading to water clarification. EC is used for the treatment of wastewater from fine-dispersed pollutants such as oil products or organic suspended compounds. The method is not suitable for the removal of dissolved substances from water. The duration of EC treatment in electrolyzers depends on the pollutant properties and amounts and usually do not exceed 5 min. Because electrodes are prone to contaminations, it is recommended to pretreat the water from coarse-dispersed contaminants before EC.

There are no standard solutions in the construction of EC reactors. However, there are typical designs. In general, EC reactors are compact and simple to operate. Electrolyzers for EC process are usually rectangular or cylindrical vessels in which an electrode system (stack of electrodes) is placed. Wastewater flows between the electrodes, which can be of flat plate or cylindrical shape. Plate electrodes are located usually vertically inside reactors, which can be explained by the greater rigidity and the invariance of the dimensions of the electrode system, as well as better conditions for removing the evolved gases and flotation process [4]. EC reactors are mounted together with exhaust ventilation systems for removing gases. When treating water containing toxic and dangerous compounds, in addition to exhaust ventilation, EC reactor should be supplied with a level gauge and leak-proof body. EC reactors should have mechanical devices for removing flotation products from the water and sediments from the lower part of the reactor. Anodes and cathodes are often made of the same material, which allows prolonging the service life of electrodes when changing their polarity periodically.

The majority of EC reactors are a free-flow plate electrolyzers of horizontal (Fig. 4.7A) or vertical (Fig. 4.7B) type. Metal plates serving as electrodes are disposed vertically at a distance of 3—20 mm from one another and held by insulating inserts. Electric current is applied to each plate.

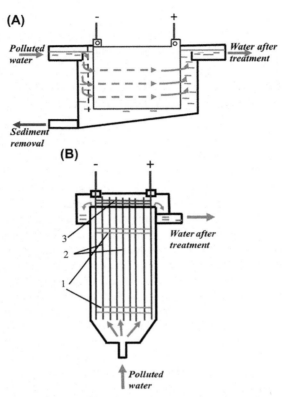

FIGURE 4.7 Free-flow electrocoagulation reactors of horizontal (A) and vertical (B) types. 1, Insulating inserts; 2, electrodes; 3, current-conducting inserts.

Depending on the disposition of electrodes and baffles, plate-type EC reactors can work in single-flow and multiflow modes (Fig. 4.8) [5]. Multiflow scheme of water movement (Fig. 4.8A) means that water flows simultaneously through all the channels formed by electrode plates. Water velocity in this type of reactors should be sufficiently small to have enough time to be mixed with the dissolved coagulant. This type of EC reactors is structurally simple; however, there is a problem of electrode passivation. To partially suppress passivation of electrodes, single-flow EC reactors should be used (Fig. 4.8B). In this type of reactor, water passes through the labyrinth formed by the electrodes or series connection of electrode channels. In single-flow EC reactors, water flow rate is $(n - 1)$ times (n is the number of electrodes) greater comparing to the multiflow reactor configuration.

The process of worn anode replacement is considerably simplified in EC electrolyzers with hanging electrodes (Fig. 4.9). The body of such reactor is made of steel and polarized cathodically. Cathodes are made of the same

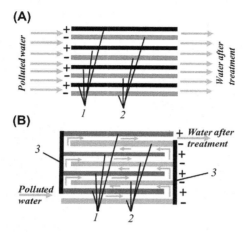

FIGURE 4.8 Multiflow (A) and single-flow (B) schemes of water movement in electro-coagulation reactor. 1, Anodes; 2, cathodes; 3, dividing plates.

material as reactor body and welded alternately to the inside left and right walls of the reactor, forming a labyrinth for water passage. To improve the process of sediment removal from the reactor, there is a gap between the electrodes and reactor bottom. Aluminum anodes are mounted on the anodic plate and secured by insulating panels and liners. Steel cathodes and the reactor vessel are protected from electrochemical corrosion by cathodic polarization and can be destroyed mainly in the idle state at power failure.

Formation of hydro complexes during EC often clogs the interelectrode space. Therefore, mechanical treatment, hydropneumatic treatment, or treatment of electrodes with abrasives is often used in EC reactors. For example, electrode cleaning can be conducted by attaching electrodes to the rotating shaft and mounting the pectinate separator, whose teeth pass between the

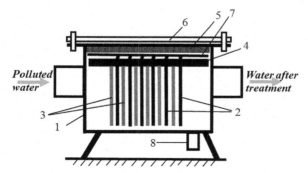

FIGURE 4.9 Electrocoagulation reactor with hanging electrodes. 1, Reactor wall; 2, anodes; 3, cathodes; 4, anodic plate; 5, insulating panel; 6, cover; 7, liner; 8, sediment discharge pipe.

FIGURE   4.10 Electrocoagulation reactor with rotating electrodes and a pectinate device for electrode cleaning. 1, Reactor wall; 2, rotating shaft; 3, electrodes; 4, bus; 5, clean water outlet; 6, pectinate separator.

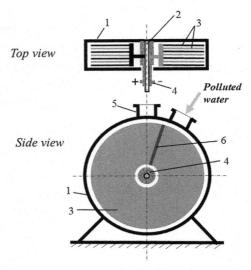

electrodes (Fig. 4.10). Rotation of electrode plates between the separator teeth provides cleaning of electrodes and prevents the mixture of treated water with the polluted one.

Circulation of treated water between the electrodes can also decrease polarization of electrodes. An example of EC reactor with water circulation induced by compressed air is shown in Fig. 4.11. Polluted water is supplied to the ejector and circulated in the space between the electrodes covering the ejector. In addition to polarization suppression, water circulation allows improving hydraulic, physical, and chemical conditions of coagulant flake formation. Clean water is withdrawn from the reactor through the pipe mounted tangentially to the reactor wall. Reactor can be emptied through the pipe mounted on the bottom of the reactor.

To simplify installation and reduce electricity consumption, bipolar connection of electrodes can be used (Fig. 4.12). An electric current in bipolar connection is applied to electrodes, which are separated by a number of plates, which are not connected to external current source. In other words, electric current is not applied to each plate but through a number of plates (Fig. 4.12A). Dissolution of intermediate plates occurs due to their polarization in electric field established between current-connected electrodes. An example of EC reactor with bipolar connection and continuous cleaning of electrodes by abrasion particles (for example, sand) circulating between electrodes and external and internal reactor body is shown in Fig. 4.12B. As it is seen in Fig. 4.12B there are four current-conductive electrodes (the first, firth, ninth, and thirteenth), which significantly simplify the reactor assembly

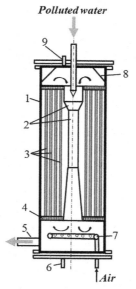

FIGURE 4.11 Electrocoagulation reactor with water circulation. 1, Reactor wall; 2, ejector system; 3, cylindrical electrodes; 4, permeable support; 5, clean water outlet; 6, pipe for reactor emptying; 7, air supply pipe; 8, cone for off-take gases; 9, pipe for discharge of gases.

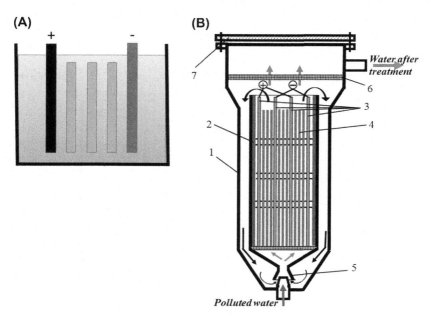

FIGURE 4.12 Bipolar connection of electrodes (A) and Electrocoagulation reactor with bipolar electrode connection and continuous treatment of electrodes (B). 1, Reactor wall; 2, lining made of the nonconductive material; 3, current-conductive electrodes; 4, intermediate plates; 5, diffuser of hydraulic water elevator; 6, mesh preventing removal of cleaning abrasion particles from reactor; 7, cover.

and replacement of worn electrode. Current-conductive electrodes are connected to the external circuit by electric conductors, and intermediate plates are placed into the reactor together with insulating gaskets fixed on the plates without worrying about electrical contacts. Current-conductive electrodes are about 1.5 times thicker than the intermediate plates. Circulation of abrasion particles in the reactor is enabled by a hydraulic water elevator. To prevent flushing of abrasion particles out of the reactor, a mesh is installed above the EC unit.

EC is often combined with other processes (for example, filtration or sedimentation) in one reactor to reduce working areas and obtain clarified water in one treatment step. Fig. 4.13 shows an example of combined EC and filtration processes in one reactor [6]. Wastewater is fed into the reactor downward through the inlet water pipe and dispensing nozzle directly to EC unit. Electrochemical destabilization of organic pollutant takes place in the part of the reactor with EC unit. When EC unit is passed, destabilized organic and inorganic compounds directly come into contact with the top coarse layer of the filtration material, where coagulation of pollutants takes place. The final filtering occurs in the second loading layer of the filtration material, after which treated water is discharged through the water outlet pipe at the bottom of the reactor. Contact filter washing is conducted in an opposite direction to water treatment, i.e., backwash water is supplied from the bottom of the reactor upward and wash water is discharged from the reactor through the outlet pipe. Operation mode is selected according to the quality of wastewater. Usually electrode polarity is switched every 10—15 min. Filtration can also be

FIGURE 4.13 Electrocoagulation unit combined with filtration process in one reactor. 1, Dispensing nozzle; 2, wash water outlet; 3, electrocoagulation unit; 4, coarse top filter load; 5, second fine filter load.

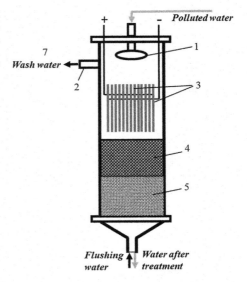

conducted prior to the EC as a pretreatment step to remove large suspended solids from wastewater. In such configuration, water will be supplied from the bottom and withdrawn from the top of the reactor. Funnel with a mesh preventing filter load from washing out of the reactor will be mounted on the water outlet pipe.

Fig. 4.14 illustrates an example of sectional EC reactor combined with the sedimentation tank. Wastewater is supplied through the water supply pipe in the upper part of the reactor and passes through the annular gap between the steel pipe (cathode) and the anodically polarized aluminum cylinder. Dissolution of aluminum anodes at applied electric field between the electrodes causes the formation of aluminum hydroxides, which precipitate with pollutants at the bottom part of the reactor in the sedimentation tank. Clarified water is discharged from the upper part of the sedimentation tank, and the sediment is withdrawn from the bottom part of the sedimentation tank.

A schematic diagram of an EC unit combined with the vertical clarifier for the treatment of oil- and grease-containing wastewaters is shown in Fig. 4.15. Wastewater is delivered to the EC unit from the bottom. Under the influence of electric field, anodes are dissolved and formed hydroxides interact with water pollutants. After passing the EC unit, water enters the chamber of light oil fraction separation, where popped up pollutant fractions are collected from the top of the reactor. Coagulant-pollutant complexes further enter the vertical clarifier, where they precipitate. The treated water is removed from the top outlet piping, and sediments are discharged from the bottom part of the reactor.

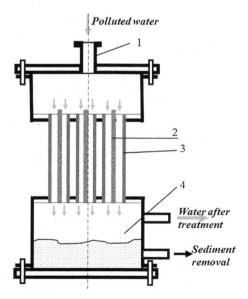

FIGURE 4.14 Scheme of sectional electrocoagulation reactor with cylindrical electrodes and the sedimentation tank. 1, Water inlet pipe; 2, sacrificial anodes; 3, stainless steel cathode; 4, sedimentation tank.

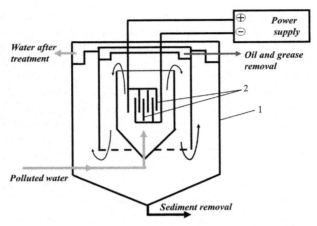

FIGURE 4.15 Electrocoagulation unit for the treatment of wastewaters containing oil and grease. 1, Vertical settler; 2, electrodes.

Another configuration allowing treating oil- and grease-containing wastewaters is EC combined with clarification in inclined plate settler (Fig. 4.16). The settler is divided into separate layers (tiers) by means of inclined parallel plates. The angle of inclination of the plates is about 50 degrees. The process of impurity precipitation occurs in a small thickness of the water layer inside each tier. This design allows the suspension to be precipitated more quickly because collected impurities creep down the slope by gravity into flocculation and sedimentation zone.

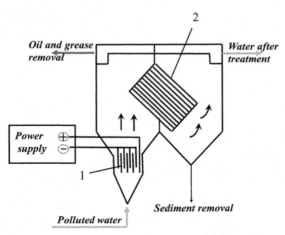

FIGURE 4.16 Electrocoagulation reactor combined with inclined plate settler. 1, Electrodes; 2, inclined plates.

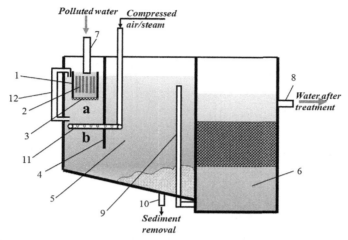

FIGURE 4.17 Electrocoagulation (EC) reactor combined with filtration and sedimentation. 1, EC unit; 2, electrodes; 3, slot; 4, EC chamber; 5, clarifier; 6, filter; 7, water inlet pipe; 8, water outlet pipe; 9, pipeline connecting the clarifier 5 with filter 6; 10, sediment discharge pipe; 11, air agitation; 12, airlift pump.

It is also possible to combine EC with filtration and sedimentation in one reactor (Fig. 4.17) [7]. As it is seen from Fig. 4.9, EC unit with electrodes is positioned at the upper part of the EC compartment, which is combined with clarifier. Right beneath the electrodes there is a slot intended for the output of treated water in the EC chamber. The EC chamber of the reactor is separated into two parts, which are mixing zone **a** and stabilization zone **b**, by an air (steam) agitation system made of perforated pipes. Air mixing system intensifies the process of coagulation in the mixing part **a**. As a result, large flakes are formed even at low temperatures below 10°C. Stabilization zone **b** provides a smooth transition of flow from the mixing zone to the clarifier, which increases the efficiency of use of the clarifier volume and, hence, the performance of the whole reactor. Moreover, direct output of water in the clarifier reduces sedimentation on the electrodes and thereby increases device efficiency. Additional mounting of airlift pump with the input located above the air mixing allows local recycling of treated water with coagulated particles and speeds up the process of coagulation and formation of denser flakes. Coagulated pollutants precipitate in the sedimentation tank and clarified water flows through the pipeline to the filter. Clean water from the filter is further removed through the water outlet pipe, and the sediment from the clarifier is periodically removed through the sediment discharge pipe.

Electrochemical filtration is one of the types of contact coagulation. A typical electrochemical filter is shown in Fig. 4.18 [8]. The filter is loaded with at least three layers of granular materials. The materials of layers 3 and 5 must

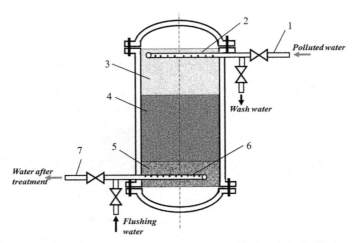

FIGURE 4.18 Electrochemical filter. 1, Polluted water inlet; 2 and 6, distribution systems; 3, layer of granular aluminum; 4, layer of filtering granular material; 5, activated carbon layer; 7, clean water outlet.

be electrically conductive, having different values of the standard potential. The material of layer 3 must be electronegative, for example granular aluminum, capable of forming an insoluble hydroxide. The material of layer 5 should be electropositive. Layers 3 and 5 are separated by layer 4 consisting of a nonconductive granular filter material. When passing water through the filter layers, electromotive force and hence galvanic current are established between layers 3 and 5. Under the influence of established electric field, electronegative material of layer 3 starts to dissolve, forming coagulant on the grains of filtering material. In addition to coagulation process occurring in layer 4, redox reaction takes place in layers 3 and 5, thus significantly affecting the efficiency of water treatment in the case when organic pollutants are present in wastewater. Moreover, activated carbon of layer 5 acts as a sorbent.

Main Calculations for Plate-Type Electrolyzer With Aluminum of Iron Electrodes

For a given quality and consumption of wastewater the main parameters, which can be calculated for electrolyzers, are the dose of aluminum or iron (II or III) and value of applied current, required to conduct the treatment, surface area and thickness of the electrodes, and dimensions of electrolyzer itself. Moreover, the system of electrode connection and electrical characteristics should be chosen.

1. The required dose of metal ions (aluminum or iron, D, g/m^3) is determined experimentally based on trial coagulation. Hourly metal ion consumption

$(G$, g/h), which is needed for water treatment is calculated using the following equation:

$$G = D \cdot Q \tag{4.4}$$

where Q is water flow rate, m³/h.

When wastewater contains only one compound, metal consumption can be determined as follows:

$$G = \frac{Q \cdot C \cdot q_{Me}}{K} \tag{4.5}$$

where q_{Me} is the specific metal consumption (in g) required to remove 1 g of pollutant compound from wastewater and K is the stock utilization ratio of the electrode material. Depending on the thickness of the electrode plates, k is taken about 0.6–0.8.

2. The value of applied current $(I$, A) needed for dissolution of anode is determined as follows:

$$I = \frac{100 \cdot G}{k \cdot CE} \tag{4.6}$$

where k is the electrochemical equivalent, $k_{Al} = 0.336$ g/(A·h), $k_{Fe(II)} = 1.04$ g/(A·h), and $k_{Fe(III)} = 0.695$ g/(A·h); CE is the current efficiency of anodic metal dissolution process.

When only one pollutant is present in wastewater, required current can be determined as follows:

$$I = Q \cdot C \cdot q_I \tag{4.7}$$

where C is the initial concentration of pollutant in wastewater (g/m³) and q_I is the specific electricity consumption required for the removal of 1 g of pollutant compound from wastewater, A·h/g.

3. The total surface area of anodes $(S$, m²) is determined based on the optimal value of current density $(j$, A/m²) for a particular anode material.

$$S = I/j \tag{4.8}$$

4. The surface area of one electrode can be determined using the following equation:

$$S' = 2b \cdot h \tag{4.9}$$

where b is the width of electrode plate, which is chosen arbitrary, and h is the working height of electrode plate (the height of the electrode plate portion immersed in a liquid).

5. Anode thickness $(\delta$, m) taking into account their utilization ratio and required service life can be found using the following equation:

$$\delta = \frac{V_a}{S \cdot k \cdot 0.5} = \frac{2 \cdot m_a}{\gamma \cdot S \cdot K} = \frac{2 \cdot Q \cdot D \cdot t \cdot 10^{-3}}{K \cdot \gamma \cdot S} \tag{4.10}$$

where V_a is the anode volume (m^3), m_a is the mass of anode (kg), γ is the density of anodic material (kg/m^3), and t is the time of continuous operation hours of anodes before their change (h).

6. Time of continuous operation hours of anodes before their change:

$$t = \frac{K \cdot \delta \cdot \gamma}{2 \cdot j \cdot k} \qquad (4.11)$$

7. The total number of electrode plates required (N_{EC}) is determined as follows:

$$N_{EC} = 2S/S' \qquad (4.12)$$

The total number of electrode plates in a stack should be below 30. If the calculated total number of electrode is greater than 30, it is necessary to make provision for few stacks.

8. The thickness of the electrode stack (b_{st}) can be determined by the following equation:

$$b_{st} = N_{EC} \cdot \delta + (N_{EC} - 1) \cdot l \qquad (4.13)$$

where l is the distance between electrodes (m).

9. Working volume of EC reactor (V_r, m^3) can be determined as follows:

$$V_{r,EC} = Q \cdot \tau_{EC} \qquad (4.14)$$

where τ_{EC} is the residence time of wastewater in EC reactor (h).

10. The width (B) of the EC reactor is $B = b_{st} + 2b'$, where b' is the gap between the electrode system and the wall of the EC reactor.

11. The length of the EC reactor (L, m) is equal to $L = b + 2b'$.

12. The area occupied by EC reactor (A, m^2) is equal to $A = B \cdot L$.

While conducting electrolysis process, hydrogen and oxygen are always generated. The mixture of these gases is explosive, and the mixture of hydrogen with air is explosive at a hydrogen concentration of more than 4 vol.%. According to the safety conditions, all ERs should be equipped with exhaust ventilation systems, which ensure dilution of the liberated hydrogen with air to a concentration of less than 0.4 vol.%, i.e., 10 times below the explosive threshold.

4.3 ELECTROFLOTATION REACTOR

When insoluble electrodes are used, bubbles of released hydrogen and oxygen gases sorb pollutants on their surface and carry them away by lifting up. This is the principle of EF process. EF is one of the most effective water treatment methods for oil products, fine particles, and dissolved organic compounds. The size of the gas bubbles depends on the composition and shape of the electrodes, as well as the conditions for electrolysis such as current density and temperature. This means that it is possible to obtain desired size distribution of gas bubbles during electrolysis. EF reactors are widely used for the removal of surfactants and petroleum pollutants from water.

There are different kinds of reactors used in EF process. Depending on the direction of water movement, EF reactors can be divided into countercount flow, when water and flotation gases move in opposite directions (Fig. 4.19A), direct flow (Fig. 4.19B), and mixed flow (Fig. 4.19C) of water and gases. In turn, direct-flow EF reactors can be divided into reactors with horizontal or vertical arrangement of electrodes. Electrodes are placed at the bottom of reactors completely covering it. This allows even distribution of formed gas bubbles throughout the cross section of the reactor. Electrodes can be in the form of plates or wire mesh usually made of copper or stainless steel. Plate

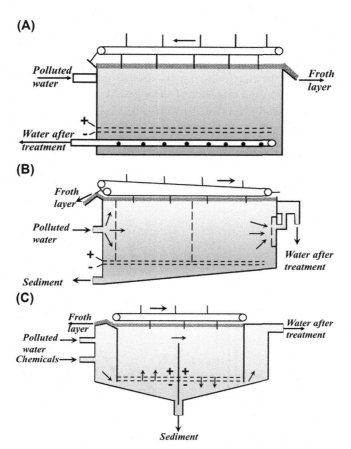

FIGURE 4.19 Electroflotation reactors with countercount (A), direct flow (B), and mixed flow (C) of water and gas bubbles.

electrodes are placed vertically in the reactor and mesh electrodes are mounted usually horizontally, thus preventing deposition of dispersed impurities on their surface. Another way to prevent deposition of solid particles on the surface of electrodes, which leads to increased power consumption, is to make anodes of triangular prism shape and locate them in a checkerboard pattern on the bottom of the reactor. An EF reactor with triangular prism anodes made of graphite or other electrochemically stable material and wire mesh cathodes made is shown in Fig. 4.20. Cathodes are bent at an angle and located above the anodes parallel to their faces. Wastewater is delivered through the inlet pipe located in the upper part of the reactor to flotation compartment, where it is saturated with gas bubbles. The gas bubbles float contaminants to the air—water interface, where the froth layer is removed from through the inclined chute. A pipe with hot water passes inside the chute. Froth in contact with the hot surface bursts, which contributes to its better removal through the chute. Purified water rises to the clean water compartment and is removed through the water outlet pipeline. Sediments, which can settle in the reactor, can be discharged though the pipeline at the bottom of the apparatus.

Because the size of floated bubbles depends on the contact angle value and surface curvature of the electrodes, it is possible to obtain desired dispersion of gas phase by changing the diameter of the wire. EF reactors can have single (Fig. 4.19A) and multiple compartments (Fig. 4.19B and C). In multicompartment EF reactors, wastewater is delivered to feeder compartment, separated from the main part of reactor by partition. Multicompartment reactors can have few partitions, which stir and direct water flow therein. When passing through the reactor, water is saturated with gas bubbles generated at the electrode surface. Pollutants stick to the floating gas bubbles, thus forming froth layer on the surface of water, which is removed by scrapers and other mechanical devices.

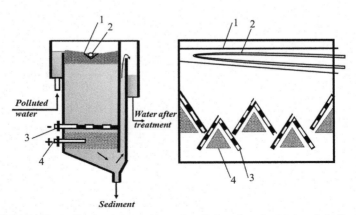

FIGURE 4.20 Electroflotation reactor with anodes of triangular prism shapes. 1, Inclined chute; 2, coolant pipeline; 3, cathode; 4, anode.

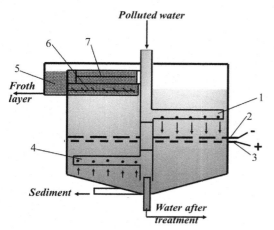

Polluted water

Froth layer

Froth layer

Sediment ←

Water after treatment

FIGURE 4.21 Electroflotation reactor with countercount flow of water and gas bubbles. 1, Rotating water feeding system; 2, mesh cathode; 3, mesh anode; 4, rotating water collecting system; 5, froth collector; 6, froth collecting pipe with auger; 7, scraper.

Another example of EF reactor with countercount flow of water and gas bubbles is shown in Fig. 4.21. Wastewater is fed to the reactor through the center pipeline and the rotating water feeding system and moves downward opposite to the floating gas bubbles. The froth is accumulated at the surface of the reactor, continuously scrapped and transported by an auger into the froth collector. Purified water is collected by a rotating device and removed from the reactor through the bottom pipe. Sediment is removed through the sediment discharge pipe at the bottom of the vessel. The main feature of the reactor is that rotating feeding and collecting devices distribute water more evenly in the volume of the apparatus, thereby increasing efficiency of wastewater treatment.

Wastewaters of high flow rates are treated usually in two-compartment EF reactors. Figs. 4.22 and 4.23 shows different constructions of two-section EF reactors. The two-section EF reactor of horizontal type (Fig. 4.22) is divided into electrode and settling compartments. When passing water inlet compartment and stabilizing screen, wastewater is delivered to electrode compartment, where it is saturated with gas bubbles. Flotation of pollutants occurs in settling part of the reactor, where froth is collected by scrapers into the frothing tray. Solid particles precipitated at the bottom of the reactor are discharged through the sediment discharge pipe.

Fig. 4.23 shows another typical configuration of EF reactor used for high flow rates of wastewater. In contrast to the EF reactor in Fig. 4.22, the EF compartment in apparatus of Fig. 4.23 is in the middle of the reactor. Wastewater is fed to the rotating water distribution system controlled by the electric motor. When entering the EF compartment, water moves down toward floating gas bubbles generated at the cathode. Floatation of pollutants occurs in

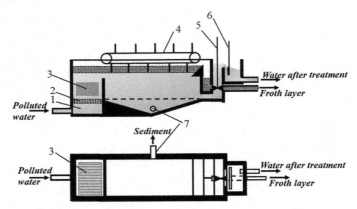

FIGURE 4.22 Electroflotation reactor of horizontal type. 1, Water inlet compartment; 2, stabilizing screen; 3, electrode system; 4, scraper; 5, slurry removal regulator; 6, level regulator at the water outlet; 7, sediment discharge pipe.

the same compartment, at the top of which it is connected to the froth collecting tray. Purified water is merged into the settling compartment through the electrode mesh and then rises along the walls of the reactor and ring partition into the ring tray, from which it is further discharged.

The highest efficiency of wastewater treatment is achieved when EF is combined with EC in one apparatus. In this case, the pollutants are exposed to both the gas bubbles as a result of water decomposition reactions and metal oxides formed during anode dissolution. Duration of EC and EF steps should be the same. The maximum total duration of EC and EF of wastewater is about 30−40 min. An example of such combined EC and EF reactor is shown in

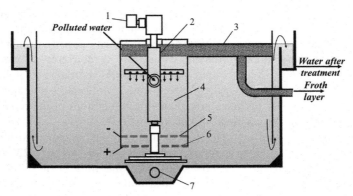

FIGURE 4.23 Two-section electroflotation (EF) reactor. 1, Electric motor; 2, combined mechanism for wastewater distribution, froth scrapping and sediment collection; 3, froth collecting tray; 4, EF compartment; 5, cathode; 6, anode; 7, sediment discharge pipe.

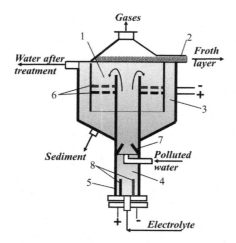

FIGURE 4.24 Combined electroflotation–electrocoagulation reactor. 1, Flotation compartment; 2, froth-collecting tray; 3, purified water collector; 4, coagulation compartment; 5, reactor wall; 6, sacrificial (soluble) electrodes; 7, reverse valves; 8, insoluble electrodes.

Fig. 4.24. Wastewater is fed to the coagulation compartment, where it reacts with the metal hydroxides formed as a result of sacrificial electrode dissolution. When passing the coagulation compartment, water enters the flotation compartment, where the formed pollutant hydroxo complexes are adhered to the electrochemically generated gas bubbles. As a result of such interactions, flotocomplexes of pollutants are accumulated on the surface of water in the flotation part of the reactor forming the froth layer, which is removed from the froth-collecting tray. Purified water is withdrawn through the purified water collector part of the reactor. When necessary, sediments are discharged trough the sediment discharge pipe.

As mentioned earlier, EF is an efficient method for the removal of oil products, fine particles, dissolved organic compounds, etc. Fig. 4.25 shows an example of schematic diagram of wastewater treatment using EF for the removal of zinc salt coming from the production of viscose. The treatment combines both chemical water treatment and EF. Before entering the EF unit, the pH of wastewater is neutralized in a mixer vessel either by a limewater or sulfuric acid. EF reactor in the scheme is a three-compartment reactor consisting of the presettling chamber for the removal of coarse impurities and two EF compartments. After EF reactor, water is first delivered to the decanter centrifuge and then to the vacuum filter for even deeper removal of pollutants from water. If necessary, the acidity of purified water is regulated in the purified water tank.

Fig. 4.26 shows a diagram of wastewater treatment from galvanic plants using EF. Wastewater from galvanic plants passes successively through the EF,

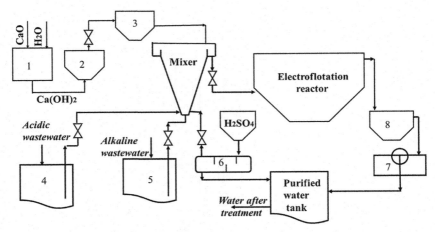

FIGURE 4.25 Schematic diagram of electroflotation treatment for zinc salt removal from wastewater of the viscose production. 1, Tank for preparation of limewater; 2, stirrer; 3, limewater dispenser; 4, tank for acidic wastewaters; 5, tank for alkaline and viscose wastewaters; 6, agitation tank; 7, vacuum filter; 8, decanter centrifuge.

flotation, and hydromechanical flotation reactors, respectively. Processed in flotation reactors, water further enters horizontal and lamella (inclined plate settler) clarifiers and is deeply treated by the filter. Purified water undergoes an electrochemical pH correction before unloading from the water treatment plant. Sediments are collected separately.

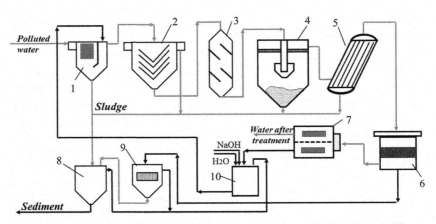

FIGURE 4.26 Schematic diagram of wastewater treatment from galvanic plants. 1, Electroflotation reactor; 2, flotation reactor; 3, hydromechanical flotation reactor; 4, clarifier; 5, inclined plate settler; 6, filter; 7, electrolyzer for pH correction; 8, sediment collector; 9, settler of wash water from the filter; 10, catholyte tank.

Main Calculations for Electroflotation Reactors

When calculating the EF reactors, the design of the electrode blocks and their location in the working compartments of the apparatus should be taken into account [9].

Example of Calculation for Electroflotation Reactor Sizing With Vertical Parallel Plate Electrodes

1. Working volume of EC reactor (V_r, m³) can be determined as follows:

$$V_{r,EF} = Q \cdot \tau_{EC} \tag{4.15}$$

2. The area of flotation compartment (A_F, m²) is calculated as follows:

$$A_F = V_{r,EF}/H_0 \tag{4.16}$$

where H_0 is the layer height of the treated liquid (m).

3. The total EF reactor height H_r is determined using the following dependence:

$$H_r = H_0 + h_1 + h_2 \tag{4.17}$$

where h_1 and h_2 are the height of the froth layer ($h_1 = 0.05-0.15$ m usually) and the side of the apparatus above the froth layer, taking into account the mounting of the device for removing the froth layer ($h_2 = 0.2-0.4$ m).

4. The length (L_F) of the flotation compartment:

$$L_F = A_F/B \tag{4.18}$$

where B is the width of the EF reactor. When the water flow rate Q is less than 10 m³/h, $B = 0.8-1$ m; when Q is up to 20 m³/h, $B = 1.5-2$ m; and when Q is up to 50 m³/h, $B_F = 3-3.5$ m.

Usually $L_F = (3-4)B_F$

5. The total length (L) of EF reactor:

$$L = L_F + L_1 \tag{4.19}$$

where L_1 is the length of the collector of treated water ($L_1 = 0.15-0.3$ m).

6. The total number of anode ($N_{a,EF}$) and cathode ($N_{c,EF}$) plates is determined as follows:

$$N_{a,EF} = (B - 2b' + \delta_c + 2l)/(2l + \delta_c + \delta_a) \tag{4.20}$$

$$N_{c,EF} = (B - 2b' - 2l)/(2l + \delta_c + \delta_a) \tag{4.21}$$

where b' is the gap between the electrode system and the wall of EF reactor, l is the distance between the electrodes ($l = 0.005-0.02$ m), δ_c and δ_a are the thickness of cathode and anode plates, respectively.

7. The active surface area of one vertical electrode can be determined using the following equation:

$$S' = 2b \cdot h \tag{4.22}$$

where b is the width of electrode plate, which is chosen equal to L_F and h is the working height of electrode plate ($h = 0.1-0.15$ m).

8. The value of applied current I is determined as follows:

$$I = j_c \cdot S' \cdot N_{c,\text{EF}} = j_a \cdot S' \cdot N_{a,\text{EF}} \tag{4.23}$$

9. The weight (W) of the electrode system is calculated using the following equation:

$$W = \rho_c \cdot S' \cdot N_{c,\text{EF}} \cdot \delta_c + \rho_a \cdot S' \cdot N_{a,\text{EF}} \cdot \delta_a \tag{4.24}$$

10. Time of continuous operation of the EF reactor in the case if graphite anodes are used:

$$\tau = W_a / (q_g \cdot I) \tag{4.25}$$

where W_a is the weight of anodes and q_g is the deterioration of graphite during electrolysis, $q_g = 85$ mg/(A·h).

Calculation for Combined Column-Type Electroflotation—Electrocoagulation Reactor

The main calculations are given for the column—type EF—EC reactor shown in Fig. 4.24

1. The duration of the process of complete coagulation (τ_{EC}) is determined using Eq. (4.10).

$$\tau_{\text{EC}} = \frac{3}{4} \cdot \frac{\eta \cdot N_A}{(R_1 + R_2) \cdot T \cdot C_0} \tag{4.26}$$

where N_A is the Avogadro's number, T is the temperature, and R_1 and R_2 are coagulating contaminant particle radii.

The duration of complete coagulation can be also determined from the experimental data when plotting the dependence of the pollutant removal on the amount of electrogenerated metal hydroxides.

2. The working volume of EC compartment ($V_{r,\text{EC}}$, m³) is calculated as follows:

$$V_{r,\text{EC}} = Q \cdot \tau_{\text{EC}} \tag{4.27}$$

3. The diameter (d_{EC}) and the height (H_{EC}) of EC compartment can be found using the following equations:

$$H_{\text{EC}} = \tau_{\text{EC}} \cdot \vartheta_{l,\text{EC}} \tag{4.28}$$

where $\vartheta_{l,EC}$ is the linear velocity of water movement in the EC compartment

$$d_{EC} = 2\sqrt{\frac{V_{r \cdot EC}}{\pi \cdot H_{EC}}} \tag{4.29}$$

4. The diameter of the water inlet pipe ($d_{in.\ pipe}$) can be calculated using the following equation:

$$d_{in.\ pipe} = \sqrt{\frac{4Q}{\pi \cdot \vartheta_{in.\ pipe}}} \tag{4.30}$$

where $\vartheta_{in.\ pipe}$ is the velocity of water movement in the inlet water pipe. When water movement is driven by hydrostatic head, $\vartheta_{in.\ pipe} = 0.8 - 1.2 \ m/s$. When water movement is driven by gravity, $\vartheta_{in.\ pipe} = 0.3 - 0.6 \ m/s$.

5. The working volume of EF compartment is equal to

$$V_{r,EF} = Q \cdot \tau_{EF} \tag{4.31}$$

where τ_{EF} is the duration of water stay in the EF compartment; $\tau_{EF} = 0.3 - 0.6 \ h$.

6. The diameter (d_{EF}) and the height (H_{EF}) of EF compartment can be found using the following equations:

$$H_{EF} = \tau_{EF} \cdot \vartheta_{l,EF} \tag{4.32}$$

where $\vartheta_{l,EF}$ is the velocity of water movement in the EF compartment; $\vartheta_{l,EF} = 0.005 - 0.2 \ m/s$

$$d_{EF} = 2\sqrt{\frac{V_{r \cdot EF}}{\pi \cdot H_{EF}}} \tag{4.33}$$

7. The area of annular insoluble anodes ($S_{a,EF}$) of the EF compartment can be calculated as follows:

$$S_{a,EF} = \frac{4}{\pi} \cdot \left(d_{EF}^2 - d_{EC}^2 \right) \tag{4.34}$$

The electrodes are mounted at a depth of $0.5-1$ m below the liquid level, depending on the pollutant concentration in the flotation zone.

8. The recommended current density at insoluble anodes in the flotation compartment is equal to $j_{EF} = 0.01-0.02 \ A/cm^2$.

9. The total anode area of the EC compartment ($S_{a,EC}$) is determined by the following equation:

$$S_{a,EC} = \frac{D \cdot Q}{CE \cdot k \cdot j_{EC}} \tag{4.35}$$

The required dose of metal ions (aluminum or iron, D, g/m^3) is determined experimentally based on trial coagulation, and anodic current density in EC unit is usually established between 0.01 and 0.05 A/cm^2.

10. The thickness of anodes (δ, m) in the EC compartment is determined by Eq. (4.10).

11. The number of soluble electrodes ($N_{a,EC}$) with the accepted square cross section of the EC compartment with side dimensions equal to a can be calculated as follows:

$$N_{a,EC} = \frac{a}{\delta + l + x + 1} \tag{4.36}$$

where x is the thickness of dielectric spacer, b is chosen depending based on technological factors and assembly conditions, and l is the distance between electrodes.

12. The surface area of one electrode of EC compartment is determined as follows:

$$\frac{S' = S_{a,EC}}{N_{a,EC}} \tag{4.37}$$

13. The height of the electrodes is $h_{el,EC} = S'/b$, where b is the width of electrode plate; $b = a - 2x$.

The upper edge of the soluble electrodes is located below the point of polluted water inlet at the distance equal to about $(2-3)a$.

4.4 EXAMPLES OF THE USE OF ELECTROCHEMICAL WATER TREATMENT METHODS IN PRACTICE

Treatment of drinking and wastewater is an integral part of modern society. Every day millions of cubic meters of various solutions from pretreated water and chemical reagents obtained from natural mineral raw materials are prepared in the world. As a result of this activity, millions of cubic meters of spent technological solutions are cleaned daily before discharging into the sewage system, trying to free them from harmful substances. Moreover, the use of water in everyday day life and agriculture leads to additional formation of the vast amount of wastewaters. The first wastewater treatment plants (WWTPs) were built in the late 1800s and the early 1900s. Due to population growth and industrial development, the need for clean water and the load on WWTPs are increasing. Moreover, the tightening of environmental legislation, such as the Urban Waste Water Treatment Directive in Europe, requires the use of improved treatment methods to achieve the required quality of water [10]. According to the US Environmental Protection Agency, only US$271 billion is needed to improve wastewater infrastructure in the United States, including wastewater-carrying pipes, treatment, and managing methods for storm water runoff [11]. Investments in the global water and wastewater treatment sector are estimated at $6 trillion over the next 20 years with major investments in South America and Asia [12]. Hence, the world market need for water treatment technologies is huge.

The intensive development of electrochemical water treatment technologies initiated their increased introduction into the market. As mentioned earlier, one of the first methods introduced into industrial water treatment processes were EC, EF, and electrodisinfection methods, which are still widely used as an independent or primary stage of wastewater treatment. For example, the American company *OriginClear* headquartered in Los Angeles successfully implements EC and EF processes in combination with advanced oxidation processes (AOx) in patented Electro Water Separation (EWS) technology [13]. It allows efficient removal of nonsoluble petroleum hydrocarbons, chemical oxygen demand (COD), biological oxygen demand (BOD), total suspended solids (TSS), ammonia, sulfur, etc., from wastewaters of pharmaceutical, oil and gas, semiconductor, municipal, and other industries. The EWS:AOx technology combines EC, EF, and electrooxidation (EO) processes sequentially into a line. Depending on the particular application, EC step can be performed using either sacrificial anodes or insoluble dimensionally stable anode system, where previously injected coagulants are mixed. During EC step suspended solids, nonsoluble organic compounds, oils, and grease are coagulated. Suspended coagulated pollutants are further removed in the following stage of EF process. EF treatment stage is followed by EO step. Wastewater pretreated by EC and EF and containing insignificant amounts of micron-sized suspended solids and dissolved pollutants passes through ER where oxidation of organic pollutants by electrogenerated oxygen reactive species, such as hydroxyl radicals, ozone, and hydrogen peroxide, takes place. Nonoxidized pollutants are remediated at the cathode. Final products of AO either precipitate in the AO reactor or dissipate as off-gas. EO reactors can be made in the tube-shape inline reactors or tanks. If the EWS:AOx system provides insufficient removal rates of pollutants, water is further posttreated by ultrafiltration and reverse osmosis. Moreover, ultrafiltration can also be incorporated after EF process and before EO step if required.

One example of a successful implementation of EWS technology as a midstage solution in decentralized wastewater treatment is the pilot scale ECOPOD System used in Bakersfield, California, United States. Bakersfield's Kern County has the largest oil-producing and fourth most productive agricultural region in the United States. So agricultural and oil companies are competing for fresh water resources. The treated oilfield-produced water is reused in crop irrigation and other industries after treatment by ECOPOD system. Moreover, the system allows wet oil recovery. It requires no chemicals and minimum operator supervision as well as treats up to 54 m^3 of wastewater per day. Key pollutant reduction counts, which were achieved in the project, were as follows:

Heterotrophic plate count—99.8% from $1.85 \cdot 10^4$ to 35 CFU/mL
Total recoverable petroleum hydrocarbons—99.8% from 590 to 1.3 mg/L
TSS—60.7% from 15 to 5.9 mg/L

Moreover, complete removal of benzene, toluene, m,p-xylene, o-xylene, and naphthalene was achieved during wastewater treatment by EWS:AOx technology. OriginClear also developed ECOPOD System 1.0 that is able to process 477 m^3/day of oil-containing wastewater into clean water suitable for crop irrigation.

Another successful example of the EWS:AOx system is the treatment of leachate from a municipal landfill of a rural northern city in China [14]. Leachate discharge is 2000 tons/day, and the site is remote with subzero temperatures in winter. The influent water quality is described by the following parameters:

COD—10,000 mg O_2/L
NH_4—3500 mg/L
pH—between 8.35 and 8.7
TDS—11,000 mg/L
Average high and low atmospheric temperatures—6 and 0°C, respectively.

The following technology setup sequence we used in the process:

Sand filter for the removal of coarse mechanical pollutant.
2000 L holding tank, where coagulant with a dose of 1000 mg/L was added and mixed to initiate flocculation process. NaOH was used to adjust the pH of solution in the holding tank to 10.

- EWS EF reactor with (mixed metal oxide) MMO electrodes at 4.6 V and 59 A
- Ultrafiltration
- EO at pH 6.5. Hydrochloric acid is used to maintain the required acidity of solution
- Ultrafiltration
- Reverse osmosis

As a result of the treatment, ammonia was reduced to 8 ppm and COD to the value below 100 mg O_2/L.

Israeli company pH$_2$O *Water Technologies* offers technical solutions for wastewater treatment using EF, EC, and electrolytic pH correction. One of the standard solutions offered by the pH$_2$O Water Technologies company is wastewater treatment at the waste truck wash plant. Let us consider the proposed solution using the example of wastewater treatment from washing trucks at the Ganei Hadass landfill, which is the largest in southern Israel [15].

Every day the landfill accepts up to 120 waste trucks. Before leaving the landfill, all trucks are washed with tap water for disinfection and to prevent odor. The landfill does not have a leachate treatment system, as it is located in an arid place. All leachate is collected from the landfill and sprinkled over it, preventing the formation of dust and improving the bacterial decomposition of wastes. Around 200–300 L of water is required to wash one truck; however, because the costs of fresh tap water are high and water quota was cut, there was a need for a new water recycling unit. As a result a water treatment facility with a capacity of 30 m^3/d, which consisted of EC and EF treatment steps, was installed. Water treatment system has a holding tank filled with 5 m^3 of clean water. Water from the tank is used for washing trucks. Polluted water after washing is supplied to the water treatment unit, after which clean water is returned to the holding tank. Every 2 weeks the tank is emptied and filled with clean water. Spent water is sprinkled on the landfill. The system reduced consumption of clean tap water from 1080 to 10 m^3 per months. Sludge (around 2%) formed during the treatment of washing water is placed in a landfill. The efficiency of the unit is 99.8% in terms of COD removal, 98% in TSS removal, and 100% in terms of *Escherichiacoli* removal.

Commercial solutions for wastewater treatment by EO are also offered by the Canadian company *Axine Water Technologies Inc.* headquartered in Vancouver [16]. Axine's EO reactors are made of electrolytic cells, which in turn consist of anode and cathode separated by an ion-exchange membrane. Cells are collected into stacks and stacks are collected into modules, the size of which depends on the influent water parameters and flow rates. Axine's commercial systems range in capacity from 30 m^3/day to several tens of thousands m^3/L. When conducting EO process, polluted water containing organic compounds flows through the anodic compartment only, where organics are by hydroxyl radicals and other reactive species. Axine's EO modules can be integrated into existing WWTPs or combined with other treatment technologies.

Many companies offer services on water treatment by electrochemical activation (EA) method. The method of EA is based on the electrolysis process of low-mineralized solutions in a two-compartment electrolytic cell or electrolyzer, where anodes and cathodes are separated by a diaphragm. It is suggested that the low solution conductivity, along with unipolar cathodic or anodic electrochemical exposure and, as a result, the high voltage of the double electric layer at the surface of the electrodes, leads to an increased formation of electrochemically generated reactive species, which in turn facilitate the efficiency of treatment process. This can also be explained by the reduced rate of recombination of the electrogenerated active species in the separated compartments of the electrolyzer. Initially, the method was used for water disinfection by hypochlorite ions generated during the electrolysis of sodium chloride solution (2.1–2.2, see Chapter 2). Disinfection of water in EA can be carried out either directly when polluted water is passed through the anode compartment of the electrolyzer or indirectly by dosing the anolyte obtained during the electrolysis of the sodium chloride solution.

TABLE 4.1 Companies Working in the Field of Electrochemical Water Treatment

Company Name	Water Treatment Technology	Headquarter
Ecologix Environmental Systems	EC	Alpharetta, Georgia, USA
Axine Water Technologies Inc.	EO	Vancouver, Canada
ADVANCED Equipment and Services	EDI	Coconut Creek, USA
Danish Clean Water A/S - Danfoss A/S	ED	Sønderborg, Denmark
Envirolyte	EA	Estonia
ECOtek Water Systems	EC, EDI	Murray Bridge, Australia
MIOX Corporation	ED	Albuquerque, USA
Novasep	Electrodialysis	Boothwyn, USA
ProMinent GmbH	ED	Heidelberg, Germany
Pure Water Group	EDI	Sprundel, Netherlands
Radical Waters	EA	Johannesburg, South Africa
PCCell	Electrodialysis	Heusweiler, Germany
SnowPure	Electrodialysis	San Clemente, USA
Titanium Tantalum Products Limited	EA	Chennai, India
Vitold Bakhir Electrochemical Systems and Technologies Institute	EA, ED, EO	Moscow, Russia
Water Vision Inc.	EC	Conroe, USA
Zap Water Technology, Inc.	EA	Richfield, USA

EA, electrochemical activation; *ED*, electrochemical disinfection; *EDI*, electrodeionization.

So far, most companies, using the term of EA in water treatment, mean the electrochemical disinfection by hypochlorite ions generated while conducting the electrolysis of sodium chloride solution. Despite this, there are many companies that integrate the processes of EC, EF, electrophoresis, and electrolysis of water under one term of EA. Therefore, it is difficult to distinguish the mechanism of water treatment used in some cases, unless additional information is provided. For example, if soluble electrodes are used in the EA process, it can be concluded that the removal of substances is mainly due to the EC process rather than disinfection. A wide range of services for the electrochemical disinfection of water by EA methods in health care, livestock and shrimp farms, swimming pools, food, oil and gas, and other industries is offered by the Estonian company *Envirolyte* [17]. In addition to water disinfection, Envirolyte provides solutions for wastewater treatment. A list of some companies providing services in the field of electrochemical water treatment is given in Table 4.1.

4.5 SUMMARY

The development and implementation of electrochemical methods is a progressive trend in water and wastewater treatment technology. Electrochemical cleaning is an effective alternative to the conventional mechanical, biological chemical, and physicochemical water treatment methods, which often provide insufficient degradation rates of pollutants or cannot be used because of a lack of working areas, the complexity of delivery and use of chemicals, or for other reasons. These methods allow correcting the physical chemical properties of the treated water, concentrating and extracting valuable chemical products and metals from water, providing deep mineralization of organic contaminants, and biological safety through the disinfection. In many cases, electrochemical methods are environmentally friendly, eliminating secondary water pollution with anionic and cationic residues, which are typical for chemical water treatment technologies.

To date, significant progress has been made in the development of anode and cathode materials, as well as reactor designs for electrochemical water treatment that allow water purification with high efficiency. For example, electrodes made of metal oxides and boron-doped diamond are widely used for the oxidation of organic compounds, and graphite and metal electrodes are effective for electrochemical reduction and recovery of metals. Moreover, the higher the surface of the electrodes, the higher the efficiency of water purification; hence, the electrodes are usually assembled in stacks or have developed in 3D surface. Electrodes can be mounted in single compartment reactors or can be separated by ion-exchange membranes or diaphragms in multiple compartments. Moreover, electrochemical water treatment can be combined in one reactor with other water purification methods, such as filtration, sedimentation, flotation, and others, when it is not possible to achieve desired degree of pollutant removal from water or to save the working areas. It is promising to use electrochemical methods in combination with other physical—chemical methods, especially when creating closed loop water supply systems. In such cases electrochemical methods are usually used as a midstage treatment followed by the polishing treatment methods such as ultra- and nanofiltration, electrodialysis, and reverse osmosis. Because electrochemical technology is characterized by advantageous mass/volume characteristics and compactness of equipment, it is of great importance in the reconstruction of existing treatment facilities and the construction of new treatment facilities in cramped conditions. Moreover, since the methods are easy to automate, insignificant operator control is required. Despite the great breakthrough in the field of electrochemical water treatment, the constant development of new purification methods put new tasks for researchers and engineers on the development of new electrode materials, constructive solution of reactors and their calculation, quantitative determination of kinetic regularities of purification processes, etc.

SELF-CONTROL QUESTIONS

1. What is the difference of single-compartment EF reactors from multi-compartment EF reactors?
2. How can electrodes be assembled in EF reactors?
3. How can EF reactors be classified?
4. How can precipitation of solid particles on the surface of electrodes be prevented during EF and EC processes? What are consequences of electrode contamination by precipitates during EF and EC?
5. What are the main size characteristics calculated for EC and EF reactors?
6. What are the most commercialized electrochemical water treatment technologies?

REFERENCES

[1] P. Brennecke, H. Ewe, E. Justi, Highly Porous Electrodes Hot Pressed from Nickel Powder for Alkaline Water Electrolyzers, March 11, 1981. USA Patent US4447302 A.
[2] S. Frank, Graphite Felt Flowthrough Electrode for Fuel Cell Use, September 1, 1978. USA Patent US4264686 A.
[3] P. Weinstock, Petrochemical Industry Wastewater Treatment by Electrochemical Methods, Ufa state Petroleum Technological University, Russian, Ufa, 2008.
[4] A. Timonin, in: Engineering and Environmental Hand Book, vol. 2, Publishing house N. Bochkareva, Kaluga, 2003, p. 884.
[5] L. Kulskii, P. Strokach, V. Slipchenko, E. Saigak, Water Treatment by Electrocoagulation, Budivelnik, Russian, Kiev, 1978.
[6] S. Breus, S. Linevich, Natural Water and Effluents Treatment Plant, November 18, 2010. Russia Patent Ru 2464235 C2.

[7] N. Anopol'skij, G.N. Fel'dshtejn, E. Fel'dsjhtejn, J. Shauer, Device for Cleaning Waste, February 15, 2006. Russian Federation Patent RU 2 317 949 C2.

[8] V. Nazarov, V. Zentsov, M. Nazarov, Water Supply in Oil Production: A Textbook, second ed., Ufa: Oil and Gas Business, Russian, 2010.

[9] S. Yakovlev, I. Krasnoborodko, V. Rogov, Technology of Electrochemical Water Treatment, Stroyizdat, Leningrad, 1987.

[10] The Council of the European Communities, Council Directive 91/271/EEC of 21 May 1991 concerning urban waste-water treatment, Off. J. L135 (1991) 40−52.

[11] R. Eckelberry, Onsite treatment recycles produced water for irrigation, World Water 39 (6) (2016).

[12] PPE Outlook in the Global Water and Wastewater Treatment Industry: Market Opportunities and Outlook 2012, 2012, 9833-00-3C-00-00.

[13] OriginClear Technologies, Technology Overview. Electro Water Separation with Advanced Oxidation, (Online). Available: http://www.originclear.com/tech/technologies.

[14] OriginClear Technologies, Landfil leachate tretament with EWS: AOxTm, (Online). Available: http://www.originclear.com/pdf/leachate-treatment-china-ews-aox.pdf.

[15] Wastewater treatment case study, Waste truck wash plant, pH_2O Water Technologies, (Online). Available: http://ph2o.net/Ganei%20Hadass%20CS.html.

[16] Axine Water Technologies, Axine TM Solutions, (Online). Available: http://www.axinewater.com/solutions.

[17] Envirolyte®, Envirolyte water disinfection systems, (Online). Available: http://www.envirolyte.com/applications.html.

Appendices

Appendix 1. Exercises for Chapter 1

EXERCISE 1.1

Draw a cell diagram and write redox reactions occurring in a galvanic cell of lead-acid battery (discharge process). Specify redox reactions occurring during recharge of lead-acid battery. Calculate the standard cell potential. Anode of charged lead battery consists of lead and cathode is of lead dioxide. The balanced chemical reaction is as follows:

$$Pb_{(s)} + PbO_{2(s)} + 2H_2SO_{4(aq.)} \underset{\text{Charging}}{\overset{\text{Discharging}}{\rightleftarrows}} 2PbSO_{4(s)} + 2H_2O$$

Solution

Anode half-cell reaction:

$$Pb + SO_4{}^{2-} \underset{\text{Charging}}{\overset{\text{Discharging}}{\rightleftarrows}} PbSO_4 + 2e^-$$

Cathode half-cell reaction:

$$PbO_2 + SO_4{}^{2-} + 4H^+ + 2e^- \underset{\text{Charging}}{\overset{\text{Discharging}}{\rightleftarrows}} PbSO_4 + 2H_2O$$

Overall cell reaction:

$$Pb + PbO_2 + 4H^+ + 2SO_4{}^{2-} \underset{\text{Charging}}{\overset{\text{Discharging}}{\rightleftarrows}} 2PbSO_4 + 2H_2O$$

Based on the half-cell reactions the cell diagram is as follows:

$$Pb_{(s)}, \ PbSO_{4(s)} \ \big| H_2SO_{4(aq)} \big| PbO_{2(s)}, \ PbSO_{4(s)}, \ Pb_{(s)}$$

$$E^0 = E^0_{\text{Red, cathode}} - E^0_{\text{Red, anode}} = 1.682 \ V - (-0.3588 \ V) = 2.04 \ V$$

EXERCISE 1.2

A cell diagram of an alkaline nickel–iron galvanic cell is as follows:

$$Fe, \ Fe(OH)_2 | KOH | \ Ni(OH)_2, \ Ni(OH)_3, \ Ni$$

Please write anode and cathode half-cell reactions and the overall cell reaction.

Solution

Anode half-cell reaction: $Fe + 2OH^- \rightarrow Fe(OH)_2 + 2e^-$

Cathode half-cell reaction: $2Ni(OH)_3 + 2e^- \rightarrow 2Ni(OH)_2 + 2OH^-$

Overall cell reaction: $Fe + 2Ni(OH)_3 \rightarrow Fe(OH)_2 + 2Ni(OH)_2$

EXERCISE 1.3

Draw a cell diagram of Ni–Cd galvanic cell containing KOH electrolyte if the balanced overall cell reaction is as follows:

$$2NiO(OH) + Cd + 2H_2O \rightarrow 2Ni(OH)_2 + Cd(OH)_2$$

Solution

Anode half-cell reaction: $Cd + 2OH^- \rightarrow Cd(OH)_2 + 2e^-$

Cathode half-cell reaction: $2NiO(OH) + 2H_2O + 2e^- \rightarrow 2Ni(OH)_2 + 2OH^-$

Based on the half-cell reactions the cell diagram is as follows:

$$Cd, \ Cd(OH)_2 | KOH | \ Ni(OH)_2, \ NiO(OH), \ Ni$$

EXERCISE 1.4

Calculate the constant current, which were passed through solution of $CuSO_4$ for 4 h and lead to the deposition of 346 mg of Cu at the cathode. The process is shown in, Fig. 1.6.

Solution

Cathode reaction of Cu deposition can be written as follows:

$$Cu^{2+} + 2e^- \rightarrow Cu$$

As we can see the electrons number required for transformation of Cu^{2+} ions to metal Cu is equal to 2, i.e., $z = 2$. According to the Faraday's law the mass m of the substance liberated at an electrode is directly proportional to the electric charge Q (see , Eq. 1.13). Taking into account the molar mass of Cu, which is equal to $M = 63.5$ g/mol and the fact that $Q = I \cdot t$ (1 C = 1 A s)

$$m = k \cdot Q = k \cdot I \cdot t = \left(\frac{I \cdot t}{F}\right) \cdot \left(\frac{M}{z}\right),$$

therefore

$$I = \frac{mFz}{Mt} = \frac{0.346 \text{ g} \cdot 96500 \text{ C/mol} \cdot 2}{63.5 \text{ g/mol} \cdot 4 \cdot 60 \cdot 60 \text{ s}} = 0.073 \text{ A} = 73 \text{ mA}$$

EXERCISE 1.5

While conducting the electrolysis process, a current of 0.7 A was passed through the solution of $CuSO_4$ for 3 h. What is the mass of copper deposited on the cathode?

Solution

Cathode reaction of Cu deposition can be written as follows:

$$Cu^{2+} + 2e^- \rightarrow Cu$$

According to the Faraday's law $m = k \cdot I \cdot t$ and taking into account that $M_{Cu} = 63.5$ g/mol $k = \frac{M}{F \cdot z} = \frac{63.5}{96500 \cdot 2} = 3.29 \cdot 10^{-4}$ g/C. Therefore, $m = 3.29 \cdot 10^{-4} \cdot 0.7 \cdot 3 \cdot 60 \cdot 60 = 2.48$ g.

EXERCISE 1.6

A direct current of 0.08 A was passed through two cells in series for 180 min. The first cell contained the aqueous solution of $AgNO_3$ and the second cell was filled with aqueous solution of $Au(NO_3)_3$. Calculate the amount of deposited silver in gold in each cell.

Solution

Cathode reactions of Ag and Au deposition can be written as follows:

$$Ag^+ + 1e^- \rightarrow Ag_{(s)} \quad z = 1; \ M_{Ag} = 107.9 \text{ g/mol}$$

$$Au^{3+} + 3e^- \rightarrow Au_{(s)} \quad z = 3; \ M_{Au} = 197 \text{ g/mol}$$

$$m_{Ag} = k \cdot I \cdot t = \frac{M}{Fz} \cdot I \cdot t = \frac{107.9 \cdot 0.08 \cdot 180 \cdot 60}{96500 \cdot 1} = 0.97 \text{ g}$$

$$m_{Au} = k \cdot I \cdot t = \frac{M}{Fz} \cdot I \cdot t = \frac{197 \cdot 0.08 \cdot 180 \cdot 60}{96500 \cdot 3} = 0.59 \text{ g}$$

EXERCISE 1.7

What current is required to produce 150 L of hydrogen gas at room temperature and pressure (RTP, 20°C, 1 atm) from the electrolysis of water in 2 h at neutral pH?

Solution

The formation of H_2 gas in neutral media at water electrolysis is described as follows:

$$2H^+ + 2e^- \rightarrow H_2, \quad z = 2.$$

It is known that 1 mol of any gas at RTP conditions occupies 24 L of volume. Consequently, the amount of H_2 gas ($M_{H_2} = 2$ g/mol) moles required to be produced at water electrolysis is $150/24 = 6.25$ mol and the amount of H_2 gas grams required to be produced at water electrolysis is 6.25 mol $\cdot 2$ g/mol $= 12.5$ g.

$$m = k \cdot I \cdot t = \frac{M \cdot I \cdot t}{Fz},$$

therefore

$$I = \frac{mFz}{Mt} = \frac{12.5 \text{ g} \cdot 96500 \text{ } C/\text{mol} \cdot 2}{2 \text{ } g/\text{mol} \cdot 2 \cdot 60 \cdot 60 \text{ s}} = 167.5 \text{ A}.$$

EXERCISE 1.8

What is the work done by a battery (galvanic cell) with a nominal cell voltage of 1.5 V when 3 mol of electrons participated in the chemical transformation therein?

Solution

Because $W = z \cdot F \cdot E_{cell}$ and F is the amount of charge in 1 mol of electrons, the work done by the battery is $W = 3$ mol $\cdot 96500$ C/mol $\cdot 1.5$ V $= 434$ kJ

EXERCISE 1.9

What is the $Pt|H_2|H^+\|Cl_2|Cl^-|Pt$ battery potential? Calculate the Gibbs energy of the battery.

Solution

Anode half-cell reaction: $H_2 \rightarrow 2H^+ + 2e^-$ $E^0 = 0$ V
Cathode half-cell reaction: $Cl_2 + 2e^- \rightarrow 2Cl^-$ $E^0 = 1.36$ V

$$E^0_{cell} = E^0_{red, cathode} - E^0_{red, anode} = 1.36 - 0 = 1.36 \text{ V}$$

$$\Delta G^0 = -z \cdot F \cdot E^0_{cell} = -2 \cdot 96500 \cdot 1.36 = -262480 \text{ J} = -262.5 \text{ kJ}$$

EXERCISE 1.10

Write a galvanic cell diagram, half-cell reaction, choose a positive electrode and calculate the value of the cell potential if the negative electrode of the cell is made of Cd. Half-cells are separated by a salt bridge. How will the cell potential change if the Cd^{2+} concentration at the anode is equal to 0.001 M?

Solution

Anode is a negative electrode in galvanic cell, so anode is Cd. According to , Table 1.2, $E^0_{Cd^{2+}/Cd} = -0.403$ V. Cathode of galvanic cell should have higher electrode potential than Cd anode, that is, it should be from the right-hand side of the reactive species, for example Ag with $E^0_{Ag^+/Ag} = 0.799$ V. The galvanic cell diagram and half-cell reactions can be written as follows:

$$Cd|Cd^{2+}(1 \text{ M})||Ag^+(1 \text{ M})|Ag$$

Anode half-cell reaction: $Cd \rightarrow Cd^{2+} + 2e^-$ $E^0_{Cd^{2+}/Cd} = -0.403$ V
Cathode half-cell reaction: $Ag^+ + 1e^- \rightarrow Ag$ $E^0_{Ag^+/Ag} = 0.799$ V

$$E^0_{cell} = E^0_{red, cathode} - E^0_{red, anode} = 0.7991 - (-0.403) = 1.202 \text{ V}$$

According to the Nernst equation $E = E^0 - \frac{0.05916}{z} \lg \frac{[a_{ox}]}{[a_{red}]}$. While displacing activities with the concentration, let us calculate the potential of the Cd anode then $[Cd^{2+}] = 0.001$ M.

$$E = E^0 - \frac{0.05916}{z} \lg [Cd^{2+}] = -0.403 - \frac{0.05916}{2} \lg 0.001 = -0.314 \text{ V}.$$

The new cell potential will be equal to $E_{cell} = E^0_{red, cathode} - E_{red, anode} = 0.799 - (-0.314) = 1.113$ V. In this regard, while lowering the concentration of Cd^{2+} anions from 1 to 0.001 M, the cell potential decreases by $1.202 - 1.113 = 0.089$ V.

EXERCISE 1.11

Calculate the potential of the following galvanic cell $Sn|SnSO_4|Cr_2(SO_4)_3|Cr$, if $[SnSO_4] = 0.2$ M, $[Cr_2(SO_4)_3] = 0.05$ M, $E^0_{Sn^{2+}/Sn} = -0.136$ V and $E^0_{Cr^{3+}/Cr} = -0.744$ V. Check if the given galvanic cell diagram is correct. Write half-cell and overall cell reactions.

Solution

$$[Sn^{2+}] = [SnSO_4] = 0.2 \text{ M}; \; [Cr^{3+}] = 2 \cdot [Cr_2(SO_4)_3] = 0.1 \text{ M}.$$

According to the Nernst equation $E = E^0 - \dfrac{0.05916}{z} \lg C$

$$E_{Sn^{2+}/Sn} = -0.136 - \frac{0.05916}{2} \lg 0.2 = -0.115 \text{ V}.$$

$$E_{Cr^{3+}/Cr} = -0.744 - \frac{0.05916}{3} \lg 0.1 = -0.724 \text{ V}.$$

$E_{Sn^{2+}/Sn} > E_{Cr^{3+}/Cr}$, that is Sn should play the role of the cathode in the galvanic cell because the oxidizing activity of cations (ability to take electrons) increases with the potential rise. In this regard the cell diagram should be written as follows:

$$Cr|Cr_2(SO_4)_3|SnSO_4|Sn$$

$(-)$ Anode half-cell reaction: $Cr \rightarrow Cr^{3+} + 3e^- \quad |2$
$(+)$ Cathode half-cell reaction: $Sn^{2+} + 2e^- \rightarrow Sn \quad |3$
The overall ionic reaction: $2Cr + 3Sn^{2+} \rightarrow 3Cr^{3+} + 2Sn$
The overall molecular reaction: $2Cr + 2SnSO_4 \rightarrow 3Cr_2(SO_4)_3 + 2Sn$

EXERCISE 1.12

Write a galvanic cell diagram, half-cell reactions and overall cell reactions for the Fe−Sn and Fe−Zn pairs. Calculate the cell reactions if $E^0_{Sn^{2+}/Sn} = -0.136$ V, $E^0_{Fe^{2+}/Fe} = -0.44$ V and $E^0_{Zn^{2+}/Zn} = -0.763$ V. Calculate the Gibbs free energy of cells.

Solution

Metal with a higher electrode potential serves as the cathode in galvanic cell and metal with a lower potential plays the role of the anode. $E^0_{Fe^{2+}/Fe} > E^0_{Zn^{2+}/Zn}$, which means that Fe electrode is the cathode and Zn is the anode in the Fe–Zn pair. $E^0_{Sn^{2+}/Sn} > E^0_{Fe^{2+}/Fe}$, which means that Sn is the cathode and Fe is the anode in the Fe–Sn galvanic pair.

Fe–Zn galvanic pair

Reactions at anode: $Zn \rightarrow Zn^{2+} + 2e^-$
Reactions at cathode: $Fe^{2+} + 2e^- \rightarrow Fe$
Overall reaction: $Zn + Fe^{2+} \rightarrow Zn^{2+} + Fe$

$$E^0_{cell} = E^0_{red, cathode} - E^0_{red, anode} = -0.44 - (-0.763) = 0.323 \text{ V.}$$

$$\Delta G^0 = -z \cdot F \cdot E^0_{cell} = -2 \cdot 96500 \cdot 0.323 = -62339 \text{ J} = -62.3 \text{ kJ.}$$

Fe–Sn galvanic pair

Reactions at anode: $Fe \rightarrow Fe^{2+} + 2e^-$
Reactions at cathode: $Sn^{2+} + 2e^- \rightarrow Sn$
Overall reaction: $Fe + Sn^{2+} \rightarrow Fe^{2+} + Sn$

$$E^0_{cell} = E^0_{red, cathode} - E^0_{red, anode} = -0.136 - (-0.44) = 0.304 \text{ V.}$$

$$\Delta G^0 = -z \cdot F \cdot E^0_{cell} = -2 \cdot 96500 \cdot 0.304 = -58672 \text{ J} = -58.7 \text{ kJ.}$$

EXERCISE 1.13

The potential of Mg electrode is 0.015 V lower than the standard electrode potential of the Mg electrode. What is the concentration of Mg^{2+} ions near the electrode? $E^0_{Mg^{2+}/Mg} = -2.37$ V

Solution

Mg electrode potential $E_{Mg^{2+}/Mg} = -2.37 - 0.015 = -2.885$ V.

$$E = E^0 - \frac{0.05916}{z} \lg C,$$

thus

$$\lg C = \frac{\left(E^0_{Mg^{2+}/M} - E_{Mg^{2+}/Mg}\right) z}{0.05916} = \frac{(-2.37 - (-2.385 \cdot 2))}{0.05916} = 0.51 \text{ mol/L}$$

EXERCISE 1.14

Calculate the potential of the $Sn^{2+} - 2e^- \rightarrow Sn^{4+}$ system if activity ratio of the oxidized and reduced forms of tin is 1:20. $E^0_{Sn^{4+}/Sn^{2+}} = -0.15$ V

Solution

$$E = E^0 - \frac{0.05916}{z} \lg \frac{[a_{ox}]}{[a_{red}]} = -0.15 - \frac{0.05916}{2} \lg \frac{1}{20} = -0.11 \text{ V.}$$

EXERCISE 1.15

Calculate the galvanic cell potential formed by Ag electrode immersed into solution of 0.02 M $AgNO_3$ and Pt electrode immersed into solution of 0.2 M HNO_3. Electrodes are separated by a salt bridge. Write the cell diagram and electrodes half-reactions.

Solution

Let us assume that Pt electrode is the anode in the galvanic cell, then in the near electrode distance reaction of hydrogen gas oxidation to H^+ ions should occur, i.e., $2H^+ + 2e^- \rightarrow H_2$. However, this reaction is possible only with the continuous flow of H_2 to the platinum electrode. Because nitrogen in NO_3^- ion is in the highest degree of oxidation, it can only accept electrons. Therefore, Pt is the cathode on the surface of which reduction of NO_3^- anions takes place and Ag is the anode, which is oxidized to Ag^+ cations.

At anode: $Ag \rightarrow Ag^+ + 1e^-$ $\quad E^0_{Ag^+/Ag} = 0.799$ V

At cathode: $NO_3^- + 4H^+ + 3e^- \rightarrow NO + 2H_2O$ $\quad E^0_{Ag^+/Ag} = 0.96$ V

The cell diagram is as follows: $Ag|AgNO_3 \ (0.02 \text{ M})||HNO_3 \ (0.2 \text{ M})|NO|Pt$

$$E_{anode} = E^0 - \frac{0.05916}{z} \lg [Ag^+] = 0.799 - \frac{0.05916}{1} \cdot \lg 0.02 = 0.9 \text{ V.}$$

$$E_{cathode} = E^0 - \frac{0.05916}{z} \lg [H^+]^4 [NO_3^-] = 0.96 - \frac{0.05916}{3} \lg \left(0.2 \cdot 0.2^4\right)$$
$$= 1.03 \text{ V.}$$

$$E = E_{cathode} - E_{anode} = 1.03 - 0.9 = 0.13 \text{ V}$$

EXERCISE 1.16

Paddle mixer is made of structural steel composition of Fe—Mn—Co—Zn and operated in an aqueous KCl solution under standard conditions. Describe the process of electrochemical corrosion of the product under specified conditions. Pick the anodic and cathodic protection coatings for the alloy. Which corrosion reactions will occur with the picked anodic and cathodic coatings?

Solution

According to the reactive series by comparing the values of the standard electrode potentials of metals, let us determine which component of the alloy will corrode in the first place.

$$E^0_{Mn^{2+}/Mn} = -1.1 \text{ V} < E^0_{Zn^{2+}/Zn} = -0.763 \text{ V} < E^0_{Fe^{2+}/Fe}$$
$$= -0.44 \text{ V} < E^0_{Co^{2+}/Co} = -0.3 \text{ V}.$$

As it is seen Mn has the lowest electrode potential that means it is the most active metal, which is prone to corrosion at first. Therefore, the half-reaction of anodic Mn dissolution can be written as follows:

$(-)$ At anode: $Mn \rightarrow Mn^{2+} + 2e^-$

Because KCl does not undergo hydrolysis, the medium of the electrolyte solution stays neutral and cathodic half-reaction can be written as follows:

$(+)$ At cathode: $\frac{1}{2} O_2 + H_2O + 2e^- \rightarrow 2OH^-$

Any metal with the electrode potential lower than $E^0_{Mn^{2+}/Mn}$ can serve as anodic coating for the alloy, for example Al with $E^0_{Al^{3+}/Al} = -1.6$ V. Consequently, any metal with the electrode potential greater than $E^0_{Co^{2+}/Co} = -0.3$ V can serve as cathodic coating, for example Cu with $E^0_{Cu^{2+}/Cu} = 0.3$ V.

If anodic and cathodic coatings are damaged, then the most active metal will undergo corrosion at the first place. Thus, in the case of aluminum anodic coating, half-reactions are as follows:

$(-)$ At anode: $Al \rightarrow Al^{3+} + 3e^-$
$(+)$ At cathode: $\frac{1}{2} O_2 + H_2O + 2e^- \rightarrow 2OH^-$.

In the case of cathodic Cu coating, half-reactions are as follows:

$(-)$ At anode: $Mn \rightarrow Mn^{2+} + 2e^-$
$(+)$ At cathode: $\frac{1}{2} O_2 + H_2O + 2e^- \rightarrow 2OH^-$.

EXERCISE 1.17

Fastener is made of bronze composition of the Cu–Al–Ni–Zn. Write reactions of corrosion electrode processes in solution of dilute sulfuric acid under standard conditions. Pick the anodic and cathodic corrosion protection coating for the given alloy. Which corrosion reactions will occur with the picked anodic and cathodic coatings?

Solution

According to the reactive series by comparing the values of the standard electrode potentials of metals, let us determine which component of the alloy will corrode in the first place.

$$E^0_{Al^{3+}/Al} = -1.6 \text{ V} < E^0_{Zn^{2+}/Zn} = -0.763 \text{ V} < E^0_{Ni^{2+}/Ni}$$
$$= -0.2 \text{ V} < E^0_{Cu^{2+}/Cu} = 0.3 \text{ V}.$$

As it is seen Al has the lowest electrode potential that means it is the most active metal, which is prone to corrosion at first. Therefore the half-reaction of anodic Al dissolution can be written as follows:

(−) At anode: $Al \rightarrow Al^{3+} + 3e^-$

In the dilute solution of sulfuric acid, H^+ ions are the oxidants; therefore,

(+) At cathode: $H^+ + 2e^- \rightarrow H_2$

Any metal with the electrode potential lower than $E^0_{Al^{3+}/Al}$ can serve as anodic coating for the alloy, for example Mg with $E^0_{Mg^{2+}/Mg} = -2.3$ V. Consequently, any metal with the electrode potential greater than $E^0_{Cu^{2+}/Cu}$ can serve as cathodic coating, for example Ag with $E^0_{Ag^+/Ag} = 0.8$ V.

If anodic and cathodic coatings are damaged, then the most active metal will undergo corrosion at the first place. Thus, in the case of manganese anodic coating, half-reactions are as follows:

(−) At anode: $Mg^{2+} \rightarrow Mg + 2e^-$
(+) At cathode: $H^+ + 2e^- \rightarrow H_2$

In the case of cathodic Ag coating, half-reactions are as follows:

(−) At anode: $Al \rightarrow Al^{3+} + 3e^-$
(+) At cathode: $H^+ + 2e^- \rightarrow H_2$

EXERCISE 1.18

Calculate the cell current density of a single-electron electrode reaction if overpotentials of activation polarization are 150 and 4 mV; $j_0 = 5$ mA; $\alpha = 0.5$; and $T = 298$K and concentration polarization is absent. Compare the results with the limiting conditions of the Butler–Volmer equation.

Solution

According to the Butler–Volmer and Tafel equations the cell current density can be calculated as follows.

$$j = j_0 \cdot \left\{ \exp\left[-\frac{\alpha \cdot z \cdot F \cdot \Delta E}{R \cdot T} \right] - \exp\left[\frac{(1-\alpha) \cdot z \cdot F \cdot \Delta E}{R \cdot T} \right] \right\}$$

$$= 0.005 \cdot \left\{ \exp\left[\frac{-0.5 \cdot 1 \cdot 96500 \cdot 0.15}{8.314 \cdot 298} \right] - \exp\left[\frac{(1-0.5) \cdot 1 \cdot 96500 \cdot 0.15}{8.314 \cdot 298} \right] \right\}$$

$$= -0.093 \ \text{A/cm}^2 = -93 \ \text{mA/cm}^2$$

Let us consider the limiting conditions $|\Delta E| \ll \frac{RT}{zF}$ and $|\Delta E| \gg \frac{RT}{zF}$

$$\frac{RT}{zF} = \frac{8.314 \cdot 298}{1 \cdot 96500} = 0.026 \ \text{V} = 26 \ \text{mV}$$

If $|\Delta E| \ll \frac{RT}{zF}$, then $j = -\frac{j_0 \cdot z \cdot F \cdot \Delta E}{R \cdot T} = -\frac{0.005 \cdot 1 \cdot 96500 \cdot 0.15}{8.314 \cdot 298} = -0.029$
$\text{A/cm}^2 = -29 \ \text{mA/cm}^2$.

If $|\Delta E| \gg \frac{RT}{zF}$, then $j = -j_0 \cdot \exp\left[\frac{(1-\alpha) \cdot z \cdot F \cdot \Delta E}{R \cdot T} \right] = -0.005 \cdot \exp$
$\left[\frac{(1-0.5) \cdot 1 \cdot 96500 \cdot 0.15}{8.314 \cdot 298} \right] = -0.093 \ \text{A/cm}^2 = 93 \ \text{mA/cm}^2$.

When $|\Delta E| \ll \frac{RT}{zF}$, the calculation error while using equations of limiting conditions is high because the value of $|\Delta E| = 150$ mV is actually higher than the value of $\frac{RT}{zF} = 26$ mV. Therefore, the error at limiting condition of $|\Delta E| \gg \frac{RT}{zF}$ is absent.

The current density of the cell having electrode reactions overpotential $\Delta E = 4$ mV with regard of Butler–Volmer equation is equal to

$$j = j_0 \cdot \left\{ \exp\left[-\frac{\alpha \cdot z \cdot F \cdot \Delta E}{R \cdot T} \right] - \exp\left[\frac{(1-\alpha) \cdot z \cdot F \cdot \Delta E}{R \cdot T} \right] \right\}$$

$$= 0.005 \cdot \left\{ \exp\left[\frac{-0.5 \cdot 1 \cdot 96500 \cdot 0.004}{8.314 \cdot 298} \right] - \exp\left[\frac{(1-0.5) \cdot 1 \cdot 96500 \cdot 0.004}{8.314 \cdot 298} \right] \right\}$$

$$= 0.78 \ \text{mA/cm}^2$$

If $|\Delta E| \ll \frac{RT}{zF}$, then $j = -\frac{j_0 \cdot z \cdot F \cdot \Delta E}{R \cdot T} = -\frac{0.005 \cdot 1 \cdot 96500 \cdot 0.004}{8.314 \cdot 298} =$
$0.78 \ \text{mA/cm}^2$

If $|\Delta E| \gg \frac{RT}{zF}$, then $j = -j_0 \cdot \exp\left[\frac{(1-\alpha) \cdot z \cdot F \cdot \Delta E}{R \cdot T} \right] = -0.005 \cdot \exp$
$\left[\frac{(1-0.5) \cdot 1 \cdot 96500 \cdot 0.004}{8.314 \cdot 298} \right] = -5 \ \text{mA/cm}^2$.

Since $|\Delta E| \ll \frac{RT}{zF}$ that is 4 mV $\ll$ 26 mV, there is no error in calculations using the complete Butler–Volmer equation and its simplified version satisfying the conditions of $|\Delta E| \ll \frac{RT}{zF}$. However, when using equation for the limiting conditions of $|\Delta E| \gg \frac{RT}{zF}$, then calculation error is significant.

EXERCISE 1.19

Calculate parameters α, j_0, and k^0 of the Butler–Volmer equation. Current densities formed during $M^{2+} \rightarrow M^{3+} + e^-$ reaction at an inert anode are equal to -5 and -80 mA/cm^2 and attributed to reaction overpotentials of 0.1 and 0.4 V, respectively, formed due to the activation polarization. It is known that $[M^{2+}] = [M^{3+}] = 2$ mM, temperature is equal to 298K and concentration polarization is negligible.

Solution

Since $\quad |\Delta E| = 100 \quad$ and $\quad 400\,\text{mV} \gg \frac{RT}{zF} = 26\,\text{mV}$, then $j_i = -j_0 \cdot \exp$ $\left[\frac{(1-\alpha)\cdot z\cdot F\cdot \Delta E}{R\cdot T}\right]$. Charge transfer coefficient α can be calculated by dividing equations for current densities. Let us divide j_2 by j_1,

$$\frac{j_2}{j_1} = \frac{-j_0\cdot \exp\left[\dfrac{(1-\alpha)\cdot z\cdot F\cdot \Delta E_2}{R\cdot T}\right]}{-j_0\cdot \exp\left[\dfrac{(1-\alpha)\cdot z\cdot F\cdot \Delta E_1}{R\cdot T}\right]}$$

Hence, $\alpha = 1 - \dfrac{\ln\left(\frac{j_2}{j_1}\right)\cdot R\cdot T}{z\cdot F\cdot(\Delta E_2\cdot \Delta E_1)} = 1 - \dfrac{\ln\left(\frac{-0.08}{-0.005}\right)\cdot 8.314\cdot 298}{1\cdot 96500\cdot(0.4 - 0.1)} = 0.76.$

Since $j_i = -j_0\cdot \exp\left[\frac{(1-\alpha)\cdot z\cdot F\cdot \Delta E}{R\cdot T}\right]$, then $j_0 = 1.9$ mA/cm^2.

Heterogeneous electron transfer rate constant can be found from the following equation:

$$j_0 = z\cdot F\cdot k^0 \cdot \left[C_{ox}^{\alpha}\right]\cdot \left[C_{red}^{(1-\alpha)}\right].$$

Because the concentration of oxidized and reduced forms are equal,

$$k^0 = \frac{j_0}{z\cdot F\cdot C} = \frac{0.0019\cdot 1000}{1\cdot 96500\cdot 0.002} = 9.8\cdot 10^{-3} \text{ cm/s}.$$

EXERCISE 1.20

Determine the rate-limiting step (concentration or activation polarization) of an electrochemical double-electron electrode oxidation reaction of compound X ($C_x = 4$ mM) at an inert anode if mass transfer coefficient is equal to

0.003 cm/s and reaction occurs at stationary conditions. According to the Tafel plot, the following coefficients of the Tafel equation were obtained: $a = 1.5$ V and $b = 0.45$ V. It is known that overpotential is equal to 0.8 V and migration and convection mass transport are negligible.

Solution

To determine the rate-limiting step of anodic oxidation reaction, it is necessary to compare the limiting current density (j_L) produced at diffusion-controlled mass transport and current density generated during the charge transfer controlled kinetics in the absence of diffusion.

The limiting current density can be found using the following equation:

$$j_L = z \cdot F \cdot C_i \cdot k_m = 2 \cdot 96500 \cdot 0.004 \cdot 10^{-3} \cdot 0.003 = 0.0023 \text{ A/cm}^2 = 2.3 \text{ mA/cm}^2$$

According to the Tafel equation, taking into account the value of over-potential $\Delta E = 0.8$ V, the current density generated during the charge transfer–controlled anodic oxidation is equal to $\Delta E = a + b \cdot \lg j$, that is $0.8 = 1.5 + 0.45 \cdot \lg j$. Hence, $j = 0.0278 \text{ A/cm}^2 = 27.8 \text{ mA/cm}^2$.

Since $|j_L| \ll |j|$, that is 2.3 mA/cm$^2 \ll$ 27.8 mA/cm^2, the overpotential is caused by concentration polarization and related to the slow mass transfer of compound X. The rate-limiting reaction step is diffusion.

Appendix 2. Exercises for Chapter 2

EXERCISE 2.1

Calculate the diffuse layer thickness of particles distributed in solution containing 0.25 mM NaCl and 0.4 mM K_2SO_4. Relative dielectric permittivity of electrolyte solution (ε_r) is 78.5 and temperature is 287K. How does the thickness of diffuse layer change if the solution is diluted with water twice? How does the thickness of diffuse layer change if the temperature of solution rises to 350K and ε_r decreases to 67.5?

Solution

The thickness of diffuse layer can be calculated as follows:

$$\frac{1}{k} = \sqrt{\frac{\varepsilon_r \varepsilon_0 RT}{2F^2 \mu}},$$

where electrical permittivity of vacuum $\varepsilon_0 = 8.854 \cdot 10^{-12}$ F/m $= 8.854 \cdot 10^{-12}$ C/V m; Faraday constant $F = 96,500$ C/mol; and $R = 8.314$ J/K mol.

Taking into account, all ions present in the solution due to salt dissociation, ionic strength can be calculated as follows:

$$\mu = \frac{1}{2} \cdot \left(\sum_i C_i z_i^2 \right) = \frac{1}{2} \cdot \left([Na^+] \cdot z_{Na^+}^2 + [Cl^-] \cdot z_{Cl^-}^2 + [K^+] \cdot z_{K^+}^2 + [SO_4^{2-}] \cdot z_{SO_4^{2-}}^2 \right)$$

$$= \frac{1}{2} \cdot \left(0.25 \cdot 10^{-3} \ mol/L \cdot (1)^2 + 0.25 \cdot 10^{-3} \ mol/L \cdot (-1)^2 + 2 \cdot 0.4 \cdot 10^{-3} \ mol/L \cdot (1)^2 \right.$$

$$\left. + 0.4 \cdot 10^{-3} \ mol/L \cdot (-2)^2 \right) = 0.00145 \ mol/L = 1.45 \ mol/m^3$$

Taking into account 1 F = 1 C/V and 1 V = 1 J/C, the thickness of diffuse layer is equal to

$$\frac{1}{k} = \sqrt{\frac{78.5 \cdot 8.854 \cdot 10^{-12} \ F/m \cdot 8.314 \ J/K \ mol \cdot 287K}{2 \cdot (96,500 \ C/mol)^2 \cdot 1.45 \ mol/m^3}} = 7.84 \cdot 10^{-9} \ m = 7.84 \ nm$$

If temperature of solution increases from 287 to 350K, the diffuse layer thickness changes to

$$\frac{1}{k} = \sqrt{\frac{67.5 \cdot 8.854 \cdot 10^{-12} \ F/m \cdot 8.314 \ J/K \ mol \cdot 350K}{2 \cdot (96,500 \ C/mol)^2 \cdot 1.45 \ mol/m^3}} = 8.02 \cdot 10^{-9} \ m = 8.02 \ nm$$

If electrolyte solution is diluted twice then [NaCl] = 0.25mM/2 = 0.125 mM and [K_2SO_4] = 0.4/2 = 0.2 mM. In this regard, ionic strength of diluted solution is equal to

$$\mu = \frac{1}{2} \cdot \left(0.125 \cdot 10^{-3} \ mol/L \cdot (1)^2 + 0.125 \cdot 10^{-3} \ mol/L \cdot (-1)^2 \right.$$

$$\left. + 2 \cdot 0.2 \cdot 10^{-3} \ mol/L \cdot (1)^2 \cdot 10^{-3} \ mol/L \cdot (-2)^2 \right)$$

$$= 0.000725 \ mol/L = 0.725 \ mol/m^3$$

Consequently ionic strength of twice diluted solution is equal to

$$\frac{1}{k} = \sqrt{\frac{78.5 \cdot 8.854 \cdot 10^{-12} \ F/m \cdot 8.314 \ J/K \ mol \cdot 287K}{2 \cdot (96,500 \ C/mol)^2 \cdot 0.725 \ mol/m^3}} = 1.11 \cdot 10^{-8} \ m = 11.1 \ nm$$

As it can be concluded from the above calculations, higher temperatures and diluted solutions lead to the increase of diffuse layer thickness.

EXERCISE 2.2

Determine the efficiency of copper ions removal from the wastewater containing 100 mg/L Cu^{2+} and 0.1 M K_2SO_4 by the electrochemical reduction method, which is conducted for 5 min at applied current of 0.5 A using graphite plate anode and copper plate cathode. The working volume of electrolyzer is 15 L. Calculate time and energy consumption required for the complete removal of copper ions from wastewater at same current conditions if the voltage generated at this applied current is equal to 12 V.

Solution

The total amount of copper in the electrolyzer before treatment can be estimated taking into account the initial concentration of copper and working volume of the electrolyzer and equal to 100 mg/L · 15 L = 1500 mg = 1.5 g.

Cathodic half-reaction of copper deposition is described as follows:

$$Cu^{2+} + 2e^- \rightarrow Cu_{(s)} \quad z = 2; \quad M_{Cu} = 63.55 \text{ g/mol}$$

According to the Faraday's law the amount of deposited copper can be calculated by the following equation:

$$m_{Cu} = k \cdot I \cdot t = \frac{M}{Fz} \cdot I \cdot t = \frac{63.55 \cdot 0.5 \cdot 5 \cdot 60}{96,500 \cdot 2} = 0.049 \text{ g}$$

The efficiency of copper removal in electrochemical reduction process under given conditions is equal to

$$\alpha(\%) = \frac{0.049}{1.5} \cdot 100\% = 3.3\%$$

The time required for the complete removal of copper can be calculated as follows:

$$t = \frac{m_{Cu} \cdot F \cdot z}{M \cdot I} = \frac{1.5 \cdot 96500 \cdot 2}{63.55 \cdot 0.5} = 9110.9 \text{ s} = 151.8 \text{ min} = 2.53 \text{ h}$$

The energy consumption (EC, kWh/m^3) of the process can be estimated using the following equation:

$$EC = \frac{UIt}{V} = \frac{12 \cdot 0.5 \cdot 2.53}{15} = 1.012 \text{ kWh/m}^3$$

where U is the applied voltage (V); I is the current intensity (A); t is the time of treatment (h); and V is the volume of treated solution (L).

EXERCISE 2.3

During electrochemical removal of nickel ions (250 mg/L initial concentration) from electrolyte solution of Na_2SO_4, 500 mg of nickel was deposited onto Ti/Pt cathode within 30 min while applying 1 A current to the electrodes. Ti/IrO$_2$ electrode was used as an anode. The distance (l) between electrodes is equal to 1 cm and the surface area (A) of each electrode is $10\,\text{cm}^2$. The working volume (V) of electrolyzer is 5 L. Please determine the current efficiency (CE) and specific energy consumption (SEC) required for electrochemical deposition of 1 g nickel. Calculate the time and EC required for complete removal of nickel from electrolyte solution. Cell voltage during the electrolysis process can be determined using the following equation:

$$U_{\text{cell}} = [E_a - E_c + (1 + \alpha) \cdot I \cdot R_s] \cdot (1 + \beta),$$

where $E_c = -0.5$ V and $E_a = 2.2$ V are anodic and cathodic potentials, $\alpha = 0.1$ is the coefficient taking into account the increase in the electrolyte resistance due to the gas supply, $R_s = l/(A \cdot \rho)$ is the electrolyte resistance; $\rho = 0.5\,\Omega^{-1}\,\text{cm}^{-1}$ is the specific conductance of electrolyte solution, and $\beta = 0.03$ is the coefficient taking into account the potential drop in terminals.

Solution

Cathodic half-reaction of Ni^{2+} reduction at cathode can be written as follows:

$$Ni^{2+} + 2e^- \rightarrow Ni_{(s)} \quad z = 2;\ M_{Ni} = 58.69\ \text{g/mol}$$

CE can be determined through the ratio of the actual amount of metal precipitated on the cathode and the quantity of the metal calculated by the Faraday law (m_t).

$$m_{cu,t} = k \cdot I \cdot t = \frac{M \cdot I \cdot t}{Fz} = \frac{58.69 \cdot 1 \cdot 30 \cdot 60}{96,500 \cdot 2} = 0.55\ \text{g}$$

$$CE = \frac{ma}{m_t} \cdot 100\% = \frac{0.5}{0.55} \cdot 100\% = 90.9\%$$

Cell voltage during the electrolysis process is equal to

$$U_{\text{cell}} = \left[2.2 - (-0.5) + (1 + 0.1) \cdot 1 \cdot \frac{1}{10 \cdot 0.5}\right] \cdot (1 + 0.03) = 3\ \text{V}$$

Taking into account, current efficiency of the process SEC can be determined as follows:

$$SEC = \frac{U_{\text{cell}}}{k \cdot CE} = \frac{U_{\text{cell}} \cdot z \cdot F}{M} = \frac{3 \cdot 2 \cdot 26.8}{58.69 \cdot 0.91} = 3\ \text{W h/g}$$

where $1\ F = 26.8$ A h/mol $= 96\,500$ C/mol.

Taking into account the CE of the process time and EC required for the complete removal of nickel can be calculated as follows:

$$t = \frac{C_{Ni} \cdot V \cdot F \cdot Z}{M \cdot I \cdot CE} = \frac{0.25 \cdot 5 \cdot 96,500 \cdot 2}{58.69 \cdot 1 \cdot 0.91} = 4517\ \text{s} = 75.3\ \text{min} = 1.26\ \text{h}$$

$$EC = \frac{U \cdot I \cdot t}{V \cdot CE} = \frac{3 \cdot 1 \cdot 1.26}{5 \cdot 0.91} = 0.83\ \text{kWh/m}^3$$

EXERCISE 2.4

Concentration of surfactants in acidic wastewater is equal to 1000 mg/L and is planned to be removed by electroflotation method using graphite anode and stainless steel cathode. The distance between electrodes is equal to 1.5 cm and surface area is equal to 20 cm^2. The volume of the wastewater needed to be treated is 50 L. The preliminary electrolysis at applied current of 0.8 A conducted for 10 min allowed removal of $m_t = 10$ g of surfactant from wastewater. CE of electroflotation process was estimated equal to 85%. What is the volume of hydrogen gas and time of electrolysis required for the complete removal of surfactant from wastewater if the electrolysis is conducted at the same parameters as preliminary test? What is the energy consumption required for the allocation of 1 m^3 hydrogen if cell voltage is composed of potential drops in all parts of the circuit and can be estimated from the following equation:

$$U_{cell} = E^0_{water} + \Delta E_a + \Delta E_c + \Delta E_{sol} + \Delta E_{terminals},$$

where $E^0_{water} = 1.23$ V; $\Delta E_a = 0.1$ V; $\Delta E_c = 0.15$ V; $\Delta E_{sol} = j \cdot \rho \cdot l \cdot k_{sol}$; $\Delta E_{terminals} = 0.05$ V; $k_{sol} = 1.15$ and $\rho = 0.1$ $\Omega^{-1}cm^{-1}$.

Solution

While conducting electrolysis of water, allocation of hydrogen gas takes place at the cathode and oxygen gas at the anode. Acidic conditions lead to the following electrode reactions:

At the cathode: $2H^+ + 2e^- \rightarrow H_2$ $z = 2$
At the anode: $H_2O - 2e^- \rightarrow \frac{1}{2}O_2 + 2H^+$ $z = 2$

The amount of surfactants ($m_{complete}$) in a given volume of wastewater is equal to 1000 mg/L $\cdot$ 50 L $= 50$ g. The volume of hydrogen gas produced during the preliminary electrolysis can be calculated as follows:

$$V_{H_2,t} = k \cdot I \cdot t \cdot CE = 4.56 \cdot 10^{-4} \cdot 0.8 \cdot 10/60 \cdot 0.85 = 5.2 \cdot 10^{-5} m^3,$$

where k is electrochemical equivalent of hydrogen ($k_{H_2} = 4.56 \cdot 10^{-4}$ m^3/A h); t is the electrolysis time (h); and I is the current (A).

The amount of hydrogen gas required for complete removal of 1000 mg/L surfactant from wastewater can be determined using the following ratio:

$$\frac{m_t}{V_{H_2,t}} = \frac{m_{complete}}{V_{H_2,complete}}$$

$V_{H_2,complete} = (5.2 \cdot 10^{-5} \cdot 50)/10 = 2.6 \cdot 10^{-4}$ m^3.

Time required for the complete flotation of surfactant is equal to

$$t = \frac{V_{H_2,complete}}{k \cdot I \cdot CE} = \frac{2.6 \cdot 10^{-4}}{4.56 \cdot 10^{-4} \cdot 0.8 \cdot 0.85} = 0.84 \text{ h} = 50.4 \text{ min}$$

The cell voltage is composed of potential drops in all parts of the circuit and can be estimated from the following equation:

$$U_{cell} = 1.23 + 0.1 + 0.15 + (0.8/20 \cdot 0.1 \cdot 1.5 \cdot 1.15) + 0.05 = 1.54 \text{ V}$$

The energy consumption required for the allocation of 1 m^3 hydrogen under the normal operation can be calculated as follows:

$$EC = \frac{U_{cell} \cdot z \cdot F}{V_{mH_2} \cdot CE} = \frac{1.54 \cdot 2 \cdot 26.8}{0.0224 \cdot 0.85} = 4.33 \text{ kWh/m}^3$$

where V_{mH_2} is the molar volume of 1 mol hydrogen $\left(V_{mH_2} = 0.0224 \text{ m}^3\right)$.

EXERCISE 2.5

Please determine whether solid particles will be wetted by water of be floated to the water surface if $\gamma_{sg} = 0.0251$ J/m^2, $\gamma_{sl} = 0.0521$ J/m^2 and $\gamma_{lg} = 0.0821$ J/m^2. What will be the probability of formation of "gas bubble—particle" complexes if the concentration of gas bubbles (C_g) in the fluid containing those particles is 30 vol%, gas bubble radius (R) is equal to 50 µm, and solid particles radius (r) is equal to 25 µm?

Solution

The equilibrium between surfaces energies in the system of water droplet on a smooth homogeneous solid can be described by the Young's equation as follows:

$$\gamma_{sg} = \gamma_{sl} + \gamma_{lg} \cdot \cos \theta_c$$

$\cos \theta = (\gamma_{sg} - \gamma_{sl})/\gamma_{lg} = (0.0251 - 0.0521)/0.0821 = -0.3289$ or $\theta = 109.2°$, which means that particle has low wettability and can be floated efficiently.

The probability of formation of the "bubble—particle" complex can be determined using the following equation.

$$\omega = C_g\left[\left(1 + \frac{r}{R}\right)^3 - 1\right] = 30 \cdot \left[\left(1 + \frac{25}{50}\right)^3 - 1\right] = 71.25\%$$

As it can be seen from the result, solid particles with $\theta = 109.2°$ at given conditions have relatively high probability to form flotocomplexes.

EXERCISE 2.6

Calculate the work of adhesion for the "gas bubbles—solid particles" interaction system if $\gamma_{sg} = 0.0551$ J/m^2, $\gamma_{sl} = 0.0432$ J/m^2 and $\gamma_{lg} = 0.0921$ J/m^2. It is known that diameter of solid particles in the water is smaller than diameter of generated gas bubbles.

Solution

According to the Young's equation,

$$\cos \theta = \left(\gamma_{sg} - \gamma_{sl}\right)/\gamma_{lg} = (0.0551 - 0.0432)/0.0921 = 0.129 \text{ or } \theta = 82.6°,$$

which means that particle surface is hydrophilic and tends to wettability.

In the case when relatively small solid particles interact with big gas bubbles in aqueous medium, the work of adhesion for the system can be expressed through the following equation:

$$\Delta W_{sg} = \gamma_{lg}(1 - \cos \theta_c) = 0.0921 \cdot (1 - \cos 82.6°) = 0.08 \text{ J/m}^2.$$

EXERCISE 2.7

During the flotation of aluminum particles with density $\rho_s = 2700$ g/L by the nitrogen gas in water having density $\rho_l = 1010$ g/L "gas bubble—solid particle complexes" with diameter $d_{gs} = 130$ μm are formed. Calculate the rising velocity of the formed complexes in the laminar flow and velocities of the particles (u_s) and bubbles (u_g) movement relative to the medium.

The dynamic viscosity of water $\eta = 1002$ kg/(s m). Radii of a gas bubble and a solid particle $R = 30$ μm and $r = 5$ μm and concentrations of gas bubbles and solid particles in water $C_g = 55\%$ and $C_s = 6\%$, respectively. Gravitational acceleration $g = 9.8$ m/s^2.

Solution

If the movement of a gas bubble—particle complex is laminar, then the rising velocity of the complex can be calculated as follows:

$$u = \frac{g[\rho_l V_g - V_s(\rho_s - \rho_l)]}{3\pi\eta d_{gs}},$$

where V_g and V_s are the volumes of a gas bubble and a particle, respectively. Assuming that both gas bubbles and particles in the water have spherical shape and one bubble interacts with one particle,

$$V_g = 4/3\pi R^3 = 4/3 \cdot 3.14 \cdot (30 \cdot 10^{-6})^3 = 1.1 \cdot 10^{-13} \text{ m}^3$$

$$V_s = 4/3\pi r^3 = 4/3 \cdot 3.14 \cdot (5 \cdot 10^{-6})^3 = 5.2 \cdot 10^{-16} \text{ m}^3$$

$$u = \frac{9.8 \cdot (1010 \cdot 1.1 \cdot 10^{-13} - 5.2 \cdot 10^{-6} \cdot (2700 - 1010)}{3 \cdot 3.14 \cdot 1002 \cdot 130 \cdot 10^{-6}} = 8.8 \cdot 10^{-10} \text{ m/s}$$

According to the balance of forces, the rising velocity can be calculated as follows.

The velocity of the particles (u_s) and bubbles (u_g) movement relative to the medium is determined by the following equations:

$$
\begin{aligned}
u_s &= -\frac{2}{9}\left(\frac{gr^2\rho_l}{\eta}\right)\left[(1-C_s)\left(\frac{\rho_s}{\rho_l}-1\right)+C_g\right]\\
&= -\frac{2}{9}\cdot\left(\frac{9.8\cdot(5\cdot10^{-6})^2\cdot1010}{1002}\right)\cdot\left[(1-0.06)\cdot\left(\frac{2700}{1010}-1\right)+0.55\right]\\
&= 1.1\cdot10^{-10}\ \text{m/s}
\end{aligned}
$$

$$
\begin{aligned}
u_g &= \frac{1}{9}\left(\frac{gR^2\rho_l}{\eta}\right)\left[1+C_s\left(\frac{\rho_s}{\rho_l}-1\right)-C_g\right]\\
&= \frac{1}{9}\cdot\left(\frac{9.8\cdot(30\cdot10^{-6})^2\cdot1010}{1002}\right)\cdot\left[1+0.06\cdot\left(\frac{2700}{1010}-1\right)-0.55\right]\\
&= 5.4\cdot10^{-10}\ \text{m/s}
\end{aligned}
$$

EXERCISE 2.8

Calculate the total energy G of interaction of two spherical particles with a radius of 120 nm each and plot G, G_{el}, and G_A as a function of h using the following data.

$A_H = 4\cdot10^{-20}$ J; $k = 2\cdot10^8$ m^{-1}; $\phi_\delta = 32$ mV; $h = 1, 2, 5, 8, 10, 20, 30$ and 40 nm; $\varepsilon_0 = 8.854\cdot10^{-12}$ F/m $= 8.854\cdot10^{-12}$ C/V m and $\varepsilon_r = 80$.

Solution

The total energy of two particles interaction is equal to $G = G_A + G_{el}$.

For two identical particles ($R_1 = R_2 = R_{12}$) at close proximity $h \ll R$ ($40 \ll 120$ nm) attraction energy can be found as follows:

$$
G_A = -\frac{A_H R_{12}}{12h} = -\frac{4\cdot10^{-20}\cdot120\cdot10^{-9}}{12\cdot1\cdot10^{-9}} = -40\cdot10^{-20}\ \text{J}
$$

Repulsive forces of particles interaction can be determined using the following equation:

$$
\begin{aligned}
G_{el} &= 2\pi\varepsilon_r\varepsilon_0 R_{12}\phi_\delta^2 \exp(-kh)\\
&= 2\cdot3.14\cdot8.854\cdot10^{-12}\cdot80\cdot120\cdot10^{-9}\cdot(32\cdot10^{-3})^2\cdot\exp\left(-2\cdot10^8\cdot1\cdot10^{-9}\right)\\
&= 44.7\cdot10^{-20}\ \text{J}
\end{aligned}
$$

The total energy of two particles interaction at 1 nm distance between particles is equal to

$$G = G_A + G_{el} = -40 \cdot 10^{-20} + 44.7 \cdot 10^{-20} = 4.7 \cdot 10^{-20} \text{ J}$$

Attraction, repulsion, and total energies of particle interaction at other interparticle distances can be determined similarly to the above performed calculations and listed in the below table. Dependences of energies of particle interactions on h distance are plotted based on the obtained results from the table.

h, nm	$G_A \cdot 10^{-20}$, J	$G_{el} \cdot 10^{-20}$, J	$G \cdot 10^{-20}$, J
1	−40	44.7	4.7
2	−20	36.6	6.6
5	−8	20.1	12.1
8	−5	11	6
10	−4	7.4	3.4
20	−2	1	−1
30	−1.3	0.1	−1.2
40	−1	0.02	−0.98

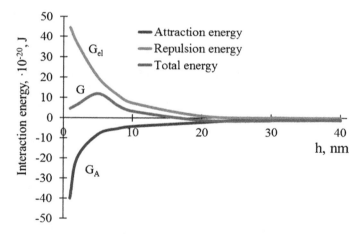

The dependence of the energy of electrostatic repulsion, molecular attraction, and the total energy of the particle interaction on the distance between particle surfaces.

EXERCISE 2.9

Calculate electrostatic repulsion, molecular attraction, and the total energies of the interaction of two particles (diameter of 300 nm each) in 0.1 M Na_2SO_4 solution when the distance between particle surfaces is 10 nm.

$\varepsilon_0 = 8.854 \cdot 10^{-12}$ F/m $= 8.854 \cdot 10^{-12}$ C/V m; $F = 96,500$ C/mol; $R = 8.314$ J/K mol; $A_H = 2.3 \cdot 10^{-20}$ J; $\varepsilon_r = 79.2$; $T = 285$K and $\phi_\delta = 25$ mV.

Solution

The thickness of diffuse layer can be calculated as follows:

$$\frac{1}{k} = \sqrt{\frac{\varepsilon_r \varepsilon_0 RT}{2F^2 \mu}},$$

where R is the gas constant and μ is the ionic strength of the solution, which can be calculated as follows:

$$\mu = \frac{1}{2} \cdot \left(\sum_i C_i z_i^2 \right) = \frac{1}{2} \cdot \left([\text{Na}^+] \cdot z_{\text{Na}^+}^2 + [\text{SO}_4^{2-}] \cdot z_{\text{SO}_4^{2-}}^2 \right)$$

$$= \frac{1}{2} \cdot \left(2 \cdot 0.1 \ \text{mol/L} \cdot (1)^2 + 0.1 \ \text{mol/L} \cdot (-2)^2 \right)$$

$$= 0.3 \ \text{mol/L} = 300 \ \text{mol/m}^3$$

$$\frac{1}{k} = \sqrt{\frac{79.2 \cdot 8.854 \cdot 10^{-12} \ \text{F/m} \cdot 8.314 \ \text{J/K mol} \cdot 285 \text{K}}{2 \cdot (96,500 \ \text{C/mol})^2 \cdot 300 \ \text{mol/m}^3}} = 5.4 \cdot 10^{-10} \ \text{m}$$

Based on the calculation on the thickness of diffuse layer $k = 1/(5.4 \cdot 10^{-10}) = 1.85 \cdot 10^9 \ \text{m}^{-1}$.

Attractive energy of two identical particles ($R_1 = R_2 = R_{12}$) interaction at close proximity $h \ll R$ ($10 \ll 150$ nm) can be found as follows:

$$G_A = -\frac{A_H R_{12}}{12h} = -\frac{2.3 \cdot 10^{-20} \cdot 150 \cdot 10^{-9}}{12 \cdot 10 \cdot 10^{-9}} = -2.85 \cdot 10^{-20} \ \text{J}$$

$$G_{el} = 2\pi \varepsilon_r \varepsilon_0 R_{12} \phi_\delta^2 \exp(-kh) = 2 \cdot 3.14 \cdot 79.2 \cdot 8.854 \cdot 10^{-12} \cdot 150 \cdot 10^{-9}$$
$$\cdot (25 \cdot 10^{-3})^2 \cdot \exp(-1.85 \cdot 10^9 \cdot 10 \cdot 10^{-9}) = 3.7 \cdot 10^{-27} \ \text{J}$$

$$G = G_A + G_{el} = -2.85 \cdot 10^{-20} + 3.7 \cdot 10^{-27} \approx -2.85 \cdot 10^{-20} \ \text{J}$$

As it is seen from the above calculations, electrostatic repulsion energy is negligible compared to the attraction energy. This means that particles can easily coagulate.

EXERCISE 2.10

Write the final products obtained in cathodic, anodic, and middle compartments of a simple electrodialyzer separated by ideally active cation-exchange and anion-exchange membranes if electrolyte solution containing NaCl is fed to the middle compartment and distilled water is fed to the electrode compartments. Draw electrodialyzer cell and schematic movement of ions within compartments of the electrodialyzer.

Solution

When potential difference is applied between electrodes of electrodialyzer, nitrate ions start to move through the anion-exchange membrane toward anode and sodium ions start to move through the cation-exchange membrane toward cathode from the middle compartment. As a result, solution in the middle compartment becomes desalinated. Water electrolysis reactions as well as interactions within compartments of electrodialyzer are shown in the following reactions and schematic illustration.

At anode: $2H_2O + 2NO_3^- - 4e^- \rightarrow O_2 + 4H^+ + 2NO_3^-$
At cathode: $2H_2O + 4e^- \rightarrow H_2 + 2OH^- + 2Na^+$

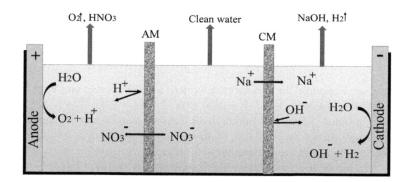

Appendix 3. Exercises for Chapter 3

EXERCISE 3.1

Please write down anodic, cathodic, and overall cell reactions occurring in a mediator-free MFC-containing acetate as a fuel source. It is known that $E^0_{red, cathode} = 0.82$ V and $E^0_{red, anode} = -0.289$ V. What would be the Gibbs free energy of the overall reaction?

Solution

Anode: $CH_3COOH + 2H_2O \rightarrow 2CO_2 + 8H^+ + 8e^-$
Cathode: $2O_2 + 8H^+ + 8e^- \rightarrow 4H_2O$
Overall combustion reaction: $CH_3COOH + 2O_2 \rightarrow 2CO_2 + 2H_2O + electricity$

$$E^0_{cell} = E^0_{red, cathode} - E^0_{red, anode} = 0.82 - (-0.289) = 1.109 \ \text{V}.$$

$$\Delta G = -z \cdot F \cdot E^0_{cell} = -8 \cdot 96500 \cdot 1.109 = -856.1 \ \text{kJ/mol}.$$

EXERCISE 3.2

Calculate the amount of energy stored in a supercapacitor of the stated capacitance equal to 4.7 F and the voltage of 12 V. What would be the minimum supercapacitor energy and voltage able to ensure the work of a motor of 20 kW nominal power for 2 min?

Solution

The maximum energy, which can be stored in capacitors, is equal to

$$W_{max} = \frac{C \cdot U_V^2}{2} = \frac{4.7 \cdot 12^2}{2} = 338.4 \ \text{J}$$

The energy required for to ensure the work of a 30-kW power motor for 2 min can be found as follows: $W = P \cdot t = 20{,}000 \ \text{W} \cdot 2 \cdot 60 \ \text{s} = 2.4 \ \text{MW s} = 2.4 \ \text{MJ}$.

The minimum voltage required to be applied across supercapacitors to obtain enough energy for the 2 min of 20-kW motor operation at full discharge of supercapacitors can be calculated as follows:

$$U_V = \sqrt{\frac{2W}{C}} = \sqrt{\frac{2 \cdot 2.4 \cdot 10^6}{4.7}} = 1010 \ \text{V}$$

EXERCISE 3.3

Calculate the wavelength of sound wave passing through water if sound frequency is equal to $5 \cdot 10^{10}$ Hz and sound speed in water is 1.48 mm/μs.

Solution

The wavelength of sound wave can be found as follows:

$$\lambda = \frac{c}{f} = \frac{1.48 \cdot 10^3}{5 \cdot 10^{10}} = 29.6 \ \text{nm}$$

EXERCISE 3.4

Please determine the energy of the photons corresponding to the longest ($\lambda = 750$ nm) and the shortest ($\lambda = 380$ nm) waves of the visible spectrum. $h = 4.13 \cdot 10^{-15}$ eV s and $c_l = 3 \times 10^8$ m/s.

Solution

The photon energy can be calculated as follows:

$$E = \frac{h \cdot c_l}{\lambda}$$

Therefore, the photon energies of the shortest and longest waves are equal to

$$E_{380} = \frac{h \cdot c_l}{\lambda} = \frac{4.13 \cdot 10^{-15} \cdot 3 \cdot 10^8}{380 \cdot 10^{-9}} = 3.26 \text{ eV}$$

$$E_{750} = \frac{h \cdot c_l}{\lambda} = \frac{4.13 \cdot 10^{-15} \cdot 3 \cdot 10^8}{750 \cdot 10^{-9}} = 1.65 \text{ eV}$$

EXERCISE 3.5

Ultrasonic transducer with diameter $d = 5$ cm is attached to the bottom of cylindrical reactor, which is filled with distilled water, and generate sound waves with the wavelength of 3 mm. At what distance from the transducer the maximum pressure of ultrasound (US) is observed and what is the power of generated US if the sound velocity $c = 1484$ m/s, sound particle velocity $v = 0.5$ mm/s, and water density is 1000 kg/m^3.

Solution

The frequency of ultrasound can be found as follows:

$$f = \frac{c}{\lambda} = \frac{1484}{3 \cdot 10^{-3}} = 495 \text{ kHz}$$

The approximate distance from the transducer to the vicinity of reactor where the maximum peak of US pressure occurs can be calculated using the following equation:

$$N = \frac{d^2 f}{4c} = \frac{0.05^2 \cdot 495 \cdot 10^3}{4 \cdot 1484} = 0.208 \text{ m} = 20.8 \text{ cm}$$

Sound intensity (acoustic intensity, I_a, W/m^2) i is the value that expresses the power (P, W) of the acoustic field at the point.

$$I = \frac{P}{A} = \frac{p \cdot v}{2} = \frac{p \cdot c \cdot v^2}{2}, \text{ hence } P = \frac{A \cdot \rho \cdot c \cdot v^2}{2}$$

$$= \frac{0.00196 \cdot 1000 \cdot 1484 \cdot 0.0005^2}{2} = 0.54 \text{ W}$$

where A is the unit area perpendicular to the direction of wave propagation. In our case, US wave is propagated through the reactor with 5 cm diameter; therefore, $A = \pi \cdot r^2 = 3.14 \cdot 0.025^2 = 0.00196$ m^2.

EXERCISE 3.6

Calculate the current efficiency and energy consumption of continuous EDI desalination process from Mg^{2+} ions with the initial concentration in the inlet water equal to 50 mg/L if the 90% of ion removal is achieved at 5 V applied constant voltage and deionized stream flow rate $Q = 10$ L/min. The process is stable, i.e., current changes in water splitting potential and solution resistance are constant. Change of resin resistance during EDI $\Delta R_{resins} = 250\ \Omega$.

Solution

Because the process is constant, it can be estimated that

$$I_{stack} = \frac{\Delta E_{cell}}{\Delta R_{resins}} = \frac{5}{250} = 0.02\ A$$

The CE (%) of the process can be determined using the following equation:

$$CE = \frac{z_i \cdot (C_0 - C_t) \cdot Q \cdot F}{1000 \cdot 60 \cdot M \cdot I} \cdot 100 = \frac{2 \cdot (0.1 - 0.1 \cdot 0.9) \cdot 10 \cdot 96,500 \cdot 100}{1000 \cdot 60 \cdot 24.3 \cdot 0.02} = 66.2\%$$

The EC (kWh/mol) of EDI process can be found using the following calculation:

$$EC = \frac{M \cdot I \cdot \overline{E_{cell}}}{(C_0 - C_t) \cdot Q} = \frac{24.3 \cdot 0.02 \cdot 5}{(0.1 - 0.1 \cdot 0.9) \cdot 10 \cdot 60} = 0.04\ kWh/mol$$

EXERCISE 3.7

Calculate the amount of Ca^{2+} ions adsorbed in a cation-exchange resin from a solution containing Na^+ and Ca^{2+} ions if concentration exchange constant $K = 0.83$ and amount of Na^+ ions adsorbed in the resin is equal to 0.8 mol/g. Initial concentrations of Na^+ and Ca^{2+} inlet solution are equal, $C_{Na^+} = C_{Ca^{2+}} = 1.5$ mol/L.

Solution

Ion exchange processes on a solid surface are described by the Nikolsky equation

$$\frac{x_1^{\frac{1}{z_1}}}{x_2^{\frac{1}{z_2}}} = K \frac{a_{e1}^{\frac{1}{z_1}}}{a_{e2}^{\frac{1}{z_2}}}$$

Because activities of cations can be replaced with the concentration of ions, the exchange capacity of cation-exchange resin toward Ca^{2+} adsorption is equal to

$$x_{Ca^{2+}} = {}^{z}Ca^{2+}\sqrt{\frac{K \cdot C_{Ca^{2+}}^{\frac{1}{z_{Ca^{2+}}}} \cdot x_{Na^+}^{\frac{1}{z_{Na^+}}}}{C_{Na^+}^{\frac{1}{z_{Na^+}}}}} = \sqrt[2]{\frac{0.83 \cdot 1.5^{\frac{1}{2}} \cdot 0.8^{\frac{1}{1}}}{1.5^{\frac{1}{1}}}} = 0.73 \ mol/g$$

EXERCISE 3.8

The rate of CO_2 generation as a result of enzymatic substrate oxidation was measured for different initial concentrations of substrate. The results of measurements are listed in the following table.

[S], mol/L	0.001	0.005	0.01	0.02	0.05
ϑ, mol/min	$1.3 \cdot 10^{-5}$	$3.7 \cdot 10^{-5}$	$5 \cdot 10^{-5}$	$5.9 \cdot 10^{-5}$	$6.7 \cdot 10^{-5}$

Solution

The dependence of rate of product formation on the substrate concentration is described by the Michaelis–Menten equation

$$\vartheta = \frac{\vartheta_{max} \cdot [S]}{K_M + [S]}.$$

The equation can be written also as follows $\frac{1}{\vartheta} = \frac{1}{\vartheta_{max}} + \frac{K_M}{\vartheta_{max}} \cdot \frac{1}{[S]}$ and graphically represented in $\frac{1}{\vartheta} = f(1/[S])$ coordinates. The intersection with the x-axes cut segments equal to $-1/K_M$. The values of $\frac{1}{\vartheta}$ and $1/[S]$ are calculated, listed in the below table and plotted.

1/[S]	1000	200	100	50	20
$1/\vartheta$	80,000	27,000	20,000	17,000	15,000

In our case $-1/K_M = -200$, hence $K_M = -1/-200 = 0.005$ mol/L.

Appendix 4. Exercises for Chapter 4

EXERCISE 4.1

EC reactor with a plate-type stainless steel electrodes is planned to be used for the treatment of wastewater containing 50 mg/L Cr(VI). Calculate the total number of electrode plates, working volume of EC reactor, hourly iron consumption, and area occupied by EC reactor. The values of following parameters are known

- flow rate of wastewater Q is equal to 50 m³/h
- specific electricity consumption required for the removal 1 g of Cr(VI) from wastewater, $q_I = 3.1$ A h/g;
- anodic current density $j = 200$ A/m²;
- width of electrode plate $b = 0.8$ m;
- working height of the electrode plate $h = 0.5$ m;
- distance between the electrodes $l = 10$ mm;
- residence time of wastewater in EC reactor, $\tau = 3$ min;
- specific metal consumption in g required to remove 1 g of Cr(VI) from wastewater, $q_{Fe} = 2$ g/g;
- stock utilization ratio of the anodes, $k = 0.8$;
- gap between the electrode system and the wall of EC reactor, $b' = 20$ mm;
- thickness of anodes, $\delta = 8$ mm

Solution

- The total number of electrode plates required (N) is determined as follows:

$N = 2S/S'$, where S is the total surface area of anodes (S, m²) and S' is the surface area of one electrode plate.

$$ S = \frac{I}{j} = \frac{Q \cdot C \cdot q_I}{j} \quad \text{and} \quad S' = b \cdot h. $$

Hence $N = \dfrac{2Q \cdot C \cdot q_I}{j \cdot b \cdot h} = \dfrac{2 \cdot 50 \cdot 50 \cdot 3.1}{200 \cdot 2 \cdot 0.8 \cdot 0.5} = 193.75 = 96.8 = 98.$

Because the total number of electrode plates in a stack should be below 30, it is necessary to make provision for 4 stacks with 26 electrode in each stack, i.e., totally $26 \cdot 4 = 104$ electrodes.

The thickness of one electrode stack (b_{st}) can be determined by the following equation:

$$ b_{st} = N \cdot \delta + (N - 1) \cdot l = 24 \cdot 8 \cdot 10^{-3} + (24 - 1) \cdot 10 \cdot 10^{-3} = 0.422 \text{ m} $$

- Working volume of EC reactor (V_r, m^3) is determined as follows:

$$V_{r,EC} = Q \cdot \tau = (50 \cdot 3)/60 = 2.5 \text{ m}^3$$

- Hourly iron consumption

$$G = \frac{Q \cdot C \cdot q_{Me}}{k} = \frac{50 \cdot 50 \cdot 2}{0.8} = 6250 \text{ g/h} = 6.25 \text{ kg/h}$$

- Area occupied by EC reactor is equal to $A = B \cdot L$, where the width of the reactor $B = b_{st} + 2b'$ and the length of the EC reactor $L = b + 2b'$. Taking into account four electrode stacks

$$B = 4 \cdot 0.422 + 2 \cdot 20 \cdot 10^{-3} = 1.728 = 1.75 \text{ m}$$
$$L = 0.8 + 2 \cdot 20 \cdot 10^{-3} = 0.84 \text{ m}$$
$$A = B \cdot L = 1.75 \cdot 0.84 = 1.47 = 1.5 \text{ m}^2.$$

Index

'*Note*: Page numbers followed by "f" indicate figures and "t" indicate tables.'

Printed in the United States
By Bookmasters